A PRACTICAL GUIDE TO THE STUDY OF GLACIAL SEDIMENTS

Edited by David J. A. Evans and Douglas I. Benn

A MEMBER OF THE HODDER HEADLINE GROUP
LONDON
Distributed in the United States of America
by Oxford University Press Inc., New York

First published in Great Britain in 2004 by
Arnold, a member of the Hodder Headline Group,
338 Euston Road, London NW1 3BH

http://www.arnoldpublishers.com

Distributed in the United States of America by
Oxford University Press Inc.
198 Madison Avenue, New York, NY10016

British Library Cataloguing in Publication Data
A catalogue record for this book is available from British Library

Library of Congress Cataloging-in-Publication Data
A catalog record for this book is available from the Library of Congress

ISBN 0 340 75959 3

1 2 3 4 5 6 7 8 9 10

Typeset in 10/13pt Caslon by Phoenix Photosetting, Chatham, Kent
Printed and bound in Malta

CONTENTS

PREFACE

'These stones are your friends' (Doug Benn, ca. 1994)

We have spent large parts of our careers trying to convince geography and earth science undergraduates that success in reconstructing landscape evolution is becoming increasingly more reliant upon the greater integration of geomorphological and sedimentological techniques. Moreover, rather than become intimidated by the variety of techniques on offer, young researchers should find confidence in the knowledge that a diverse armoury of tools is at their disposal, and mastery of the use of those tools makes them stronger scientists. The spectacular growth and evolution of glacial sedimentology over the last few decades has been paralleled by an expanding communal knowledge of sedimentary processes in modern environments. So we now have many methods for reading the sedimentary record and the contemporary analogues to guide us through the assemblage of palaeogeographical jigsaw puzzles. Teaching these methods to undergraduates has made us aware of the need for a textbook that pools the guidelines for practical glacial sedimentology, a clear and accessible guide to recording and interpreting glacial depositional successions. A practical guide can never be totally comprehensive but we have compiled here a range of complementary techniques that we find yield consistent and meaningful results in the diverse glacial environments of the world that we have been privileged to study.

David J.A. Evans & Douglas I. Benn
Mother India, Glasgow 2004

LIST OF CONTRIBUTORS

Douglas I. Benn	School of Geography and Geosciences, University of St Andrews. St Andrews, Fife KY16 9AL, Scotland, UK
Simon J. Carr	Department of Geography and Geomatics, School of Geosciences, University of Glasgow, Glasgow G12 8QQ, Scotland, UK
David J.A. Evans	Department of Geography and Geomatics, School of Geosciences, University of Glasgow, Glasgow G12 8QQ, Scotland, UK
John F. Hiemstra	Department of Geography and Geomatics, School of Geosciences, University of Glasgow, Glasgow G12 8QQ, Scotland, UK
Trevor B. Hoey	Department of Geography and Geomatics, School of Geosciences, University of Glasgow, Glasgow G12 8QQ, Scotland, UK
Emrys R. Phillips	British Geological Survey, Murchison House, West Mains Road, Edinburgh, EH9 3LA, Scotland, UK
Brice R. Rea	Department of Geography, University of Aberdeen, Elphinstone Road, Old Aberdeen, AB9 2UF, Scotland, UK
John Walden	School of Geography and Geosciences, University of St Andrews, St Andrews, Fife KY16 9AL, Scotland, UK

ACKNOWLEDGEMENTS

The gestation period of this book has been long. We are sure it has been too long for the good people at Arnold, and for that we can only apologize and hope that the product is some reward. In this respect we would like to thank Laura McKelvie, who initiated this project immediately after *Glaciers and Glaciation* had hit the bookshelves, Abigail Woodman, who took over the responsibility for the guide and provided the final push to get us to finish it, and Colin Goodlad, who took on the unenviable task of making our work look professional. Thanks to all the undergraduates who have passed through the universities of Glasgow, Aberdeen and St Andrews over the years and have invested long hours in 'stone measuring' for the good of science and the building of their character! On the home front: DJAE would like to thank Tessa, Tara and Lotte for putting up with the fall-out from another book and particularly Tara for the timeless quote '*I don't want to look at stones anymore, they don't say anything*'. Lotte, on the other hand, has a stone collection and clearly wants an early start on that character building!; DIB would like to thank Sue and Andrew for putting up with my long absences in the mountains and long hours in the office, and for appearing to show genuine interest when I launch into yet another geological lecture when we're meant to be on holiday.

FIGURE ACKNOWLEDGEMENTS

Chapter 1

1.2 – Reprinted from Walker R.G. (1992) *Facies, facies models and modern stratigraphic concepts* (in Walker R.G. and James N.P. (eds.), *Facies Models: Response to Sea Level Change*), with permission from the Geological Association of Canada.

Chapter 2

2.1 & 2.12 – Reproduced with kind permission from W. Fritz and J. Moore.
2.2 – Reprinted from Eyles N. and Miall A.D. (1984) *Glacial facies* (in Walker R.G. (ed.), *Facies Models*), with permission from the Geological Association of Canada.
2.3 – Reproduced from Harms J.C. et al. (1982) *Structures and Sequences in Clastic Rocks*, Society of Economic Paleontologists and Mineralogists, Lecture Notes for Short Course No.9.
2.5 – Reproduced from Harms J.C. (1979) Primary sedimentary structures, *Annual Review of Earth and Planetary Science*, Vol.7, p. 227–248.
2.7 – Reproduced from Lowe D.R. (1982) Sediment gravity flows II. Depositonal models with special reference to the deposits of high density turbidity currents, *Journal of Sedimentary Petrology*, Vol.52, p. 279–297, Society of Economic Paleontologists and Mineralogists.
2.9 – Reprinted from Harry D.G. and Gozdzik J. (1988) Ice wedges: growth, thaw transformation and palaeoenvironmental significance, *Journal of Quaternary Science*, Vol.3, p.39–55, © John Wiley and Sons Limited. Reproduced with permission.
2.10c – Reproduced from van der Wateren F.M. (1995) Processes of glaciotectonism (in Menzies J. (ed.), *Modern Glacial Environments: Processes, Dynamics and Sediments*), Butterworth-Heinemann, p. 309–335.
2.11 – Reproduced from Thomas G.S.P. and Connell R.J. (1985) Iceberg drop, dump and grounding structures from Pleistocene glacio-lacustrine sediments, Scotland, *Journal of Sedimentary Petrology*, Vol.55, p. 243–249, Society of Economic Paleontologists and Mineralogists.
2.17 – Reprinted from T.A. Davies et al. (1997), *Glaciated Continental Margins: An Atlas of Acoustic Images*, Figure 8, page 18 (Stoker M.S., Pheasant J.B. and Josenhans H., *Seismic methods and interpretation*, p.9–26), with kind permission of Kluwer Academic Publishers.
2.18 – Reproduced from Saunderson H.C. (1975) Sedimentology of the Brampton esker and its associated deposits: an empirical test of theory (in Jopling A.V. and McDonald B.C. (eds.), *Glaciofluvial and Glaciolacustrine Sedimentation*), Society of Economic Paleontologists and Mineralogists Special Publication 23, p. 155–176.
2.19 – Modified from Lowe J.J. and Walker M.J.C. (1997) *Reconstructing Quaternary Environments*, Longmans.
2.20 – Redrawn from Berthelsen A. (1978) The methodology of kineto-stratigraphy as applied to glacial geology, *Bulletin of the Geological Society of Denmark*, Vol.27, p. 25–38.

2.21a – Reprinted from A. Miall (1978) *Lithofacies types and vertical profile models in braided river deposits: a summary* (in Miall A. (ed.), *Fluvial Sedimentology*), with permission from the Canadian Society of Petroleum Geologists.
2.22 – Reproduced from Boyce, J.I. and Eyles, N. (2000) 'Architectural element analysis applied to glacial deposits: internal geometry of a late Pleistocene till sheet, Ontario, Canada.' *Geological Society of America Bulletin*, 112: 98–118.
2.23 – Reproduced from *A data chart for field description and genetic interpretation of glacial diamicts and associated sediments – with examples from Greenland, Iceland and Denmark* by J. Kruger and K.H. Kjaer from *Boreas*, www.tandf.no/boreas, 1999, Vol.28, p.386–402, by permission of Taylor & Francis.

Chapter 3
3.5 – Reproduced from Gale S.J. and Hoare P.G. (1991) *Quaternary Sediments*, Belhaven Press.
3.6i – Reprinted from Church M.A. et al. (1987) River bed gravels: sampling and analysis (in Thorne C.R. et al. (eds.), Sediment Transport in Gravel Bed Rivers, p.43–88, © John Wiley & Sons Limited. Reproduced with permission.
3.6ii and 3.16 – Reproduced by permission of the Royal Society of Edinburgh from *Transactions of the Royal Society of Edinburgh: Earth Sciences*, Vol.89, p.291–323.
3.9 – Reproduced from Prins M. and Weltje G. (1999) End-member modelling of siliclastic grain size distributions: the Late Quaternary record of eolian and fluvial sediment supply to the Arabian Sea and its paleoclimatic significance, in *Numerical Experiments in Stratigraphy: recent advances in stratigraphic and sedimentologic computer simulation*, Society of Economic Paleontologists and Mineralogists Special Publication No.63, p. 91–111.
3.10a – Reproduced from Benn, D.I. and Gemmel, A.M.D. (2002) *Geological Society of America Bulletin*, 114: 528–32.
3.10b – Reprinted from the *Journal of Glaciology* with permission of the International Glaciological Society.
3.12 – Reprinted from *Geomorphology*, Vol.6, T. Sugai, *River terrace development by concurrent fluvial processes and climatic changes*, 243–252, 1993, with permission from Elsevier.
3.13 – Reprinted from *Quaternary International*, Vol.56, R. Hindson and C. Andrade, *Sedimentation and hydrodynamic processes associated with the tsunami generated by the 1755 Lisbon earthquake*, 27–38, 1999, with permission from Elsevier.
3.14 – Reproduced from Hoey T.B. and Bluck B.J. (1999) Identifying the controls over downstream fining of river gravels, *Journal of Sedimentary Research*, Vol.A64, p. 916–920, Society of Economic Paleontologists and Mineralogists.
3.15 – Reproduced from Toyoshima M. (1987) Low downstream decrease rate of particle size in latest Pleistocene fluvial terrace deposits in the Dewa Mountains, NW Japan, *Science Reports of the Tohoku University*, Vol.37, p. 174–186.
3.16 – Reproduced from Bluck B.J. (1999) 'Clast assembling, bed-forms and structure in gravel beaches.' *Transactions of the Royal Society of Edinburgh*, 89: 291–323.

Chapter 4
4.2 – Reproduced from Krumbein W.C. (1941) Measurement and geological significance of shape and roundness of sedimentary particles, *Journal of Sedimentary Petrology*, Vol.11, p. 64–72, Society of Economic Paleontologists and Mineralogists.
4.4, 4.5, 4.6 and 4.7 – Reprinted from *Sedimentary Geology*, Vol.91, D.I. Benn and Ballantyne C.K., *Reconstructing the transport history of glacigenic sediments: a new approach based on the co-variance of clast form indices*, 215–227, 1994, with permission from Elsevier.
4.6 – Reprinted from *Quaternary Science Reviews*, Vol.11, D.I. Benn, *The genesis and significance of 'hummocky moraine': evidence from the Isle of Skye, Scotland*, 781–799, 1992, with permission from Elsevier.

Chapter 5
5.3 – Figure 11.9, page 223 from *Structural Geology* by R.J. Twiss and E.M. Moores, © 1992 by W.H. Freeman and Co, used with permission.
5.7, 5.13, 5.14, 5.15, 5.16 – Reprinted from Benn D.I. and Ringrose T. (2001) Random variation of fabric eigenvalues: implications for the use of a-axis fabric data to differentiate till facies, *Earth Surface Processes and Landforms*, Vol.26, p.295–306, © John Wiley and Sons Limited. Reproduced with permission.
5.10 – Reprinted from *Sedimentary Geology*, Vol.62, J. Rose, *Glacier stress patterns and sediment transfer associated with superimposed flutes*, 151–176, 1989, with permission from Elsevier.
5.11 – Reprinted from *Sedimentology*, Vol.42, D.I. Benn, *Fabric signature of till deformation, Breidamerkurjokull, Iceland*, 735–747, 1995, with permission from Blackwell Publishing Ltd.
5.12 – Reprinted from *Sedimentology*, Vol.29, G.M. Ashley et al., *Deposition of climbing ripple beds: a flume simulation*, 67–79, 1982, with permission from Blackwell Publishing Ltd.

Chapter 6
6.2 – Adapted from Carr S.J. and Lee J.A. (1998) Thin section production of diamicts: problems and solutions, *Journal of Sedimentary Research*, Vol.68, p. 217–220, Society of Economic Paleontologists and Mineralogists.
6.5 and 6.6 – Reproduced with permission from Geological Society Publishing house.
6.7 – Reproduced with permission from the Quaternary Research Association.
6.8 – Reprinted from the *Journal of Glaciology* with permission of the International Glaciological Society.
6.10 and 6.18 – Reproduced with permission from Royal Swets and Zeitlinger.
6.11 and 6.12 – Reproduced from *Comparative scanning electron microscopy study of oriented till blocks, glacial grains and Devonian sands in Estonia and Latvia* by W.C. Mahaney and V. Kalm from *Boreas*, www.tandf.no/boreas, 2000, Vol.29, p.35–51, and from *Quantification of SEM microtextures useful in sedimentary environmental discrimination* by W.C. Mahaney et al. from *Boreas*, www.tandf.no/boreas, 2001, Vol.30, p.165–171, by permission of Taylor & Francis.

6.15 – Reproduced from *Micromorphological evidence supporting Late Weichselian glaciation of the Northern North Sea* by S.J. Carr et al. from *Boreas*, www.tandf.no/boreas, 2000, Vol.29, p.315–328, by permission of Taylor & Francis.
6.16 – Reprinted from *Quaternary Science Reviews*, Vol.12, J.J.M. van der Meer, *Microscopic evidence of subglacial deformation*, 553–587, 1993, with permission from Elsevier.
6.17 – Reproduced from Mahaney W.C. (1995) Glacial crushing, weathering and diagenetic histories of quartz grains inferred from scanning electron microscopy (in Menzies J. (ed.), *Modern Glacial Environments: Processes, Dynamics and Sediments*), Butterworth-Heinemann, p. 487–506.
6.18 – Reproduced from (1994) *The Formation and Deformation of Glacial Deposits*, with permission from Chantel van Werkhoven.

Chapter 7
7.1 – Reproduced from Fichter L.S. and Poché D.J. (1993) *Ancient Environments and the Interpretation of Geologic History*, second edition, Macmillan Publishing Company.
7.2, 7.3 (part) – Reprinted from Lindholm R.C. (1987) *A Practical Approach to Sedimentology*, Allen & Unwin.
7.5, 7.6 and 7.7 – Reproduced with permission from the National Research Council of Canada.
7.8 – Reproduced from S. Snall (1985) *Weathering in till indicated by clay mineral distribution*, *Geologiska Foreningens i Stockholm Forhandlingar*, Vol.107, p.315–322, with permission from Geologiska Foreningens i Stockholm Forhandlingar and the Geological Society of Sweden.
7.9 – Reprinted from Walden J. et al. (1992) Mineral magnetic analyses as a means of lithostratigraphic correlation and provenance indication of glacial diamicts: intra- and inter-unit variation, *Journal of Quaternary Science*, Vol.7, p.257–270, © John Wiley & Sons Limited. Reproduced with permission.

Chapter 8
8.14 – Reprinted from *Quaternary Science Reviews*, Vol.16, N.R. Iverson et al., *A ring shear device for the study of till deformation: tests on tills with contrasting clay contents*, 1057–66, 1997, with permission from Elsevier.

Chapter 9
9.3 – Reprinted from *Quaternary Science Reviews*, Vol.15, D.I. Benn and D.J.A. Evans, *The interpretation and classification of subglacially-deformed materials*, 23–52, 1996, with permission from Elsevier.

All best efforts have been made to contact respective copyright holders, however this has not always been possible in all cases; any omissions brought to our attention will be corrected in future printings.

1 Introduction and rationale

Douglas I. Benn & David J.A. Evans

1.1 INTRODUCTION

Physical evidence for past Earth surface processes is preserved in two basic forms: sediments and erosional forms. Unlike erosion, which progressively removes traces of earlier processes and environments, sedimentation has the potential to build up an archive of events, recording a sequence of processes operating at a locality. Thus, careful examination of sediments has the potential to reveal shifting patterns of surface processes in space and time, and provides a window into the past, allowing us to reconstruct past environments and environmental change. In recent decades, earth scientists have learned to read many forms of sediment archive, including deep-ocean muds, desert loess and annual layers in glacier ice. This book is concerned with the techniques used to read one particular archive, the sedimentary record of former glaciers and ice sheets. Although frequently more complex and fragmentary than other sedimentary records, glacial successions can, when examined systematically and rationally, provide detailed insights into former environments and climates in places where no other evidence is available.

The last few decades have seen the spectacular growth and evolution of glacial sedimentology as a discipline. Knowledge of sedimentary processes in modern environments has increased alongside the rapid development of new methods for reading the sedimentary record. The literature on glaciers and glacial geologic processes is now vast, and covers a potentially bewildering array of techniques. We have become increasingly aware of the need for a modern overview of the subject, to provide students and researchers with a clear and accessible guide to recording and interpreting glacial successions. This book does not provide a comprehensive review of all approaches and methods currently in use. Rather, it aims to describe a range of complementary techniques that we have found to yield consistent and meaningful results in a range of contexts throughout the world, from the high Arctic to the Himalaya.

1.2 ORDER FROM CONFUSION

It is clearly impossible to record everything preserved in a sediment exposure. Constraints of time and sanity mean that we must be highly selective in what we observe and measure, and choices must be made as to what is important and relevant. This immediately raises fundamental methodological questions: How do we know what is important and relevant before we measure it? How can we be sure that the choices we make will not bias the investigation and prejudice the conclusions? It is important to address these questions because otherwise there is a very real danger that we only see what we already believe to be there. There are many unfortunate examples of sedimentological studies in which the conclusions clearly predate collection of the evidence. This is one of the side effects of the methodologies we follow as earth scientists. In the largely inductive and deductive approaches that we take when studying sediments, the types of observations we make are necessarily guided by pre-existing theory or previously reported observations. We usually pursue some form of hypothesis testing, the hypotheses emerging from the body of research that pre-dates our own efforts. In order to avoid going down scientific cul-de-sacs, we need to be armed with several, alternative working hypotheses (Chamberlin 1897). All but one of the competing hypotheses can then be gradually eliminated as research continues, following the principle of *falsification* (Popper 1972). This critical rationalist approach has become increasingly popular amongst earth scientists (Haines-Young and Petch 1986). Problems arise when a researcher chooses only one hypothesis and then, often subconciously, protects it rather than tests it. This inevitably entails the erection of '*ad hoc* protection devices', designed to stop a hypothesis being tested by others (Chalmers 1982), and often results in the selective use of real world observations. Science attempts to be objective, unbiased and reproducible. To achieve this goal, we need to find principles that will guide research design, while making sure that we are as objective as possible.

The best way to do this is to adopt a general approach that is applicable to all sediment successions, regardless of their precise origin. All depositional systems, including glacial systems, exhibit order on many different levels, either as a series of steps in *time*, or on a range of scales in *space* (Benn and Evans 1998). The temporal and spatial dimensions provide a powerful basis for structuring efficient and objective research, without presupposing any particular origin for the sediments. The *time dimension* is important, because sediment properties develop at different points of the erosional, transport and depositional history of debris as it travels towards the site of final deposition. The sequence of steps can be regarded as a type of *debris cascade system* (Chorley *et al.* 1984), and can used as a basis for sediment classification schemes (e.g. Boulton and Deynoux 1981; Dreimanis 1989). The *spatial dimension* allows the *context* of a sediment to be established with respect to adjacent sediments, the wider environment and the landscape as a whole. Study of the spatial arrangements of sediments at a hierarchy of scales allows the development of *facies models*, or summaries of how individual deposits are nested together to form depositional systems. Used correctly, these provide a structured basis for deciphering the environmental meaning of sediment successions.

1.3 STEPS IN TIME: THE DEBRIS CASCADE

Sedimentary deposits can be viewed as outcomes of a series of processes extending through time from the initial release of particles from their source, through to deposition. In Figure 1.1, this *debris cascade system* is broken down into four stages: (1) debris source, (2) position of entrainment, (3) transport path, (4) position and process of deposition. The *debris source* is the primary input to the system. For glacial sediments, this may be subglacial (e.g. plucked and abraded bedrock, or over-ridden sediments) or extraglacial (e.g. rock walls, debris-mantled slopes, wind-blown dust). The *position of entainment* is most usually subglacial or supraglacial. Debris is then carried along one or more *transport paths*, including active or passive glacial transport, rivers, suspension in lake- or sea-water, iceberg rafting and the wind. Debris may pass between transport paths many times prior to final deposition. *Depositional processes* refer to the mechanisms which lay down the final deposit, and include subglacial, fluvial, gravitational, subaqueous and aeolian processes, all of which may operate in different glacial environments.

Different properties of sediments are acquired at different points along the debris cascade:

1| The *debris source* controls the lithology of the particles in a sediment, but can also influence the shape and size of particles (particle morphology and grain-size distribution), through factors such as rock strength, joint spacing, etc.

2| The *transport path* determines the erosional processes experienced by particles as they move through the system. As such, transport exerts a strong influence on particle morphology. Grain-size distributions are also modified during transport as the result of progressive wear and the preferential transport of particular size grades by water, wind and gravitational processes. The lithology of debris can also be influenced by transport, because of the varying durability of different rock types. For glacially and gravitationally deformed sediment, aspects of deformational history can be reflected in sediment fabric and structure.

3| *Depositional processes* generally exert the strongest influence on sediment properties, and can control the geometry and lateral extent of beds, internal sedimentary structures

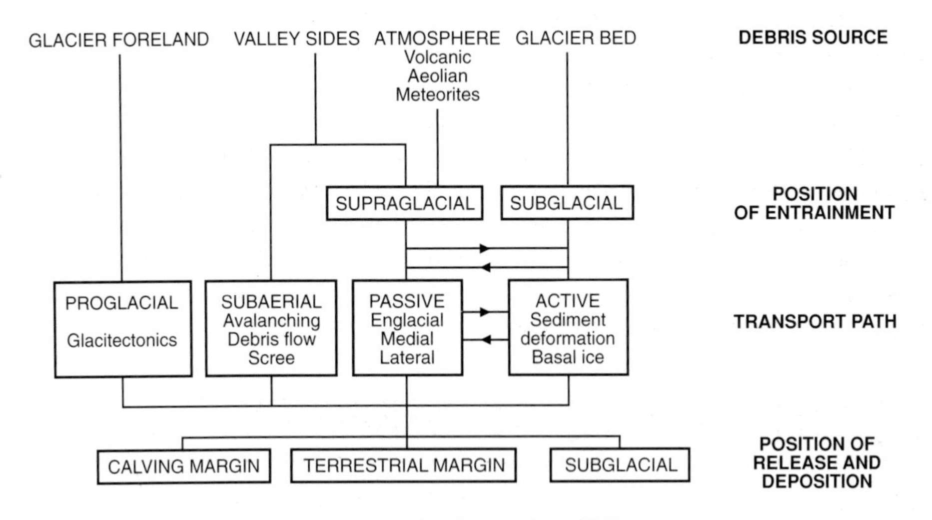

FIG 1.1 Debris cascade system for glacial environments (from Benn and Evans 1998).

(e.g. lamination, cross-bedding and grading), grain-size distributions, particle morphology, fabric, geotechnical properties (e.g. porosity, shear strength and permeability) and others.

4| *Post-depositional processes* include diagenesis, glacitectonic deformation, frost-heaving and winnowing and leaching of fine material by wind and water. They can therefore influence structures (i.e. the superimposition of deformational structures such as folds or faults), grain-size distribution and geotechnical properties.

This temporal framework therefore provides a clear rationale for the choice of techniques employed in a sedimentological study: techniques are chosen to provide information on one or more of the stages of the debris cascade, allowing aspects of the broader sedimentary system to be reconstructed.

1.4 HIERARCHIES IN SPACE: FACIES MODELS

In some cases, the characteristics of individual sediment units are sufficient to define the environment of deposition. Examples include deposits containing fossils with known environmental ranges, or certain types of till. More usually, however, a given sediment type may be formed in many different environments. A debris flow deposit, for example, may look much the same whether it is deposited in front of a glacier, in semi-arid badlands or on a coast. Therefore, the environmental significance of many sediments can only be fully understood *in context;* that is, with reference to the assemblage of adjacent sediments.

In modern environments, sediments are rarely deposited in isolation, but are laid down as part of assemblages that reflect the range of processes active in that environment. Such assemblages can be recognized at a wide range of scales, from the very small to that of a whole depositional basin. For example, cross-bedded sands may be part of an assemblage of sand and gravel infilling a fluvial channel; in turn, the channel fill could be part of an assemblage of channel and bar deposits in a braided river system; and the braided river system part of a yet larger assemblage of deposits laid down along a continental margin. The environmental context of a sediment can therefore be defined at different levels of a spatial hierarchy, beginning with the immediate locality and panning out to wider and wider horizons. At each successive level, the controls on the sedimentary system become larger in scale and longer-lasting in effect. For our example of fluvial sands, at the local level the main controls on deposition are the shape of the immediate river bed and the short-term flow conditions as determined by rainfall, glacier melt and release of stored water. At the largest scale, the formation, location and extent of a braided river system is controlled by global factors such as long-term climatic cycles, relative sea-level fluctuations and tectonics. In marine geology, this approach is formalized in *sequence stratigraphy*, which views depositional events within the context of eustatically controlled marine transgressions and regressions (van Wagoner *et al.* 1988; Posamentier *et al.* 1993). In glaciated basins, sequence stratigraphy is complicated by glacioisostatic effects on local sea level, although the value of the approach has been successfully demonstrated in some studies (e.g. Boulton 1990; Eyles and Eyles 1992; Martini and Brookfield 1995).

A hierarchical approach to sedimentology provides a powerful means of describing how sediments, landforms and landscapes fit together, and of determining how organization in the landscape reflects the organization of depositional processes and external controls in the environment. It forms the basis of *facies models,* which are descriptive and predictive models of relationships between different deposits (Reading 1986; Walker 1992). Somewhat arbitrarily, we can define four levels of organization, at increasing scales (Fig. 1.2): (1) facies, or individual deposits; (2) facies associations; (3) depositional systems, or landsystems; and (4) systems tracts, or large-scale linkages of depositional systems. *Facies* (from the Latin word meaning 'aspect' or 'appearance of' something) refers to a body of sediment with a distinctive combination of properties that distinguish it from neighbouring sediments (Reading 1986; Walker 1992). Facies are formed by single processes or groups of processes acting in close association. In most cases, they are not unique to any particular environment. Rather, depositional environments are characterized by *combinations of processes,* preserved as facies associations sediment–landform associations and, at progressively larger scales, landsystems and systems tracts. At each successive scale, we see deposits in an even wider context.

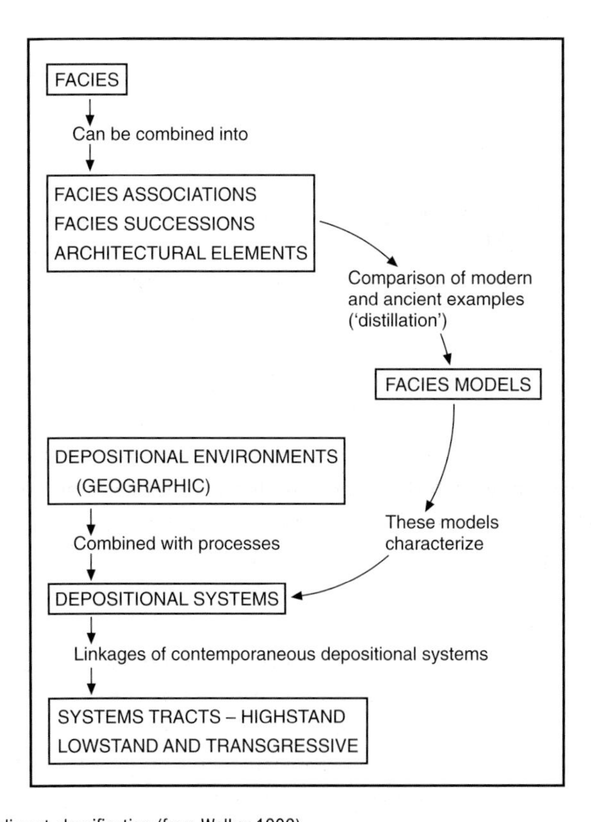

FIG 1.2 Hierarchical sediment classification (from Walker 1992).

While simple in principle, the facies model concept is not always straightforward to apply in practice. As increasingly larger scales are considered, examination of the evidence must necessarily become less and less detailed, particularly for landsystems and systems tracts that may extend over hundreds or thousands of square kilometres. This brings a very real danger that evidence is narrowly interpreted in terms of a preconceived model, magnifying any errors in interpretation made at small scales, leading to misleading conclusions at the largest scales. Basin-wide depositional models may be very neat and intellectually satisfying, but cannot be hurriedly constructed and need to be rigorously tested against new field evidence. Nevertheless, when used critically and carefully, the facies model concept provides a robust yet versatile framework for determining the nested spatial context of deposits. When used in combination with the temporal dimension of the debris cascade, this approach has the potential to yield a dynamic four-dimensional view of events preserved in the sedimentary record.

The focus of this book is the measurement and interpretation of sediments from the outcrop to the microscopic scale, which generally means the scale of facies associations and facies. Most of the techniques are applicable to individual facies and, in combination, can be used to determine their origin in terms of processes. Environmental interpretation, as noted above, requires consideration of associations of facies, in terms of both the range of adjacent facies and the geometric relationships between them. It is therefore very important when logging a section to pay close attention to the geometry of sediment bodies and the nature of their contacts. This topic is discussed in detail in Chapter 2, and we return to it again in a detailed case study in Chapter 9.

1.5 MODERN ANALOGUES

It is common practice to interpret sediments and landforms with reference to *modern analogues*, that is, by invoking similarities to modern examples whose origin and environmental significance are well known. This is a powerful approach, and has become increasingly so in recent years as the literature encompasses a growing range of modern examples. Most importantly, the use of modern analogues replaces unrestrained speculation about the meaning of sediments and landforms, which unfortunately characterized much of the English language glacial literature from the First World War to the 1960s (Evans, in press). The use of analogues underlies Lyell's famous principle of uniformitarianism, that 'The Present is the Key to the Past'. This does not imply that the past was just like the present, but that it is reasonable to suppose that the formation of sediments (and other rocks) was subject to the same physical and chemical principles operating today. From this, it follows that we can use knowledge of modern sedimentary processes and products to guide and constrain our interpretation of the sedimentary record.

There is a very large and expanding literature on modern sediment facies and facies associations, and their significance in terms of processes and environment. In addition, laboratory experiments have expanded our knowledge of process-form relationships. There is a long tradition of flume and wind tunnel experiments into fluvial and aeolian systems, and recently shear-ring and centrifuge experiments have extended this approach to subglacial processes (e.g. Hooyer and Iverson 2000a). Such experiments are important, as they provide knowledge of processes that are seldom directly observable.

It has been objected that the range of past conditions (e.g. during a glacial–interglacial cycle) was far wider than that experienced in modern times, and that certain types of deposits may have no modern analogues (e.g. Kemmis 1996; Shaw 1996, 2002). It is certainly true that some geological events were on scales vastly different to those of today, but the same physical principles should apply. The important point is that the analogue should guide interpretation, not provide a rigid template which exactly replicates past examples. Perhaps a more important limitation than the range of modern environments, is the experience and outlook of individual researchers. One's interpretation of the sedimentary record may be profoundly influenced by personal experience, which may be relatively limited. In such circumstances, it is tempting to give undue emphasis to some aspects of the evidence that fit with one's own experience of glacial systems. It is now clear that glacial environments are very variable (Evans, 2003), and one's repertoire of 'analogues' should reflect this.

1.6 HYPOTHESIS TESTING

As we discussed briefly above, scientific methods predominantly involve the complementary processes of induction and deduction. Induction refers to the formulation of hypotheses or theories from data, whereas deduction is the process of deducing facts from theoretical principles. In the classical sciences such as physics and chemistry, deduction is used to predict the outcome of experiments, and if the prediction agrees with the results, the theory is considered to have passed that particular test (see Chalmers 1982). Good theories are those that have not yet been contradicted by any reliable experimental results or, more positively, those that yield many verifiable predictions. This methodological framework provides a rigorous basis for testing theories, and for weeding out those that are incomplete, inaccurate or just plain wrong.

If we follow the progress of scientific enquiry based upon an inductive philosophy we can identify a series of steps. The first step is to collect data based upon real world observations. Even at this early stage we have, as discussed above, already made decisions about what we choose to observe and what we ignore. The second step involves the ordering of facts by way of measurement, classification, definition and analyses. This is followed by generalizations about the data, the process of induction. We are then ready to create laws or theories and to provide explanation or to predict via the process of deduction. Deduction is a difficult business but we must be bound by the strict rules of the game. This was perfectly illustrated by Sir Arthur Conan Doyle's character, Sherlock Holmes, when he decreed to the bewildered Dr Watson that 'when you have eliminated the impossible, whatever remains, however improbable, must be the truth'. The 'truth' in our practise is the most appropriate interpretation or the one with the highest probability of being correct. Holmes' principle of eliminating the impossible relies on finding evidence that clearly rules out certain interpretations. In other words, making observations that will definitively show, one way or another, whether a hypothesis is false. Eliminating possibilities also requires that hypotheses have clear, unambiguous implications that can be compared with the evidence. The best hypotheses are often the simplest. Hypotheses formulation should be guided by the principle of Occam's Razor, named after the philosopher, William of Occam (1300–49), which states that 'explanations should not assume any superfluous elements not required by the evidence'.

The interpretation of sediments is an art, but it should also be a science. Is the deductive–inductive model, considered by many to be an essential component of the scientific method, applicable in this case? Reaching an interpretation about a particular sediment association certainly involves inductive reasoning, but how can interpretations be tested? The classic experimental approach is seldom applicable in studies that attempt to reconstruct the past, but this does not mean that predictions cannot be tested in any way. Instead of predicting the outcome of experiments, sedimentologists can, and should, test their models by predicting the results of *new observations*. In practice, this means that we should say that *if* a particular interpretation of a sediment succession is correct, *then* we should expect to find certain other evidence to support it. Experienced field workers tend to do this by habit, constantly thinking through the implications of their ideas, and checking new lines of evidence to test them. This approach can also form part of multi-proxy studies, in which several lines of evidence are used to converge upon the most likely interpretation. Finding one's predictions confirmed can be one of the most exciting aspects of field geology. When the opposite happens, and the new evidence contradicts the model, you should think again, and either modify the model or replace it entirely. This should be seen not as a failure and disappointment, but as an opportunity: if the evidence does not fit with existing interpretations, it may point the way towards something entirely new. This opportunity is lost if you are so sure the model is right that you decide that the apparently contradictory observations were actually predicted by the model all along. In effect, this means that the model is *unfalsifiable*, that is, no matter what observations are made, the model is still believed to be correct. As such, the model is useless as science: it has become a tenet of belief. To have value, a model must make firm predictions that are testable against the immutable testimony of the rocks. Learning to follow these rules is ultimately far more rewarding than trying to maintain the illusion that one is omniscient.

1.7 ARE YOU THINKING CRITICALLY?

If you are to avoid the pitfalls of scientific endeavour outlined above, it is crucial that you understand the processes of objectivity and analytical thought and consciously develop the habit of critical thinking. Unfortunately for teachers and students alike, there is no one dominant theoretical recipe for critical thinking. However, Reiter (1991) identifies two major levels to critical thinking that may prove instructive. First, higher order thinking involves the ability to use analysis, synthesis and evaluation skills. Specifically, analytical skills enable the individual to break problems down into managable parts and select the relevant or valid information. The ability to synthesize allows a person to combine information in a meaningful way from a variety of sources. Evaluation skills empower the individual with the ability to make judgements based on the available evidence, data or observations.

The second level of critical thinking is multi-logical thinking, or the ability to reason objectively from multiple viewpoints. This requires the individual to exercise an open-minded approach (openness), an intellectual curiosity and a commitment to think through all the possible connotations (inquisitiveness), and a non-defensive standpoint that acknowledges weaknesses in their own interpretations (objectivity).

Of course, scientists of all denominations are human beings – they often find it difficult to accept that their interpretation of the facts, because it was so hard won, may not be the most appropriate. We must remember, however, that everyone can provide an 'explanation' of a particular collection of observations but only one explanation is strictly correct, and we may never find it! As Wolpert (1992) points out in his book *The Unnatural Nature of Science*, '... a theory that fits all the facts is bound to be wrong, as some of the facts are themselves bound to be in error ...'.

1.8 SCOPE OF THIS BOOK

We have included in this field guide overviews of the techniques presently regarded as fundamental to the description, analysis and interpretation of glacigenic sediments. Although in-depth explanations of the procedures involved in analytical techniques, especially those conducted in the laboratory, are beyond the scope of a field guide, we do provide preliminary coverage in the following chapters in order to enable an assessment of the suitability of a technique during the fieldwork stage. Each chapter concludes with short case studies that demonstrate the applications of the techniques covered therein. Because a field guide is designed for use on the job, it should get the message across succintly and practicably. Therefore, we have chosen to take an illustrative approach, whereby information is delivered through figures and photographs wherever possible.

Chapter 2 introduces the reader to the first stages of sediment description, classification and recording. It covers the principles of facies description and coding and stratigraphic logging and provides explanations of specific structures and bedding in sediments. Chapter 3 covers the principles of particle size analysis, tackling sampling methods and statistical assessments from the outcrop to the laboratory. This is followed by an overview of particle form in Chapter 4. This involves the quantification of the physical changes made to individidual particles during their transport. The macrofabric of sediment is presented in Chapter 5, covering all of the directional properties of a sediment, including the orientation of particles, bedding planes, folds, faults and erosion surfaces. This is followed in Chapter 6 by an overview of the micromorphological technique. This allows us to zoom in to the smallest of scales in order to assess the sedimentary and structural signatures that are not evident at macro scale. Particle lithology is the subject of Chapter 7, covering the petrology, heavy mineral characteristics, geochemical signatures, clay minerology and magnetic properties of materials. Of necessity, this chapter refers to the appropriate laboratory procedures. Similarly, Chapter 8 on the engineering properties of glacial materials provides important information on laboratory procedures designed to yield information on the geotechnical aspects of sediments. The final chapter in the book is a case study, designed to demonstrate how a multi-faceted approach to the analysis of Quaternary glacigenic sediment sequences in a typical quarry can culminate in the reconstruction of complex depositional environments. It goes without saying that our preferred reconstruction is, at present, the one that explains the most observations.

2 Facies description and the logging of sedimentary exposures

David J.A. Evans and Douglas I. Benn

2.1 RATIONALE

The foundation for any detailed investigation of glacial sediments should be an accurate record of the range of sediment facies present at a locality. Usually, this involves recording the facies exposed in sections created by natural processes (e.g. landslide scars, river cuttings or coastal cliffs) or by human activities (e.g. quarries, ditches, road or railway excavations), although information can also be obtained by augering, coring or geophysical methods. In this chapter, we discuss methods of facies description and section logging, with emphasis on characteristics observable in the field.

2.2 FIELD IDENTIFICATION OF FACIES AND FACIES ASSOCIATIONS

2.2.1 Introduction

A sediment *facies* can be defined as a distinctive body of sediment that forms under certain conditions of sedimentation (*cf.* Reading 1986). In other words, a sediment facies exhibits a set of physical characteristics which, collectively, reflect the processes of its formation. For example, a *turbidite* has a distinctive assemblage of bedding structures and grain-size gradations which form during the passage of a turbid underflow (turbidity current), and a *glacitectonite* exhibits distinctive deformation structures recording the over-riding and disruption of pre-existing sediments by glacier ice. However, in most cases we do not *know* the origin of a sediment, but must infer it from the sediment characteristics and our knowledge of what they mean. Thus, the identification of a genetic facies depends on an interpretation, and as such is subject to revision as ideas and perspectives change. This problem is particularly acute for glacigenic sediments, many of which form in dangerous or inaccessible environments and are consequently incompletely understood. As a result, it is often preferable to identify facies purely in terms of their physical and chemical characteristics, without reference to any inferred genesis. To make this distinction clear, the term *lithofacies* is used to refer to a sediment body in purely descriptive and objective terms. Lithofacies can then be interpreted in terms of genetic processes, independently of the description and classification stage. By clearly separating description and interpretation, the lithofacies concept ensures a more objective approach less prone to bias or error, at least in theory.

Lithofacies are identified in outcrops or boreholes from characteristics observable in the field, although definitions can be refined following laboratory analyses. Important defining characteristics are:

1| *grain size*;

2| *depositional structures*;

3| *deformation structures*;

4| *inclusions*;

5| *fossils* and *trace fossils*;

6| *bed thickness*;

7| *bed geometry*;

8| *contacts between lithofacies.*

2.2.2 Grain size

Particle size distributions of sediment facies can be determined by laboratory analysis (Chapter 3). It is useful, however, to estimate modal grain size and sorting in the field using the definitions of grain-size categories shown inside the back cover. For the larger sizes, grain size and degree of sorting can be easily estimated or measured, but for sand-sized and smaller particles, estimation must be done with the aid of charts (inside front and back cover) and rules of thumb. Silt can be distinguished from sand because wet silt can be rolled into a thin sausage shape, whereas sand cannot. Silt feels gritty when gently bitten between the teeth, whereas clay feels smooth. *Diamictons* are poorly sorted sediments, usually with a wide range of grain sizes. The term is often used somewhat loosely, although precise definitions and subdivisions can be adopted if required (e.g. Hambrey 1994). The general subdivisions normally employed are *stratified* and *massive* diamictons, whereby a stratified diamicton is one that displays some stratification at outcrop scale but is still poorly sorted (Figs. G19–23, 25 and 36). (Please note: Figures prefixed with a 'G' refer to the glossary section.)

Grading refers to systematic changes in grain size vertically through a bed (Fig. 2.1). Upward fining, resulting from the varying settling velocities of different grain sizes, is

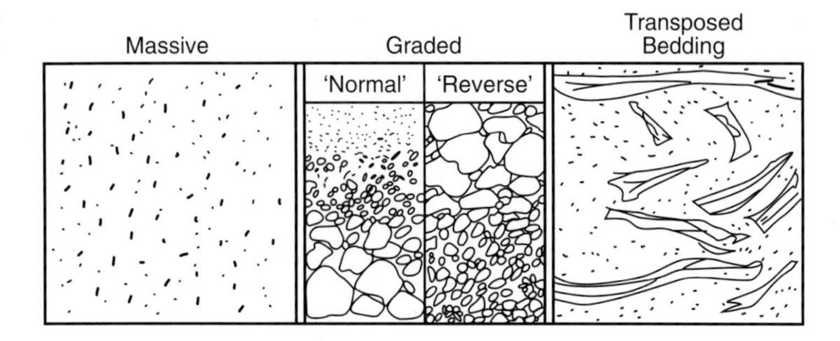

FIG 2.1 Massive and graded bedding (after Fritz and Moore 1988).

termed *normal grading* (Figs. G3 and 5), whereas upward coarsening, due to mechanical sorting in certain types of mass flow (Figs. G7, 10, and 19–23), is known as *inverse grading*. Sometimes grading affects only the coarsest grain sizes in a sediment, in which case it is known as *coarse-tail grading*. A common form of grading in diamictons is a downward increase in the concentration of coarse clasts, formed by sinking of clasts through weak matrix (Fig. G19, Lawson 1982).

2.2.3 Depositional structures

Many lithofacies exhibit internal bedding, consisting of *beds* (>1cm thick) or *laminae* (<1cm thick; Table 2.1). Internal bedding is highlighted by vertical changes in grain characteristics, such as grain size or mineralogy, and may be *bed-parallel* or *inclined*. Beds of similar character form *sets*, which can be further grouped into *cosets* on the basis of physical or genetic affinities. Fine-grained sediments exhibiting cyclic grain-size variations are referred to as *rhythmites* (Figs. G7, 11 and 12) and reflect variations in sediment supply and depositional conditions on a range of timescales (Smith and Ashley 1985; Benn and Evans 1998; Tiljander *et al.* 2001; Fig. 2.2). Rhythmites deposited by the settling out of turbid plumes in glacimarine environments display a vertical transition from silt to mud (*cyclopels*) or sand to mud (*cyclopsams*), thought to be controlled by tidal influences (Fig. G43). Bed-parallel lamination in sandy facies reflects deposition in thin,

Table 2.1 Terminology for describing laminae and internal bedding (Ingram 1954).

Term	Thickness (cm)
Thinly laminated and wisps	0.1-0.3
Thickly laminated	0.3-1.0
Very thinly bedded	1-3
Thinly bedded	3-10
Medium bedded	10-30
Thickly bedded	30-100
Very thickly bedded	100-1000

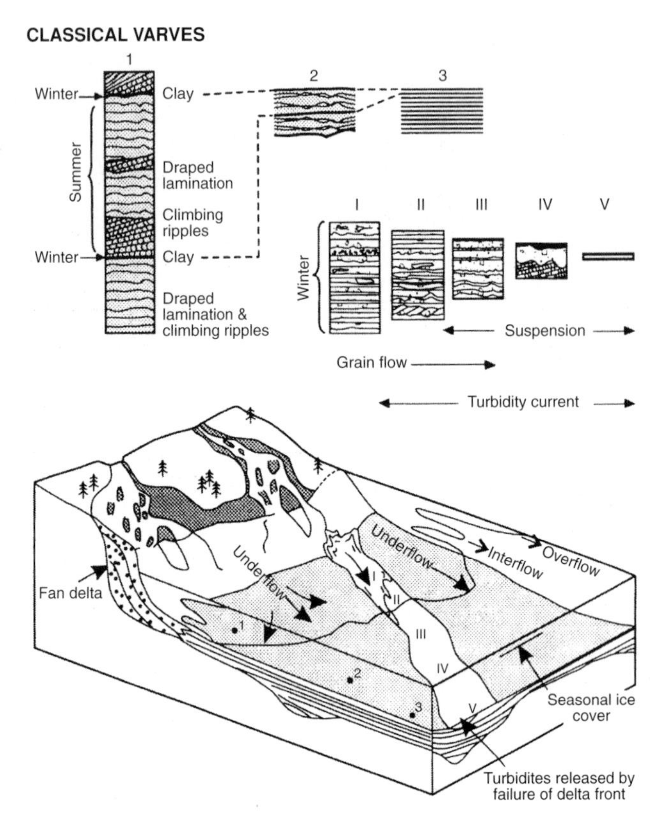

FIG 2.2 Rhythmite deposition in lake environments (modified from Eyles and Miall 1984).

horizontal sheets (*plane bed* forms) under either *lower* or *upper flow regime* conditions (i.e. slowly moving shallow water or rapidly flowing deeper water). Lamination is generally best developed in lower flow regime plane beds due to low flow velocities and efficient grain-size sorting; upper flow regime plane beds can be distinguished by the presence of thin, linear grooves and ridges on the bedding surfaces (parting lineations) parallel to the former flow direction. These are easily seen on bedding planes of lithified sandstones, but can be difficult to observe in unconsolidated sands.

Inclined internal bedding comprises various types of *cross-bedding* and *cross-lamination* (Fig. 2.3), which reflect the migration of bedforms under flowing water or wind. Bedforms can be subdivided into *microforms* (ripples) and *mesoforms* (dunes). The sequential deposition of ripples results in the production of *ripple cross-lamination* (*climbing ripples*) (Figs. G7, 8 and 11). Under unidirectional currents, facies characteristics are determined by the balance between downstream ripple migration and the settling of sediment from suspension (Fig. 2.4). Further variations in structural style are introduced by multi-directional or oscillatory currents, for example in tidal settings (Fig. 2.5). Larger scale cross-bedding structures exhibit a wide range of geometries, reflecting the form and evolution of different dune forms (Fig. 2.6). Sophisticated classifications of cross-bedding

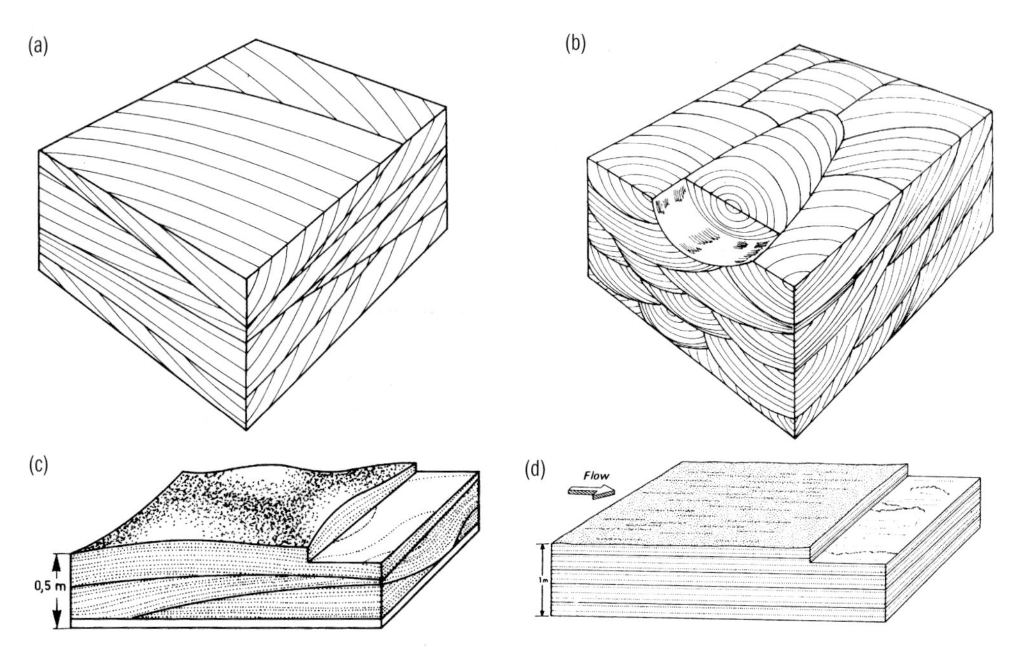

FIG 2.3 Sand and gravel cross-bedding with reactivation surfaces marked in bold: a) planar cross-bedding; b) trough cross-bedding; c) hummocky cross-stratification (HCS); d) plane beds (after Harms *et al.* 1982).

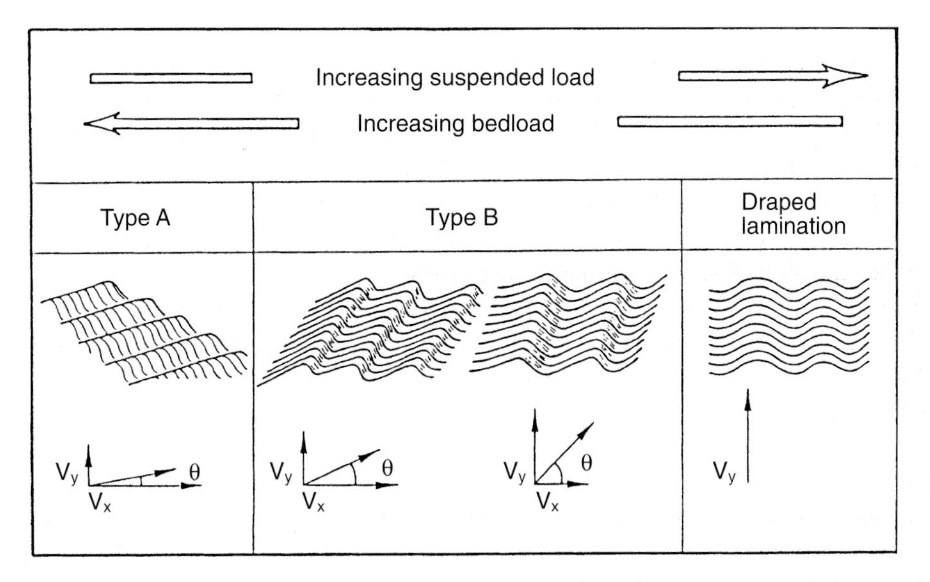

FIG 2.4 The production of ripple cross-lamination. Ripples climb at angle θ whose tangent is the mean aggradation rate V_y divided by the downstream migration rate V_x (from Ashley *et al.* 1982).

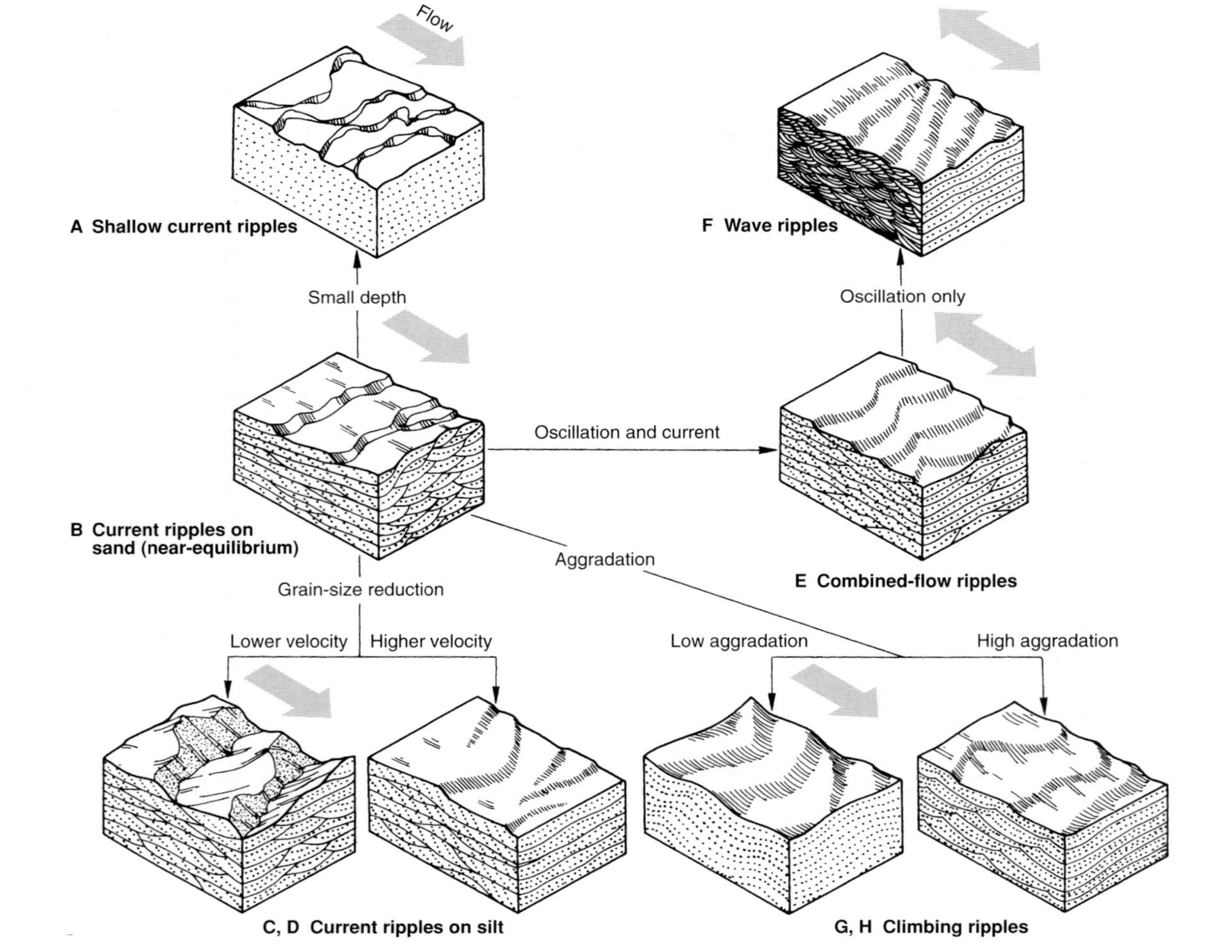

FIG 2.5 Continuum of ripple forms and stratification, organized according to changes in flow velocity, depth and direction, grain size and sediment supply (from Harms 1979).

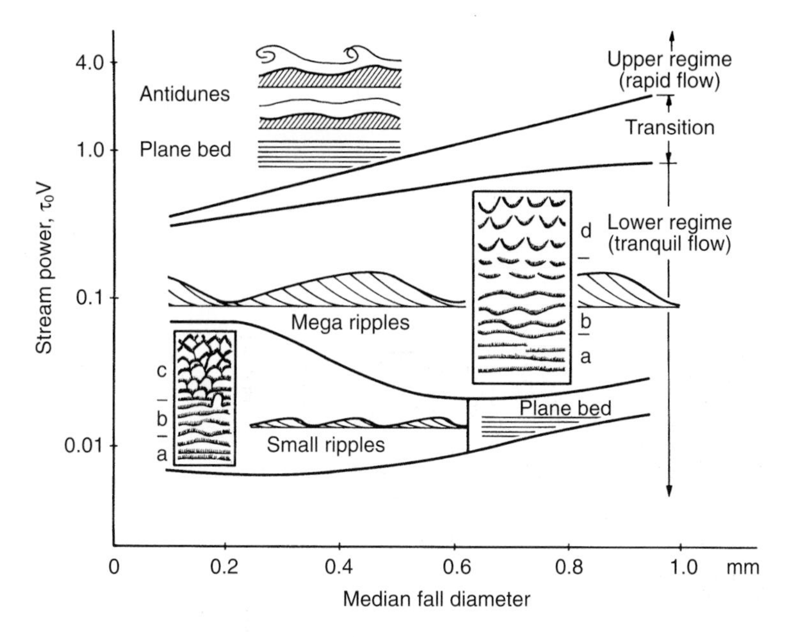

FIG 2.6 Phase diagram showing the relationship between bedforms and flow velocity or stream power and grain size or fall diameter. Ripple/dune crest morphologies are: a) straight-crested; b) undulatory; c) lingoid; and d) lunate.

types are available (e.g. Allen 1963), although for many purposes simpler classifications may suffice. Cross-bedded or cross-laminated facies often display abrupt changes in the pattern and attitude of the bedding; these changes represent *reactivation surfaces* that record switches between local erosion and deposition (Fig. G4). Laboratory experiments have demonstrated systematic relationships between flow conditions and sedimentary structures and grain size. These relationships are summarized in *phase diagrams* (Fig. 2.6), which provide a basis for reconstructing former hydraulic conditions from field data.

Large-scale *clinoforms* (Fig. G4–6 and 24) constitute an important category of bedding structure, consisting of sub-parallel sand or gravel beds dipping at angles of up to ~35° and representing the progradation of a delta front into a water body. They are commonly referred to by the genetic term *foresets*, although strictly this should only be done at the interpretation stage.

Bedding and lamination can also be developed to varying degrees in diamictons (Figs. G19, 21–23, 26, 27, 31, 32, 36, 37, 38). Crude bedding (picked out by subtle variations in grain size) is common in gravitationally reworked sediments (Fig. 2.7), and laminae of fine-grained material within diamictons can provide evidence for a water-lain origin. However, laminae in tills can also result from glacitectonic attenuation and repetition of beds, and great care must be taken to distinguish laminae of sedimentary and tectonic origin (see following section). Many diamictons may appear structureless (massive) at first sight (Figs. G34, 40). On close inspection, however, most display a surprising number of distinctive characteristics, which can give important clues about their genesis. Particular points to note are:

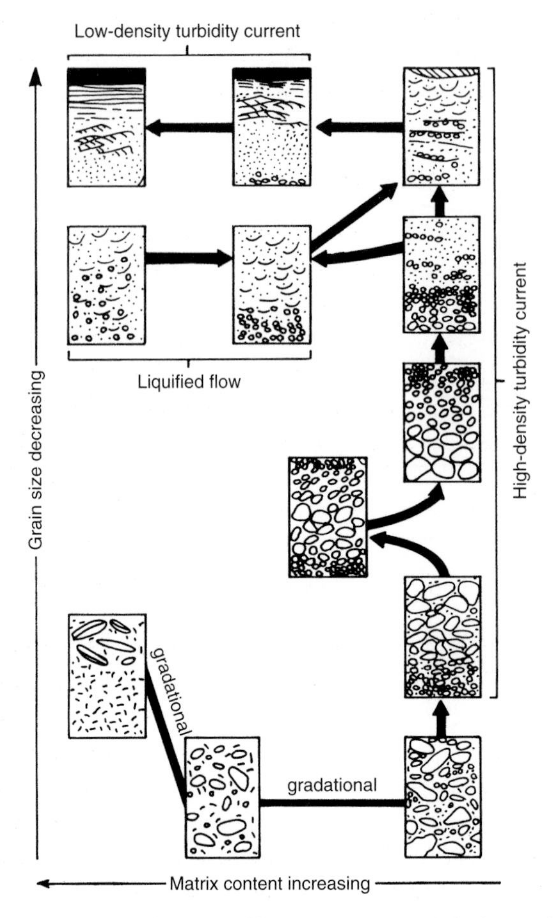

FIG 2.7 Various forms of mass flow deposits (after Lowe 1982). Arrows represent evolution into another type.

- whether the sediment framework is clast- or matrix-supported (i.e. do clasts rest on other clasts or are they enveloped in matrix?);
- variations in clast size and matrix grain size (Figs. G19 and 23);
- degree of consolidation;
- the presence and character of inclusions (e.g. rafts, pods, wisps or stringers; Figs. G25, 27, 28 and 36);
- the occurrence of clast/boulder lags that document the removal of fine matrix (Fig. G18);
- deformation structures and fissility (Figs. G25–28, 30, 34 and 35).

In most cases, however, classification and genetic interpretation of diamictons requires additional detailed information on fabric, clast shape, micromorphology and engineering properties, as described in later chapters of this book.

2.2.4 Deformation structures

Deformation structures provide important information on former environments, and should be recorded in detail on sediment logs. Deformation can take place either during deposition (*syn-depositional*) or after deposition (*post-depositional*), and the timing of deformation can be determined by the extent to which deformation structures cut across or are truncated by contacts between facies. *Water-escape structures* result from the loss of pore-fluid from sediments (*dewatering*) during consolidation (Fig. G11). The amount of disturbance can vary widely, from the *liquefaction* or *fluidization* of whole units (producing *convoluted beds, load-, flame-* and *ball and pillow structures;* Figs. G14, 16 and 30) to more localized deformation (e.g. *clastic dykes, hydrofracture fills, injection features* and *pipes*; Figs. 2.8 and G29 and 39). Loading structures resulting from density instabilities can also result from seasonal or longer-term thawing of frozen ground (*cryoturbation structures*; Fig. 2.9; Vandenberghe 1988).

Syn- and post-depositional deformation of sediments is very common in glacial environments, and can result from the removal of ice support during deglaciation, proglacial or subglacial glacitectonics, disturbance by drifting icebergs and failure induced by high pore-water pressures during or following deposition. It is useful to distinguish *compressional* and *extensional* deformation structures. Compressional structures (e.g. thrust faults and many types of folds; Fig. G30) record horizontal shortening of sediment piles by glacitectonic deformation or at the base of slope failures (Fig. 2.10). Extensional structures such as normal faults (Fig. G13) and sag folds are associated with foundering of sediments following melt out of supporting ice. Subglacial glacitectonic deformation can be either compressional or extensional (Hart *et al.* 1990). Patterns of subglacial deformation may be difficult to distinguish in massive diamictons (e.g. *deformation tills*) but are picked out clearly where pre-existing sedimentary structures provide recognizable strain markers (*glacitectonites*). Glacitectonized sediments commonly display highly variable degrees of deformation. Fine-grained sediments generally exhibit the greatest deformation, whereas coarser (sand and gravel) sediments tend to be stiffer, and may form streamlined inclusions (boudins or augen) within a more highly deformed matrix (Figs. G25–28). Some glacitectonites are strikingly laminated as a result of the attenuation, folding and repetition of pre-existing sediment facies (Figs. G31, 32, 37 and 38), and may be mistaken for waterlain sediments (e.g. Eyles 1993). In particular, tectonic laminae wrapped around clasts may be mistaken for dropstone structures, but can be distinguished by the presence of 'pressure shadow' fold patterns (Fig. G27) and the characteristics of the laminae (Hart and Roberts 1994). Structures associated with *dropstones* and *iceberg dump* and *grounding structures* are shown in Figs 2.11 and G9 and 10).

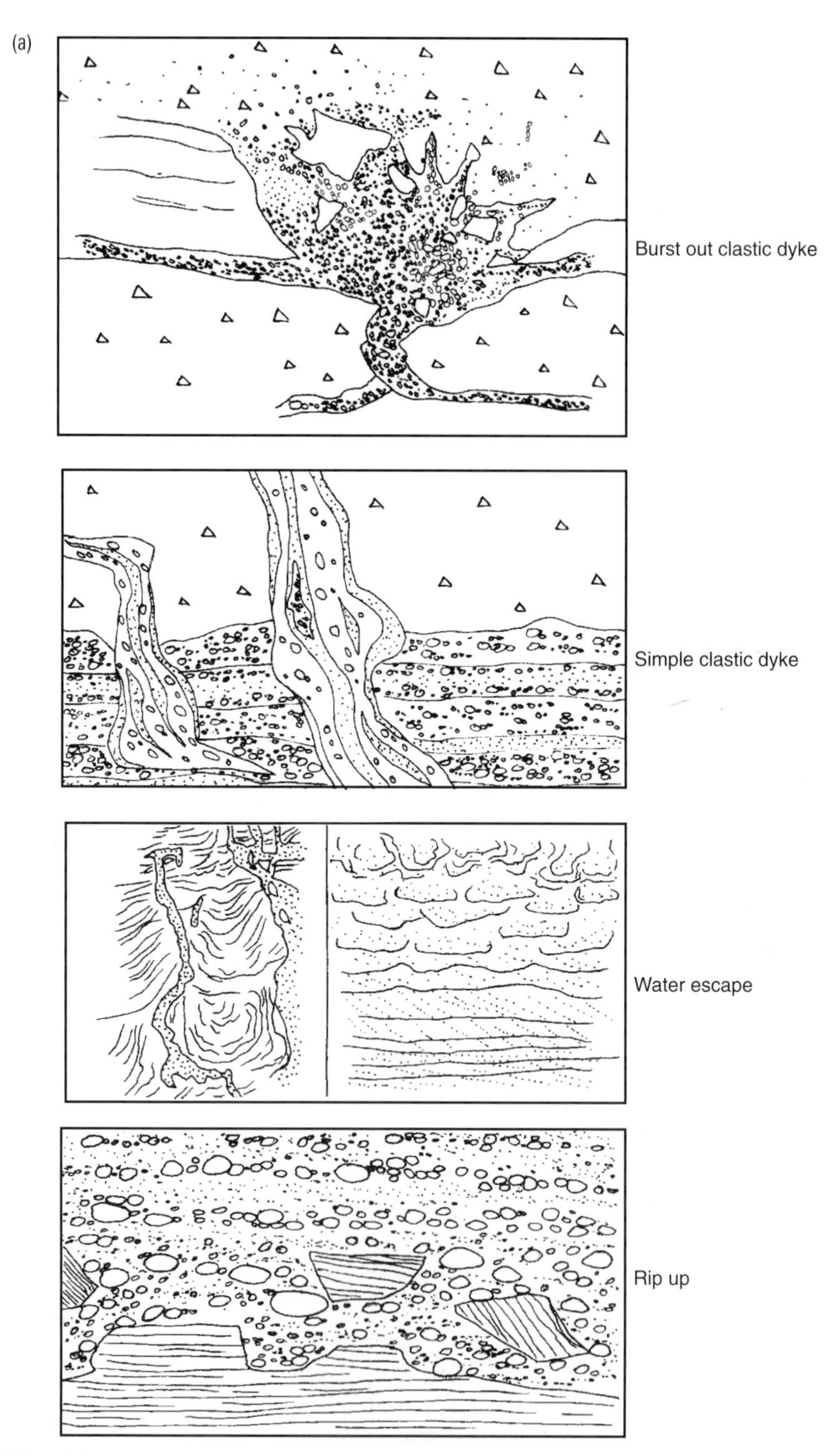

FIG 2.8 (a) Soft sediment deformation structures.

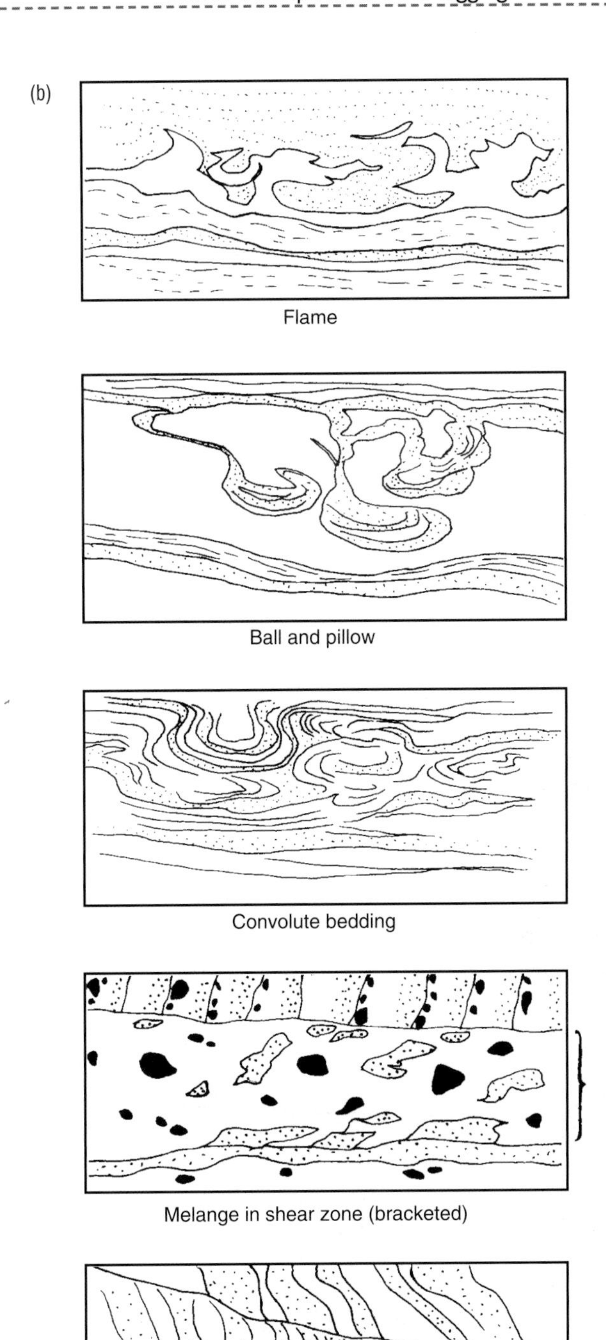

FIG 2.8 (b) Soft sediment deformation structures.

2.2.5 Inclusions, fossils and trace fossils

Glacigenic sediments may contain *inclusions* or *intraclasts* of pre-existing facies (Figs. G4 and 15). Inclusions in basal tills range widely in size from centimetre-scale rafts to kilometre-scale megablocks, and record the rip-up and mobilization of part of the glacier bed, below either ice or deforming till. Mass-flow deposits can also contain inclusions, such as fine-grained *rip-up clasts* derived from cohesive bed or bank materials (Fig. 2.8). Intraclasts of non-cohesive materials such as sand may indicate that the inclusion was frozen during transport and deposition.

Fossils (both macrofossils and microfossils) can provide valuable information on former depositional environments, climate, salinity and other factors. The classification and interpretation of fossils is too wide a subject to review here, and the reader is referred to specialist texts (e.g. Lowe and Walker 1997 and references therein). When fossils are encountered in the field, it is important to note whether they are in *life position* (i.e. the organisms are found where they lived and died) or are *derived* (i.e. underwent some degree of transport after death). Finally, a diverse range of *trace fossils* (burrows, trails, etc.) are produced by organisms living in glacial depositional environments, particularly marine settings. An excellent review is available in Reineck and Singh (1980) and a summary of the main types of trace fossil is reproduced in Figure 2.12.

2.2.6 Bed thickness and geometry

The thickness and geometry (shape) of sediment units reflect the processes and environment of deposition and should be recorded in detail (Fig. 2.13). *Tabular beds* are

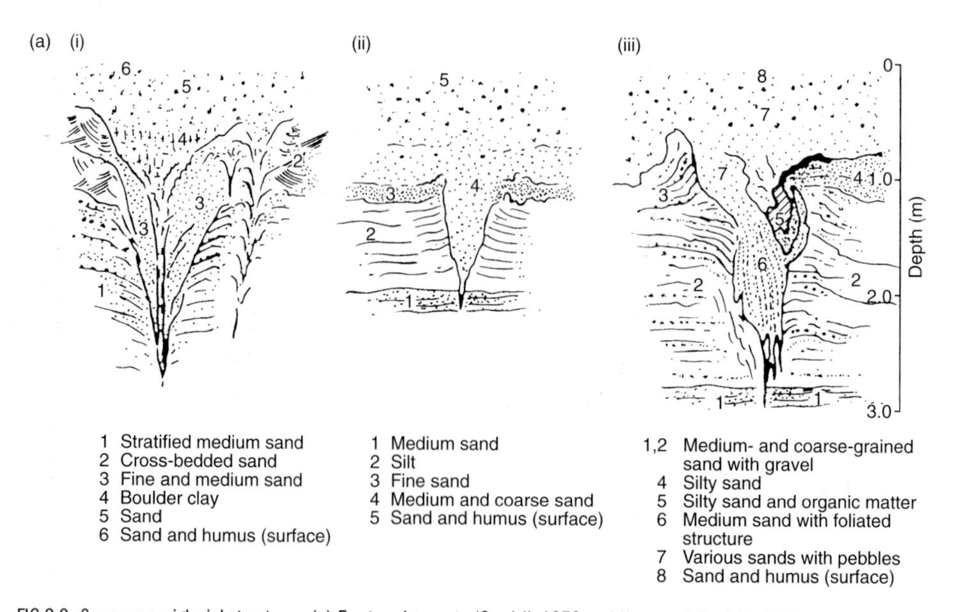

FIG 2.9 Common periglacial structures. (a) Frost wedge casts (Gozdzik 1973 and Harry and Gozdzik 1988).

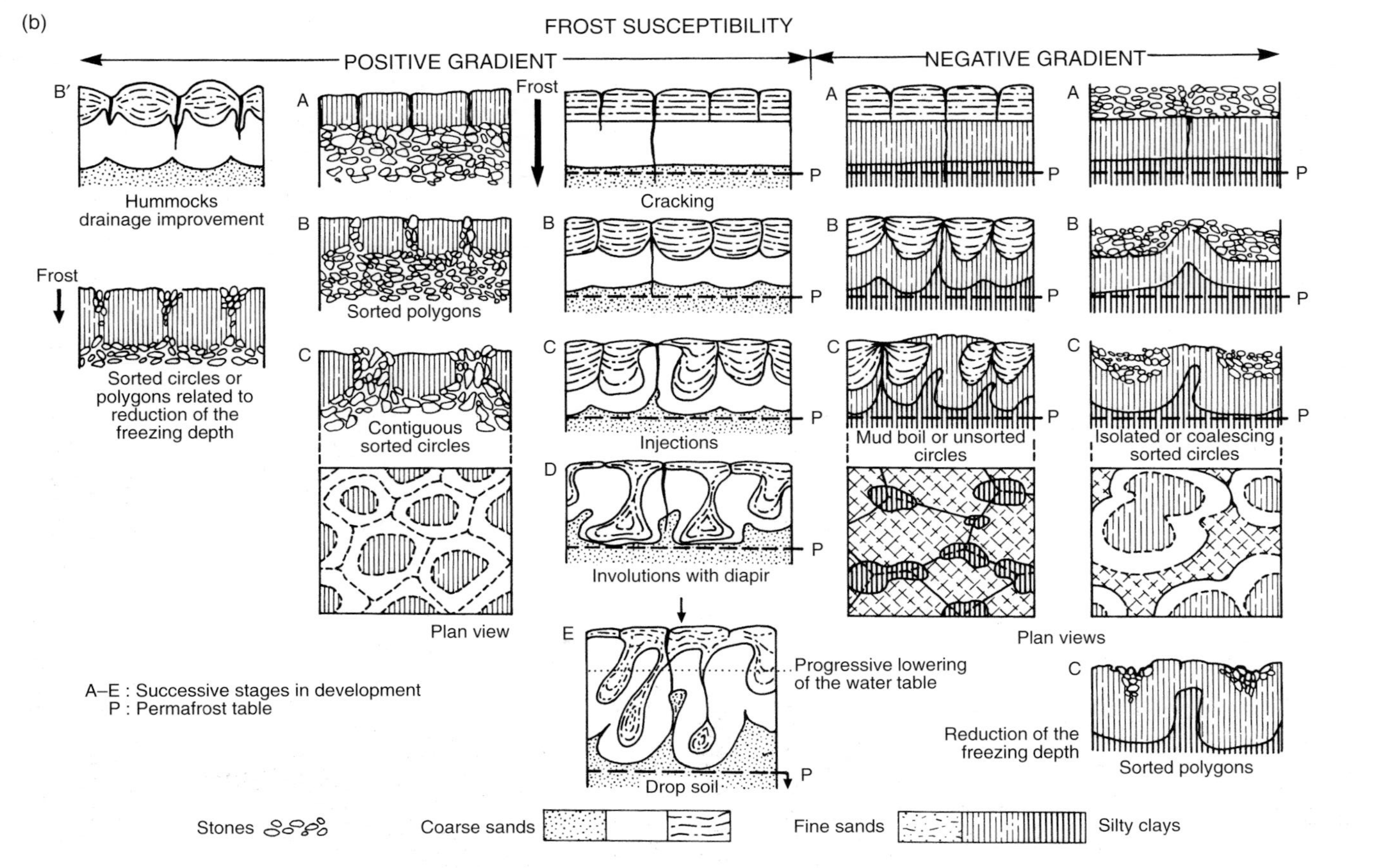

FIG 2.9 (b) Evolutionary stages in patterned ground and resulting structures in relation to frost susceptibility, differential frost heave, thermal gradient and drainage (after van Vliet-Lanoë 1988).

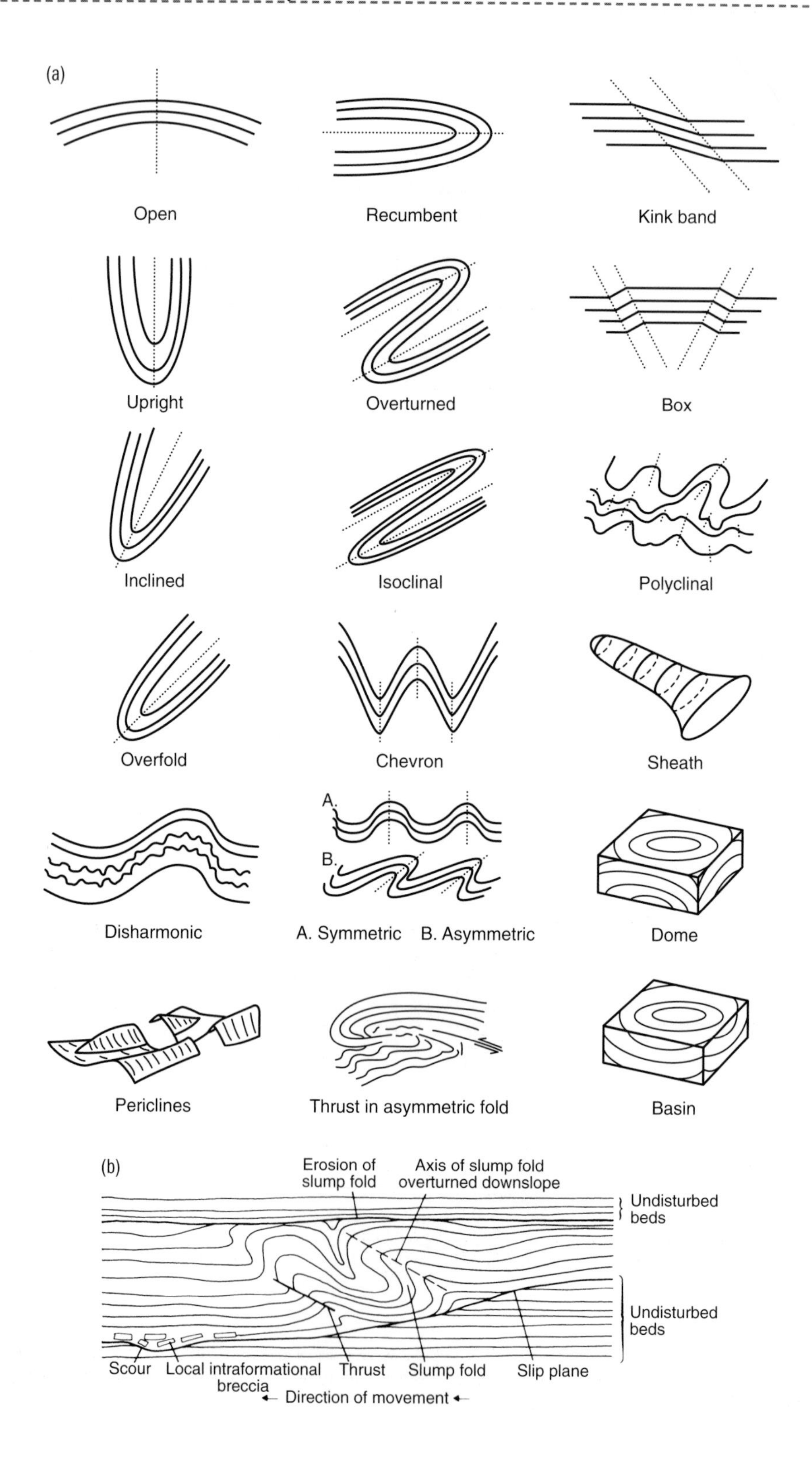
(a)
Open
Recumbent
Kink band
Upright
Overturned
Box
Inclined
Isoclinal
Polyclinal
Overfold
Chevron
Sheath
A.
B.
Disharmonic
A. Symmetric B. Asymmetric
Dome
Periclines
Thrust in asymmetric fold
Basin
(b)
Erosion of slump fold
Axis of slump fold overturned downslope
Undisturbed beds
Undisturbed beds
Scour
Local intraformational breccia
Thrust
Slump fold
Slip plane
Direction of movement

(c)

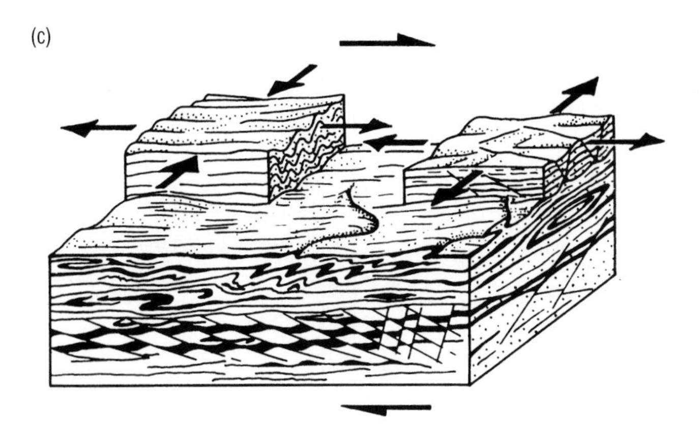

(d)

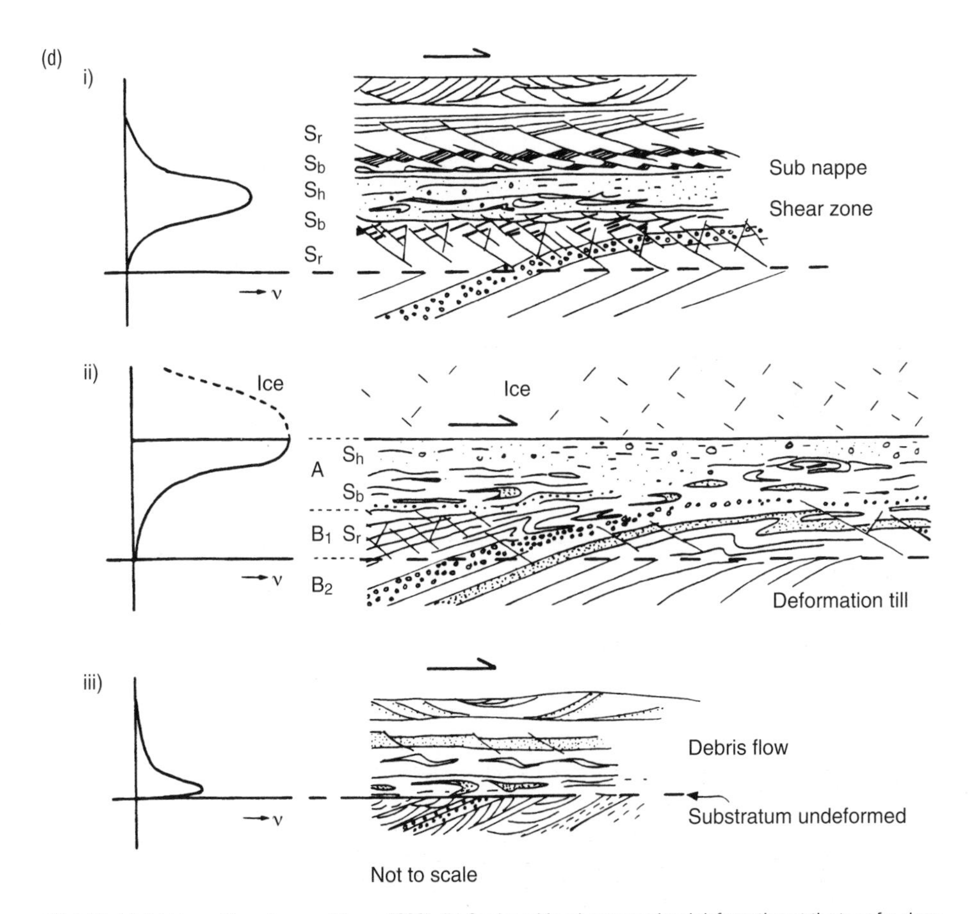

FIG 2.10 (a) Fold types (from Benn and Evans 1998); (b) Syndepositional compressional deformation at the toe of a slope failure (Tucker 1982); (c) Typical shear zone structures; d) Details of shear zones in thrust moraines [i] subglacial sediments [ii] and debris flows [iii] (van der Wateren 1995).

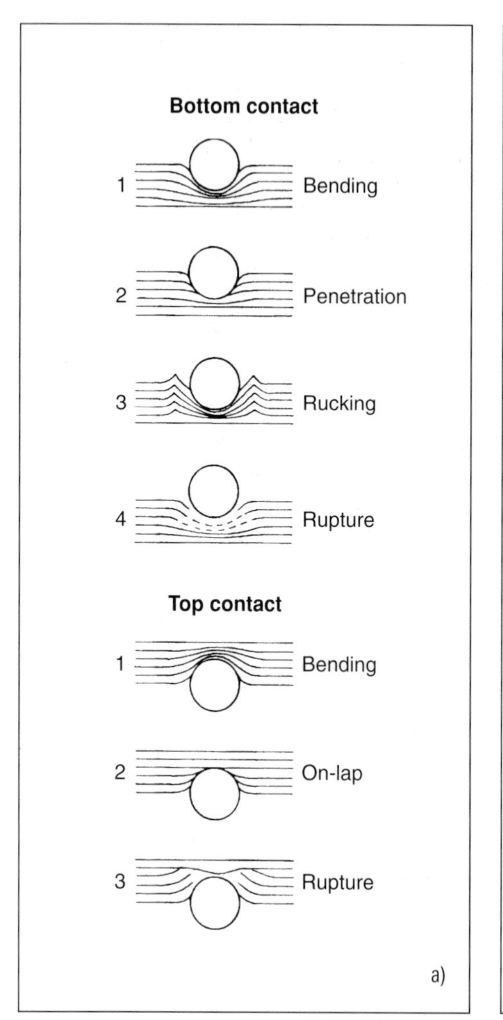

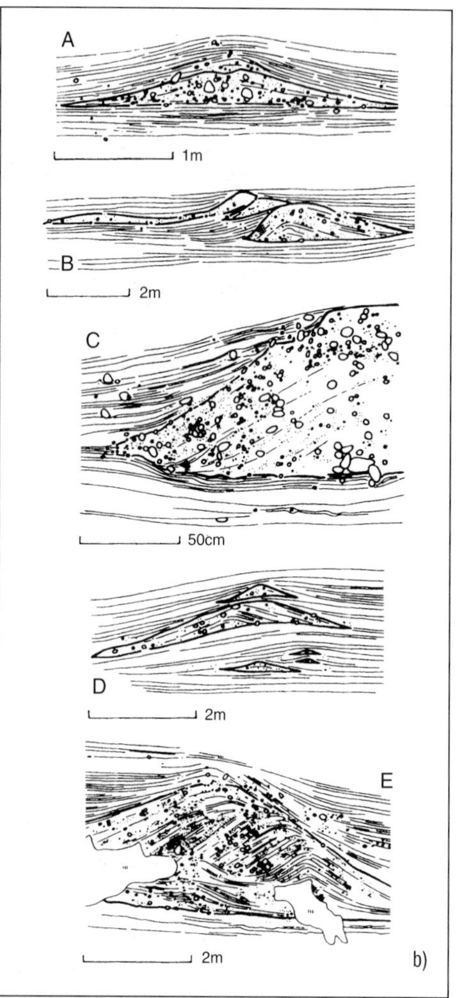

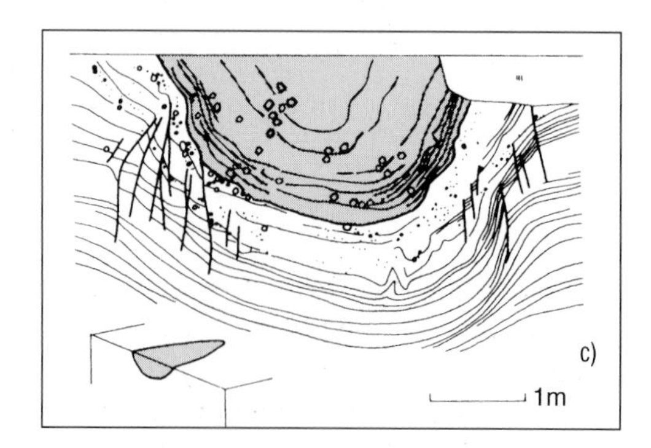

FIG 2.11 a) Dropstone relationships with associated bedding; b) Iceberg dump and grounding structures including A – single symmetric, B – stacked asymmetric/laterally overlapping, C – margin details, D – vertically stacked, symmetric, E – compound, stacked asymmetric; c) iceberg grounding structure (from Thomas and Connell 1985).

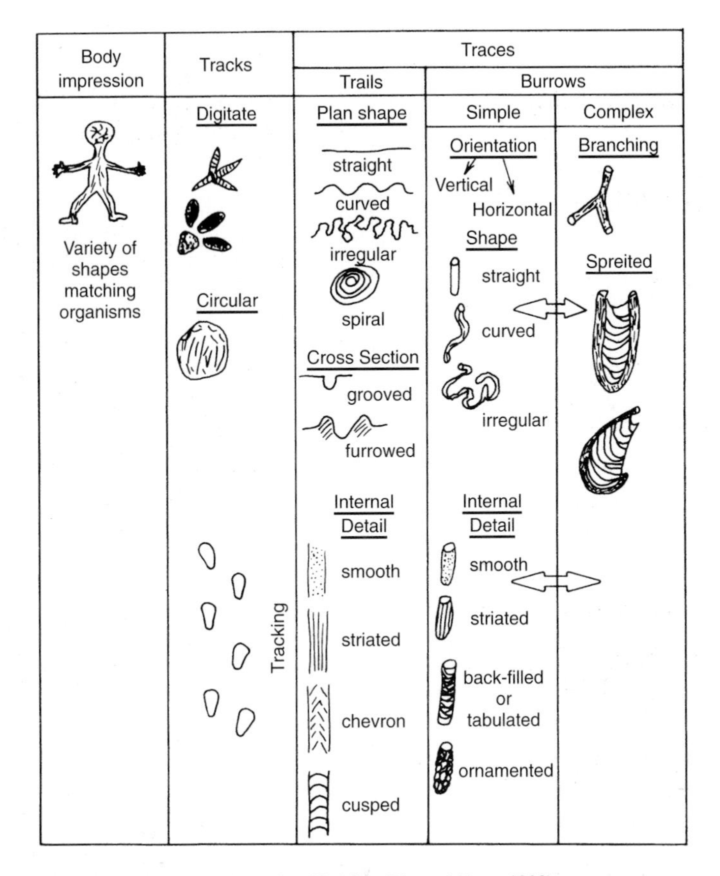

FIG 2.12 Summary of the main types of trace fossil (modified from Fritz and Moore 1988).

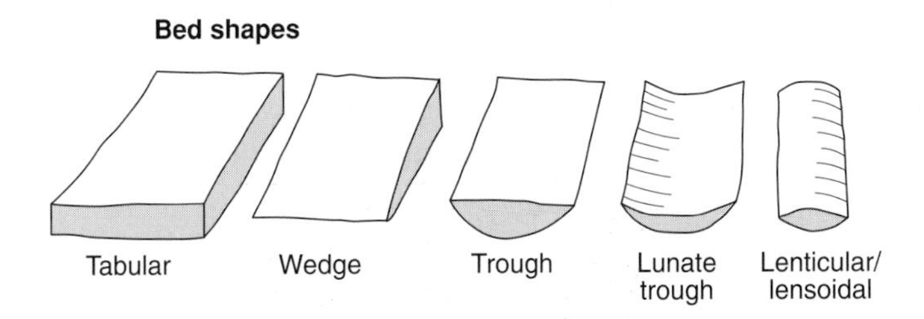

FIG 2.13 Bed geometry.

laterally extensive with parallel upper and lower surfaces. Examples include gravel or sand sheets and till beds. *Lenticular* (lens-shaped) beds can take a variety of forms. Units with concave-up bases and horizontal upper surfaces are likely to represent channel fills (Figs. G1 and 25), whereas units with convex-up upper surfaces and horizontal bases may be isolated bedforms (Reineck and Singh 1980) or streamlined inclusions in till (e.g. Benn and Evans 1996). *Irregular* sediment bodies may indicate post-depositional disturbance and/or partial reworking.

2.2.7 Contacts between lithofacies

In many ways the contacts between lithofacies contain as much information as the sediments themselves (Fig. 2.14). *Gradational* and *conformable* contacts indicate that deposition was continuous or uninterrupted by erosional episodes. *Erosional* (or *unconformable*) contacts can exhibit a range of geometries reflecting the shape of former erosion surfaces. For example, concave-up erosion surfaces record scour or channel forms (*scour and fill*; Figs. G1 and 2), and can be used to reconstruct channel dimensions, geometry and patterns of migration (e.g. Bristow 1996). Some sedimentary units can be separated by clast or boulder lags (palimpsest lags in glacimarine environments) that document periods of erosion of finer sediment (Figs. G17 and 18). Finally, *deformed (loaded)* or *faulted* contacts indicate episodes of post depositional deformation.

2.2.8 Facies codes

Lithofacies codes have been devised to enable rapid and effective sediment descriptions in the field (e.g. Miall 1977; Martin 1980; Eyles *et al.* 1983; Ghibaudo 1992; Benn and Evans 1998; Kruger and Kjaer 1999). Such codes consist of a series of letters that convey grain size characteristics (D = diamicton; B = boulders; G = gravels; GR = granules; S = sands; F = fines or silts and clays) followed by lower case letters denoting internal structures (Fig. 2.15). Lithofacies codes and associated symbols are particularly effective when annotating section drawings and logs (see below). Lithofacies codes have evolved to the point where they can convey considerable detailed information about specific sedimentary sequences. However, on their own they may not incorporate enough information to form the basis of detailed interpretations, and should be used as part of a multi-faceted approach.

2.3 FACIES ASSOCIATIONS

Facies may be grouped into *facies associations* (FA) based upon inferred genetic relationships. Similarly, lithofacies are grouped into *lithofacies associations* (LFA) on the basis of physical similarities and the nature of the contacts between individual lithofacies. The underlying principle behind facies associations is expressed by *Walther's Law of Facies* (Walther 1894; Reading 1986), which states that sediments found in an uninterrupted vertical sequence record *lateral* shifts of depositional processes that once operated side by side. For example, in a gravel-bed river, bars can migrate downstream and bury channel deposits, forming a vertical succession of trough-bedded gravels overlain by tabular gravel sheets. In some cases, the boundaries between lithofacies associations are easily defined on the basis of clear erosional unconformities or sharp changes in sediment properties, but in other cases their definition is more arbitrary, and lithofacies associations can be lumped together or split according to the purpose of the investigation.

The sedimentary *architecture*, or large scale geometry of lithofacies and LFAs, provides important information on the overall shape of the original depo-centre or depositional landform, in addition to the spatial relationships between lithofacies. Lithofacies

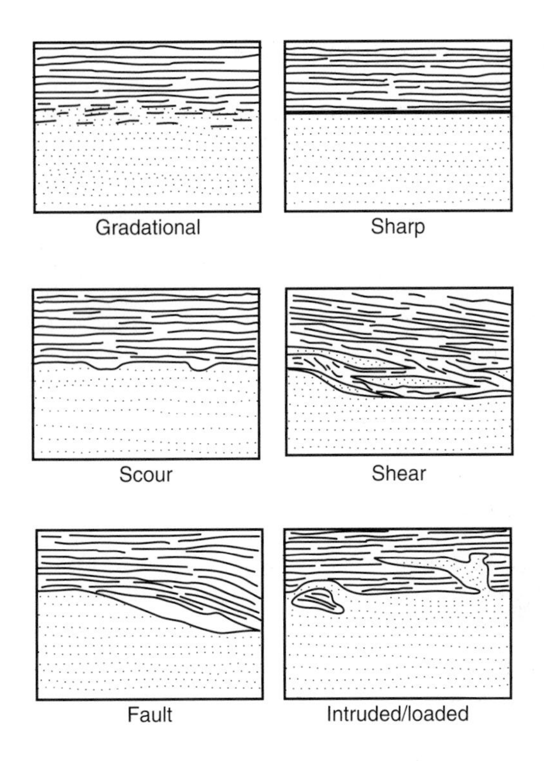

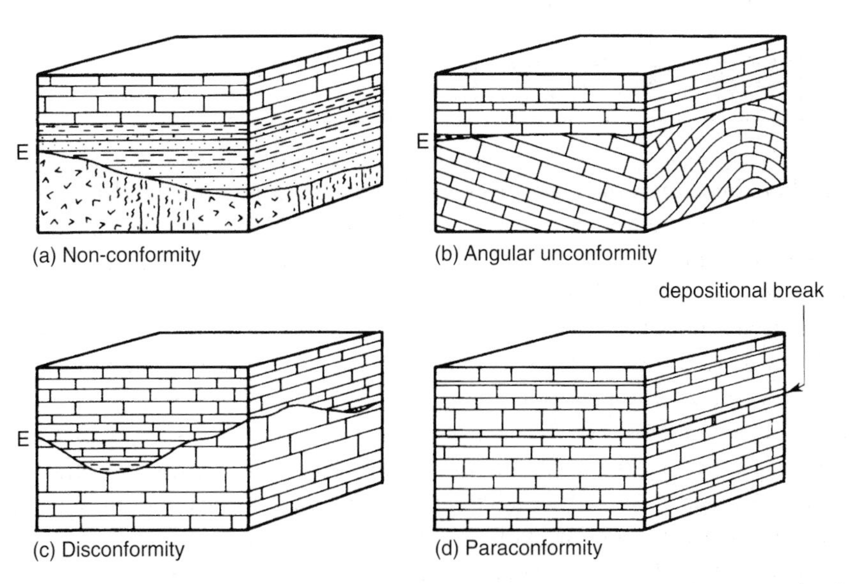

FIG 2.14 Various types of contacts between sedimentary units, including four types of unconformity. E = erosion surface (in part after Dunbar and Rodgers 1957).

Code	Description
Diamictons	*Very poorly sorted admixture of wide grain-size range*
Dmm	Matrix-supported, massive
Dcm	Clast-supported, massive
Dcs	Clast-supported, stratified
Dms	Matrix-supported, massive
Dml	Matrix-supported, laminated
- - - (c)	Evidence of current reworking
- - - (r)	Evidence of re-sedimentation
- - - (s)	Sheared
- - - (p)	Includes clast pavement(s)
Boulders	*Particles > 256mm (b-axis)*
Bms	Matrix-supported, massive
Bmg	Matrix-supported, graded
Bcm	Clast-supported, massive
Bcg	Clast-supported, graded
Bfo	Deltaic foresets
BL	Boulder lag or pavement
Gravels	*Particles of 8–256mm*
Gms	Matrix-supported, massive
Gm	Clast-supported, massive
Gsi	Matrix-supported, imbricated
Gmi	Clast-supported, massive (imbricated)
Gfo	Deltaic foresets
Gh	Horizontally bedded
Gt	Trough cross-bedded
Gp	Planar cross-bedded
Gfu	Upward-fining (normal grading)
Gcu	Upward-coarsening (inverse grading)
Go	Openwork gravels
Gd	Deformed bedding
Glg	Palimpsest (marine) or bedload lag
Granules	*Particles of 2–8mm*
GRcl	Massive with clay laminae
GRch	Massive and infilling channels
GRh	Horizontally bedded
GRm	Massive and homogeneous
GRmb	Massive and pseudo-bedded
GRmc	Massive with isolated outsize clasts
GRmi	Massive with isolated, imbricated clasts
GRmp	Massive with pebble stringers
GRo	Open-work structure
GRruc	Repeating upward-coarsening cycles
GRruf	Repeating upward-fining cycles
GRt	Trough cross-bedded
GRcu	Upward coarsening
GRfu	Upward fining
GRp	Cross-bedded
GRfo	Deltaic foresets

Code	Description
Sands	*Particles of 0.063–2mm*
St	Medium to very coarse and trough cross-bedded
Sp	Medium to very coarse and planar cross-bedded
St (A)	Ripple cross-laminated (type A)
Sr (B)	Ripple cross-laminated (type B)
Sr (S)	Ripple cross-laminated (type S)
Scr	Climbing ripples
Ssr	Starved ripples
Sh	Very fine to very coarse and horizontally/plane bedded or low angle cross-lamination
Sl	horizontal and draped lamination
Sfo	Deltaic foresets
Sfl	Flasar bedded
Se	Erosional scours with intraclasts and crudely cross-bedded
Su	Fine to coarse with broad shallow scours and cross-stratification
Sm	Massive
Sc	Steeply dipping planar cross-bedding (non deltaic foresets)
Sd	Deformed bedding
Suc	Upward coarsening
Suf	Upward fining
Srg	Graded cross-laminations
SB	Bouma sequence
Scps	Cyclopsams
- - - (d)	With dropstones
- - - (w)	Wtih dewatering structures
Silts & clays	*Particles of <0.063mm*
Fl	Fine lamination often with minor fine sand and very small ripples
Flv	Fine lamination with rhythmites or varves
Fm	Massive
Frg	Graded and climbing ripple cross-laminations
Fcpl	Cyclopels
Fp	Intraclast or lens
- - - (d)	With dropstones
- - - (w)	Wtih dewatering structures

FIG 2.15 Lithofacies codes from Benn and Evans (1998), utilizing various earlier schemes especially those of Miall (1978), Eyles *et al.* (1983) and Maizels (1993).

associations with distinctive three-dimensional geometries are termed *architectural elements*. Miall (1985) proposed a comprehensive classification of fluvial architectural elements (see Section 2.6.1). Similarly, *sediment-landform associations* constitute facies associations that possess a surface expression as a landform that is related genetically to those facies (e.g. an esker or a delta; Walker 1992).

2.4 SECTION LOGS

Information on the type, distribution and configuration of sediment facies can be recorded in one, two or three dimensions. The choice of log type is partly constrained by the degree of exposure and the resources available, but should also be guided by the nature of the sediments and the aims of the investigation.

2.4.1 Section preparation

Sections in unconsolidated sediments are commonly mantled by a veneer of loose material, which must be removed to expose the undisturbed sediment beneath prior to beginning logging. Preliminary cleaning or excavation can be done with a spade or entrenching tool, but final cleaning is best achieved with a trowel or sharp knife. In some cases cleaning may actually obscure detail, such as where stony, laminated sediments have been etched out by differential weathering, so care must be taken not to 'over clean'. The cleaning process is valuable, as it encourages a detailed examination of all accessible parts of a section. The formulation of hypotheses – or at least critical thinking about the origin of the sediments – can usefully begin at this stage. In addition to examining the fine detail, it is important to obtain a good understanding of the overall character of the exposed sediments, including the full range of facies present and the nature of the contacts between them.

The extent of exposure varies considerably between areas. In many places (e.g. in moraines in upland Britain) sections are frustratingly rare, and samples may have to be obtained from small hand-made excavations. In other areas (such as badlands in semi-arid environments, or actively eroding coasts) there may be so much exposure that choosing representative areas to work on becomes a problem. Additionally, it may not be possible to access all parts of a section if exposures are more than 2–3m high. Those with rock climbing experience may consider abseiling as a means of gaining access to the upper parts of large sections, although the unconsolidated nature of most Quaternary sediments can make this a risky business. Whenever working on unconsolidated sediments, extreme care must be taken to avoid dislodging boulders onto oneself or others. Hard hats are strongly recommended, and may be required by landowners or university safety regulations.

2.4.2 Vertical profiles

Vertical profiles are one-dimensional logs of sediment successions, which can be used to convey the detail of particular locations or to summarize extensive cliff sections. In the latter case, vertical profiles may be composite logs built up from the best-exposed parts of several areas of section. Vertical profiles communicate a wide range of field information clearly and effectively, but are of limited usefulness where sediments are structurally complex, or exhibit rapid lateral changes. As a result, vertical profiles are most useful where sediments have relatively simple 'layer-cake' form (e.g. glacilacustrine or glacifluvial successions), and are least useful for complex ice-marginal or glacitectonized sediments.

Some of the conventions of vertical profiles are shown in Figure 2.16. Unit thickness is shown on the vertical scale and modal grain size is represented by column width. The horizontal scale bar contains gradations for clays, silts, sands, gravels and diamicton (C, Si, S, G, D); further gradations for fine sands, coarse sands and granules (fS, cS, Gr) can be added if needed. Details of sedimentary structures, lithology and bounding surfaces may be portrayed using various patterns or symbols. Those used in Figure 2.16 are conventional but by no means exhaustive, and alternative schemes have been devised by field researchers to suit the nature and scale of their study. Lithofacies codes can be shown

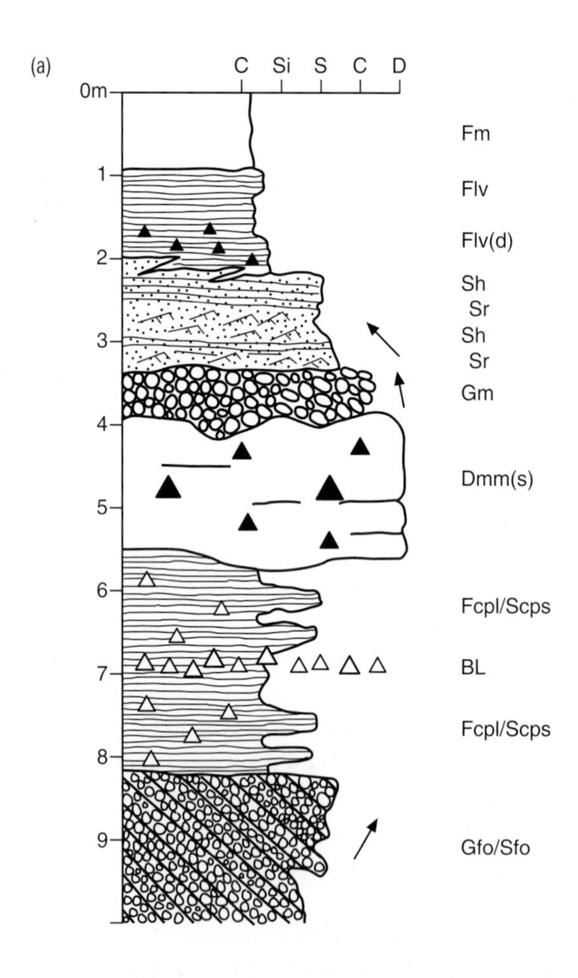

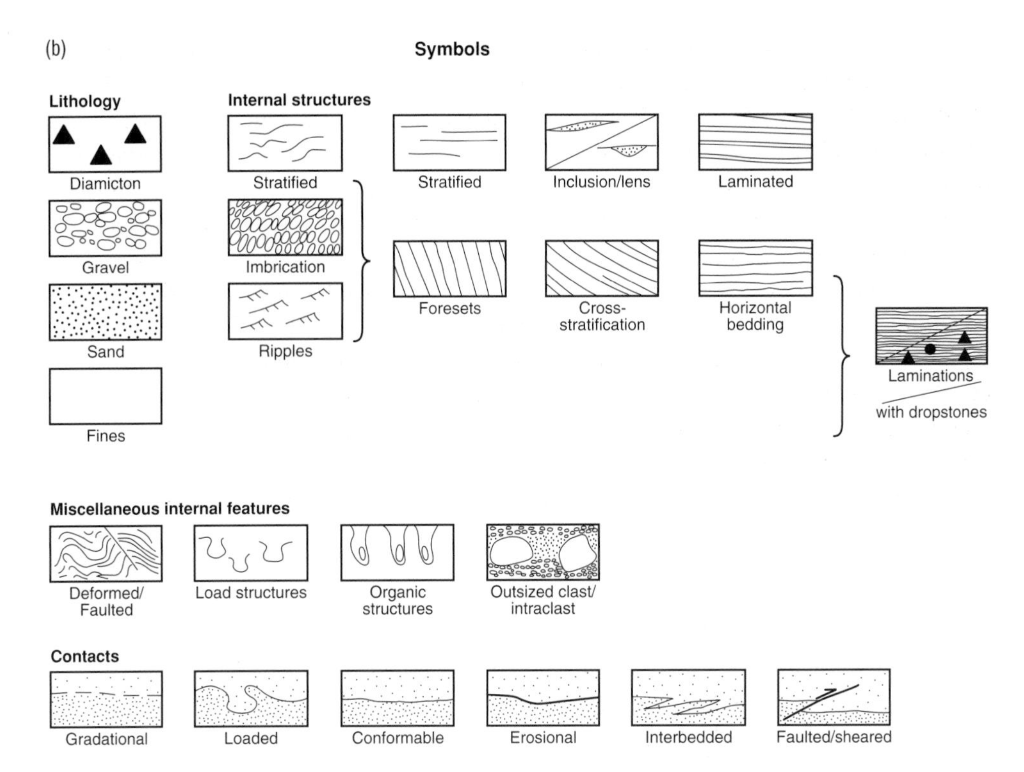

FIG 2.16 Conventional symbols for recording lithological information and sedimentary structures and bedding contacts in stratigraphic cross-sections and vertical profile logs.

alongside each facies, and additional information, such as clast lithology and fabric rose diagrams, can be added later (e.g. Benn and Evans 1993; Krüger and Kjaer 1999; Section 2.6.4).

2.4.3 Borehole logs

In the absence of surface outcrops, sedimentary sequences can be studied from *borehole* samples or *well logs*. Although the information on sedimentary structures and architecture is particularly limited, the lateral extent of buried lithofacies can often be determined from borehole data, using various physical characteristics. In borehole samples this includes geochemistry, microstructures, grain size and magnetic susceptibility (reviewed in later chapters of this guide). Well log information can be derived from various logging techniques that require specialist field teams and equipment and have been designed primarily for the gas and oil industry. However, regional correlations of lithofacies in Quaternary sedimentary sequences have been achieved through standard well logging techniques.

Hand augering is a quick and cheap method of retrieving subsurface samples where exposure is poor. Auger sampling works best for sands and silts, but may be impossible for coarse-grained or diamictic sediments. Under favourable circumstances, basic stratigraphies can be established by retrieving successive auger samples from measured depths. Sedimentary structures are destroyed in auger samples, but much useful lithological information is retained (e.g. grain size, sorting, clast lithology, heavy mineral composition).

2.4.4 Two-dimensional logging

Two-dimensional logs can be constructed from several vertical profiles, between which lines of lithostratigraphic correlation can be drawn and any lateral relationships between facies (e.g. interdigitation, pinch and swell, unconformities) can be recorded (Fig. 2.16). However, the best approach is to record detail in an accurate drawing. The ease with which high resolution images can be published in scientific journals means that many researchers now rely on photographs to convey 2-D information about sediment exposures. However, photographs rarely convey as much information as a good drawing, because much of the fine detail – easily observable by eye in the field – may be hard to distinguish in a photograph, particularly at publication scale and in monochrome. Shadows and loose surface material are often much more prominent in photographs than in the field, to the extent that the features of interest are lost or obscured. Additionally, the act of making a detailed section log is worthwhile in itself, as it forces the observer to scrutinize the section with great care, often revealing detail that may otherwise have been missed.

For these reasons, it is well worth developing the gentle art of section drawing. Although some degree of artistic ability is helpful, with practice everyone should be able to achieve a good level of competence. The keys to success are a patient, systematic approach and attention to detail, with the aim of accurately transferring the scale and proportions of the exposed sediment units to the page. The best way to achieve this is to work on graph paper, and to place conspicuous scale markers (e.g. ranging poles or survey staffs) against the section. A useful tip is to build prominent cairns at measured distances along the section, to act as horizontal scale markers. Then locate the important unit boundaries relative to the scale markers, using the old artists' trick of holding a pencil or thumb at arms length, and mark them on the drawing to scale. It is important to stand far enough back from the section to achieve an undistorted view. Once the main boundaries have been drawn, additional features can be added, making repeated visits to the face to check on details when necessary.

If it is possible to make a repeat visit to a section, a photograph mosaic of the exposure can be used as the basis for a section drawing. Unit boundaries and lithofacies details can be drawn on a clear acetate overlay. This saves a lot of time and frustration, results in planimetrically accurate logs and is thoroughly recommended. If a repeat visit is not possible, or if film processing or image printing facilities are unavailable, the best procedure is to take photographs and make a measured drawing. The photographs can be used later as the basis for accurate drawings, using the records made in the field. Although two-dimensional logging allows for an appreciation of large-scale architecture, it nearly always sacrifices some important sedimentological detail which needs to be communicated via vertical logs and/or annotated drawings of selected areas in detail.

2.4.5 Geophysical methods

Geophysical methods are increasingly being used to image subsurface sediments where exposure is poor. A comprehensive discussion of geophysical methods is beyond the scope of this book, and the reader is referred to Reynolds (1997) for details of particular techniques. *Seismic stratigraphy* is used routinely in oil and gas exploration, and has been applied with considerable success to imaging offshore sediment sequences (see Prothero and Schwab 1996 and Davies *et al.* 1997 for further details). This involves reflection seismology, whereby a loud bang is produced at the ground surface or underwater and then echoes from the subsurface are recorded. This produces a *seismic profile*, a surrogate for a stratigraphic cross section. The resulting stratigraphy is based upon seismic reflections, which are the product of abrupt changes in seismic velocity that occur at horizons of sharp contrast in density or acoustic impedance. Therefore, unit thicknesses on the profile represent two-way travel times, not actual bedding thickness. Most bedding produces a reflection providing that individual sedimentary units are sufficiently different in density (Fig. 2.17). In addition to bedding, structural features such as folds and faults are also usually clearly represented. However, there are many factors that may produce spurious reflectors and the interpretation of seismic profiles requires considerable experience.

Financial or logistical constraints generally preclude the use of seismic methods for non-commercial purposes. A cheaper and more portable alternative is ground-penetrating radar (GPR). This is similar in principle to seismic profiling, but is based on the transmission of electromagnetic waves and analysis of the return signal. GPR has been used for many years in studies of glaciers (e.g. Robin *et al.* 1969; Bogorodski *et al.* 1985), but so far has proved less useful in studies of glacigenic sediments. This is because large clasts act as prominent reflectors, which show up as arcs (parabolic reflectors) on GPR profiles. Bouldery sediments (i.e. many glacigenic diamicts) may appear as little more than a confusing mass of parabolic reflectors, obscuring other detail in the succession. Recently, however, good results have been obtained for glacifluvial and paraglacial fan sediments (e.g. Ekes and Hickin 2001; Cassidy *et al.* 2003), and it is possible that future developments may allow similar imaging of a wider range of glacigenic sediment-landform associations.

2.4.6 Three-dimensional reconstructions

Where more than one exposure is available and each is orientated in a different direction, as is often the case in quarries, a three-dimensional representation of lithofacies can be produced. Detailed borehole stratigraphies may also prove valuable in this respect. Three-dimensional data can be represented on *fence diagrams*, which clearly convey the large-scale geometry of facies, aiding the interpretation of landforms, depo-centres, former

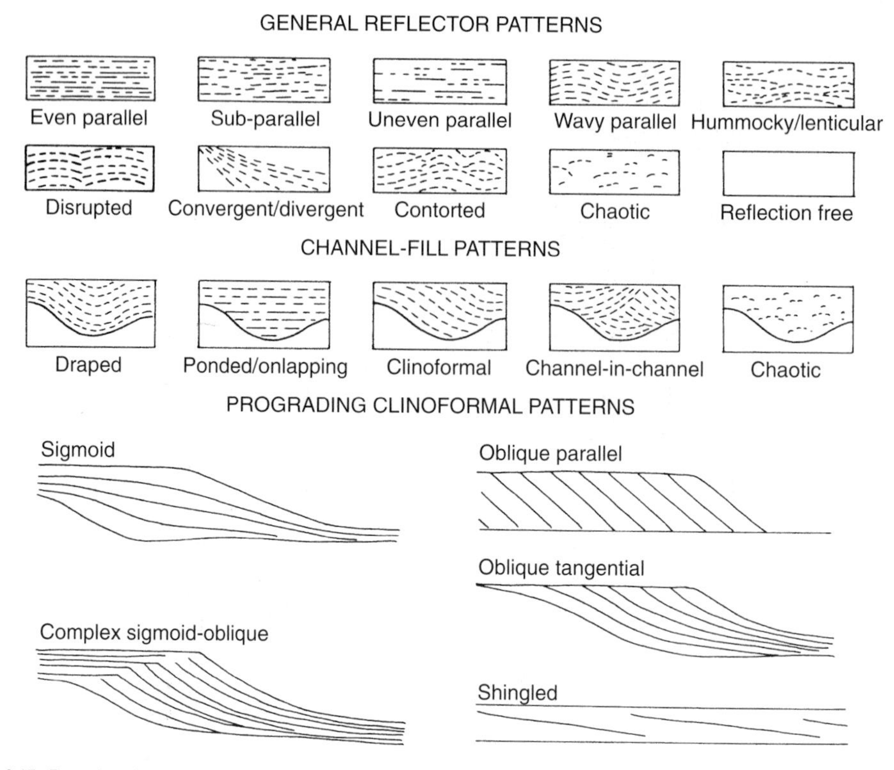

FIG 2.17 Examples of general reflector patterns used in the logging of seismic profiles (modified from Davies *et al.* 1997 and Harris *et al.* 1997).

glacier margins and changing sediment influx points (Fig. 2.18). Before reconstructions of this type are possible, it is necessary to accurately survey all lithofacies boundaries and thicknesses in order to facilitate lateral correlation of units. Given the difficulty of obtaining extensive stratigraphic information, some extrapolation is nearly always necessary when reconstructing fence diagrams.

2.5 INTERPRETATION

Interpretation of sediment sections usually focuses on reconstructing (1) depositional processes and environments, and/or (2) sequences of events (e.g. sea-level or climatic changes). When interpreting data, it is important to adopt a critical approach and to avoid merely constructing a narrative that 'explains' the observations. Wherever possible, interpretations should be arrived at by eliminating possibilities through rigorous testing of alternative hypotheses. This will usually involve the collection of additional quantitative data (fabrics, grain size, clast morphology, etc.) over and above field observations of general lithofacies characteristics (Chapter 9).

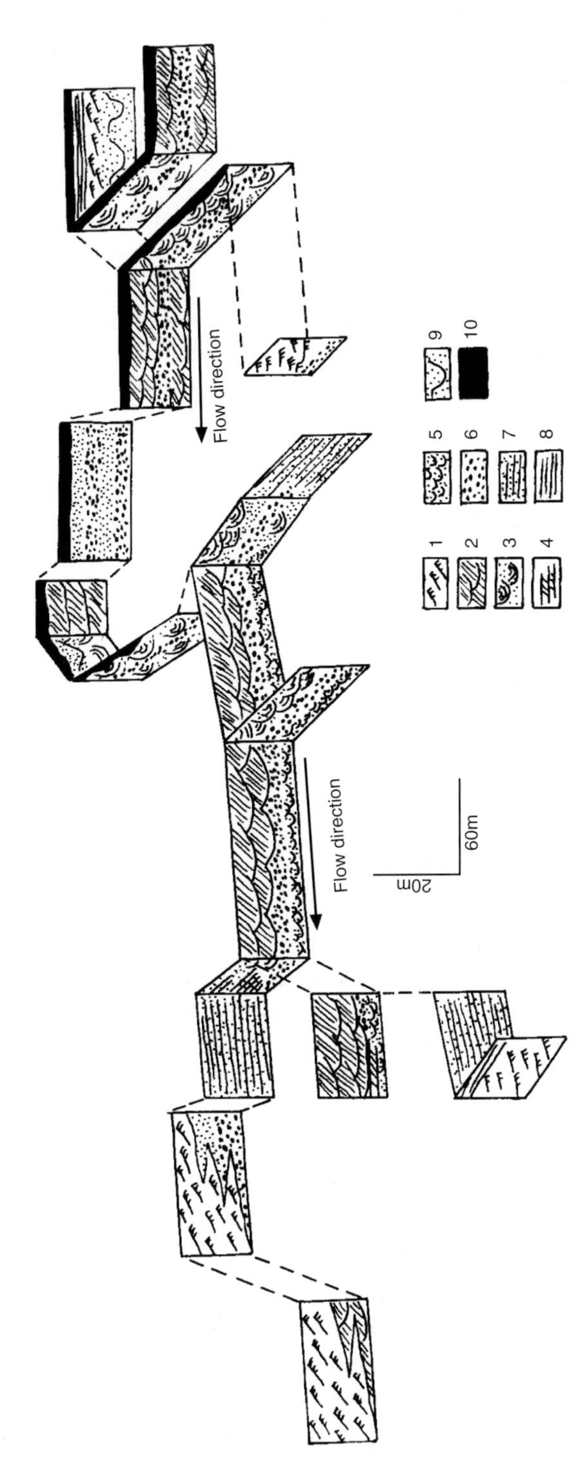

FIG 2.18 A fence diagram depicting the facies architecture of the Brampton esker (modified from Saunderson 1975). 1 = climbing ripples; 2 = trough cross-bedding in sand and gravel (longitudinal section); 3 = trough cross-bedding in sand and gravel (transverse section); 4 = tabular cross-bedded sand and gravel; 5 = large scale (deltaic) cross-bedded gravel; 6 = parallel-bedded and cross-bedded gravel; 7 = proximal rhythmites; 8 = distal rhythmites; 9 = massive sand with channel traces; 10 = till.

2.5.1 Processes and environments

Reconstruction of depositional processes and environments relies upon knowledge of the relationships between such processes and observable sediment characteristics. Sedimentologists commonly make use of *analogues* drawn from studies of modern environments or laboratory experiments (e.g. flume tanks). That is, sediments of known origin are used as benchmarks against which other lithofacies are compared. Although this approach cannot *prove* the origin of a particular facies, it can usually rule out many possibilities and identify the origin with a high degree of confidence. Interpretations based on analogues are strongest if (1) they are based on many observed characteristics and (2) all likely alternatives are ruled out. Researchers should be familiar with the range of modern glacial processes and environments, either through the literature or (preferably) through personal experience. Detailed reviews of sedimentary processes and their depositional products are available in a number of modern texts (e.g. Allen 1982, 1985; Leeder 1982; Fritz and Moore 1988; Prothero and Schwab 1996; Benn and Evans 1998; Evans 2003). Additionally, a wide range of facies models exist for glacial sediment sequences and these should be consulted when making interpretations of field data (e.g. Miall 1978; Powell 1984, 1990; Reading 1986; Levson and Rutter 1988; Eyles and Eyles 1992; Powell and Domack 1995; Benn and Evans 1998; Evans 2003).

2.5.2 Stratigraphic interpretation

Once lithofacies and lithofacies associations have been identified and interpreted, their relationships in vertical and lateral sequences can be used to reconstruct a sequence of deposition or a *stratigraphy*. Once a *lithostratigraphic unit* is identified (e.g. a till) it may be traced and often correlated over large distances. However, lithostratigraphic units are not always (and tills usually are not) *time-stratigraphic* units, but *time-transgressive* ones; they may have been produced by the same medium in response to the same climatic trigger, but they are unlikely to have been deposited at the same time everywhere. Although this is a complex subject, the details of which are outside the scope of this guide (see Hallam 1981; Fritz and Moore 1988; Doyle *et al.* 1994; Rose and Menzies 1996; Lowe and Walker 1997; Benn and Evans 1998 for further information), we provide information here on some of the more common stratigraphic units and relationships encountered in glaciated basins. All the international standard stratigraphic procedures and terminology are presented in the North American Stratigraphic Code produced by the North American Commission on Stratigraphic Nomenclature (1983) and Whittaker *et al.* (1991).

The establishment of a lithostratigraphy requires: (1) detailed descriptions of lithofacies; (2) the establishment of lithostratigraphic units (Table 2.2); and (3) an evaluation of their distribution in space (e.g. correlation). The primary and fundamental unit within the hierarchy is the *formation*, which is a mappable and easily recognized unit, based upon a type section and possessing a formal name that is non-genetic (e.g. Holderness Formation, eastern England and Scarborough Formation, Ontario, Canada). Members and beds are then subdivisions of formations, although the sub-division of sedimentary sequences can often be approached in different ways (Fig. 2.19). After the designation of units, correlation can then be made from one stratigraphic succession to another either by *tie lines* (lithological correlation) or *time lines* (correlation using time equivalent markers), noting that time lines can cross-cut tie lines.

Facies relationships in time and space provide us with models of evolving depositional environments. Relationships between lithofacies can be described qualitatively or defined quantitatively by determining the probability of transitions between particular facies (e.g. in Markov Chain Analysis), which then provide the basis for reconstructing changing environmental conditions. *Coarsening-up* or *fining-up* trends reflect either a shift in position of highest flow velocity (e.g. shifting bars on a river bed, change in efflux point on a delta), a change in proximity of the depositional site to the sediment source (e.g. glacier recession from a lake margin) or a drop in the power of the transporting medium (e.g. varves; Fig. G12). In some depositional environments (e.g. sandar) complex changes of this type are cyclical, associated with changing sedimentary structures and often incompletely recorded due to erosional episodes. In marine environments changing sea levels are recorded by *onlap* (*transgressive*) or *offlap* (*regressive* or *stationary*) sequences.

Sediments can also be directly employed in placing a chronological control on lithostratigraphy, especially when a relatively instantaneous depositional event (e.g. volcanic ash deposition) can be identified in the sediment sequence across a large area or region, thereby producing an *event stratigraphy*. Finally, glacitectonic disturbance in sedimentary sequences can be used to determine phases of disturbance by glacier ice, thereby producing a *kinetostratigraphy* or *tectono-stratigraphy* (Fig. 2.20).

2.6 CASE STUDIES

A variety of case studies using one or, more appropriately, a combination of the above techniques exist. We here reproduce summary data from some comprehensive assessments

Table 2.2 Heirachy of lithostratigraphic units (after Doyle et al. 1994; Rose and Menzies 1996).

Lithostratigraphic unit	Definition	Facies description
Supergroup	Clustering of groups based on their lithological characteristics or on their mode of formation.	
Group	Clustering of formations based on lithological characteristics or mode of formation.	Facies group
Formation	Mappable unit of homogeneous lithology.	Facies association
Member	Subdivision of a formation.	Facies
Bed	Lithologically distinct horizon or layer.	Facies or subfacies

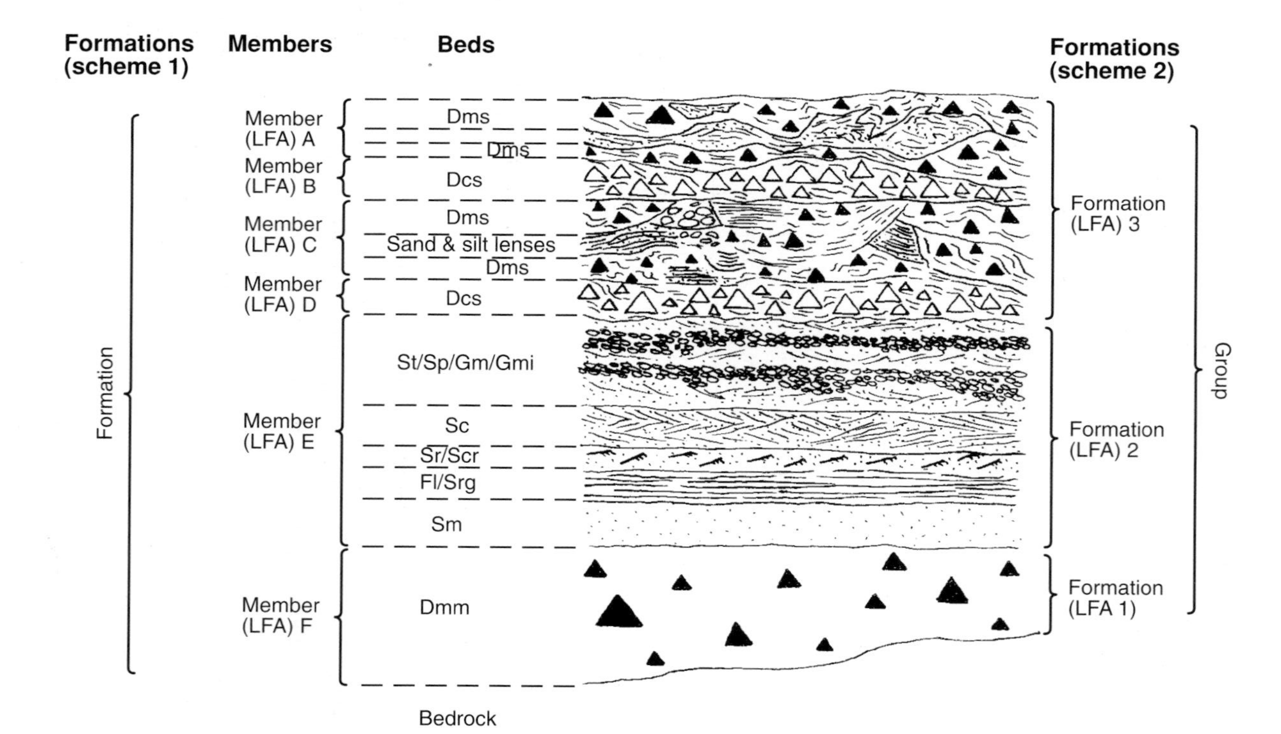

FIG 2.19 Lithostratigraphic subdivision of a glacigenic sedimentary sequence with alternative schemes (modified from Lowe and Walker 1997). Note that lithofacies associations could translate into either formations or some single members in this example.

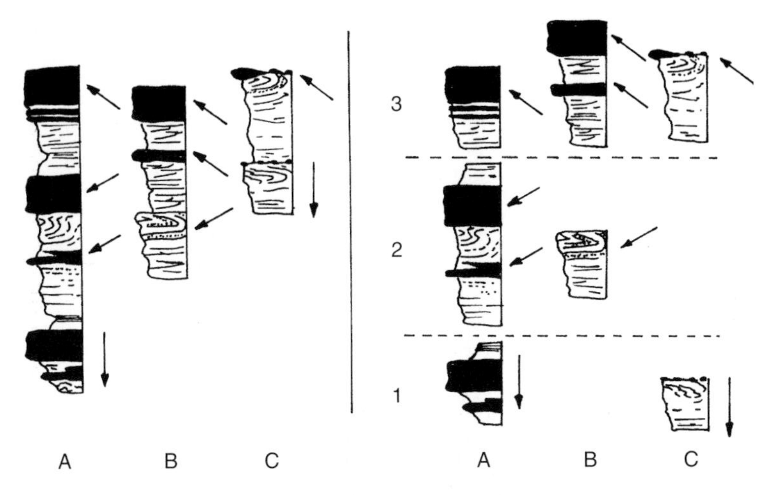

FIG 2.20 Hypothetical kineto-stratigraphic units at sites A, B and C. The former ice-flow direction is recorded by glacitectonic deformation, which can be used to isolate three kinteo-stratigraphic units (redrawn from Berthelsen 1978).

of specific depositional environments. Although they are not exhaustive, such studies provide us with working templates for the interpretation and genetic labelling of sedimentary sequences.

2.6.1 Vertical profile models for braided stream deposits and architectural elements in fluvial facies

The use of vertical profiles to characterize braided river deposits is illustrated in Fig. 2.21a (Miall 1977, 1978). Fluvial sequences are well suited to be represented by vertical profiles because individual beds are commonly laterally extensive, and vertical variation is typically much greater than horizontal variation. Each profile in Fig. 2.21a represents a 'type' example based on well-known fluvial sequences. The examples most representative of glacifluvial sediments are the Trollheim, Scott, Donjek and Platte Type assemblages, named after the North American rivers where they were first recognized. The models do not represent distinct and separate styles of deposition, but rather points on a continuum of variation. Nevertheless, they are useful points of reference which exemplify contrasting styles of sedimentation in a range of sandur settings.

A more flexible approach to characterising fluvial sediments was developed by Miall (1985, 1992), which identifies eight basic *architectural elements* or structural components of fluvial systems, such as longitudinal bars and channel fills (Fig. 2.21b). Architectural elements combine in various ways in different fluvial depositional systems, and can be used as the basis of a wide range of fluvial facies models.

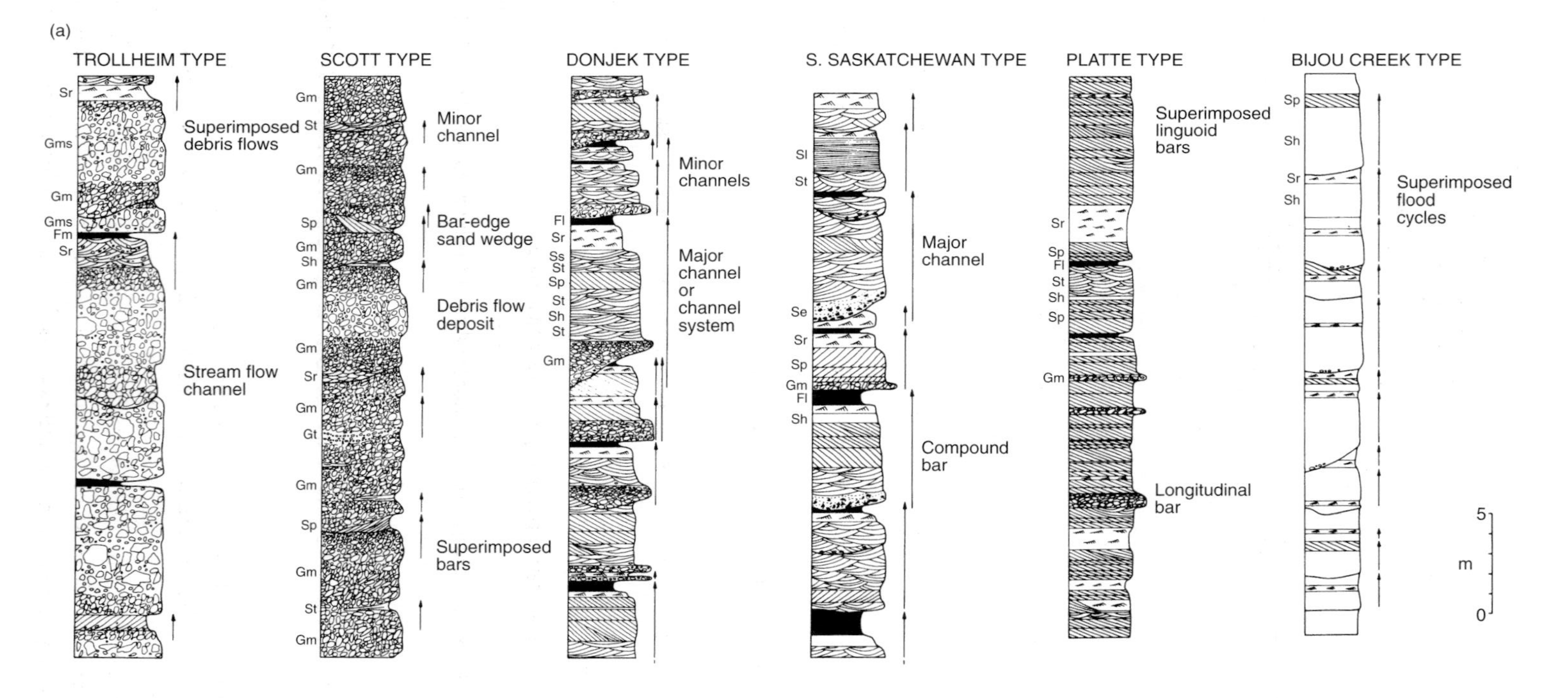

FIG 2.21 a) Vertical profile models for braided stream deposits. Coarser systems are towards the left. Arrows represent small-scale cycles. (Miall 1978).

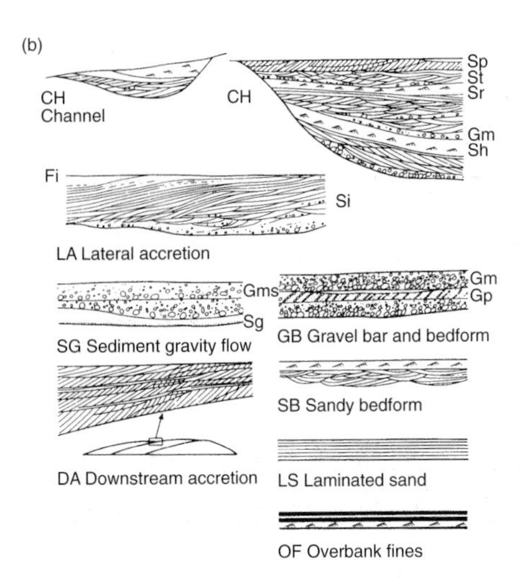

RANK	DEFINITION
1st order	A *set* (McKee and Weir, 1953) boundary, separating similar lithofacies; minor erosion, does not cross-cut underlying set.
2nd order	Defines boundary between cosets (McKee and Weir, 1953); typically erosional surfaces, separating cosets of dissimilar lithofacies.
3rd order	Minor erosional scours which cross-cut sets and cosets.
4th order	Non-erosional surface, representing boundary between stacked depositional units.
5th order	Erosional surface marking the base of an assemblage of genetically-related architectural elements.
6th order	Laterally extensive surfaces marking boundaries of mappable stratigraphic units such as *members or sub-members*.
7th order	Surfaces bounding major depositional systems in a basin fill complex.

FIG 2.21 b) Architectural elements and bounding surfaces applied to glacifluvial sediment sequences (Miall 1988).

2.6.2 Architectural element analysis applied to multiple till sequences

Boyce and Eyles (2000) adapted the architectural element approach in a study of Pleistocene subglacial tills, illustrating how a systematic classification scheme can aid in interpretation of heterogeneous sediment sequences. They identified five recurrent architectural elements (Fig. 2.22a) with characteristic internal structures (Fig. 2.22b), and separated by different orders of bounding surface (Fig. 2.22c). Repetitions in the sequence of diamictic elements, sorted interbeds and deformed zones were interpreted as evidence

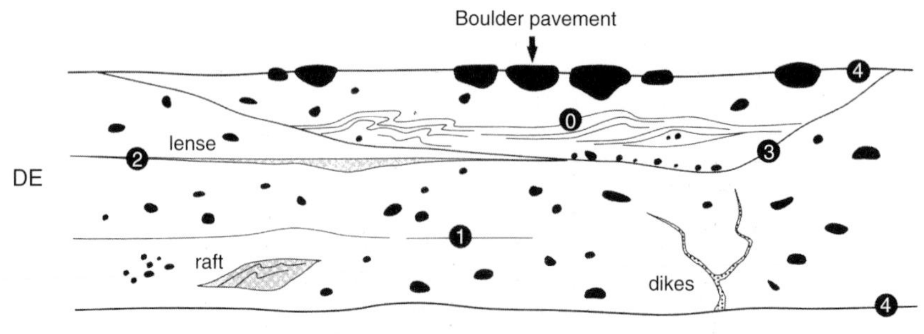

Diamict element (Dmm, Dms, Sm, Sr)

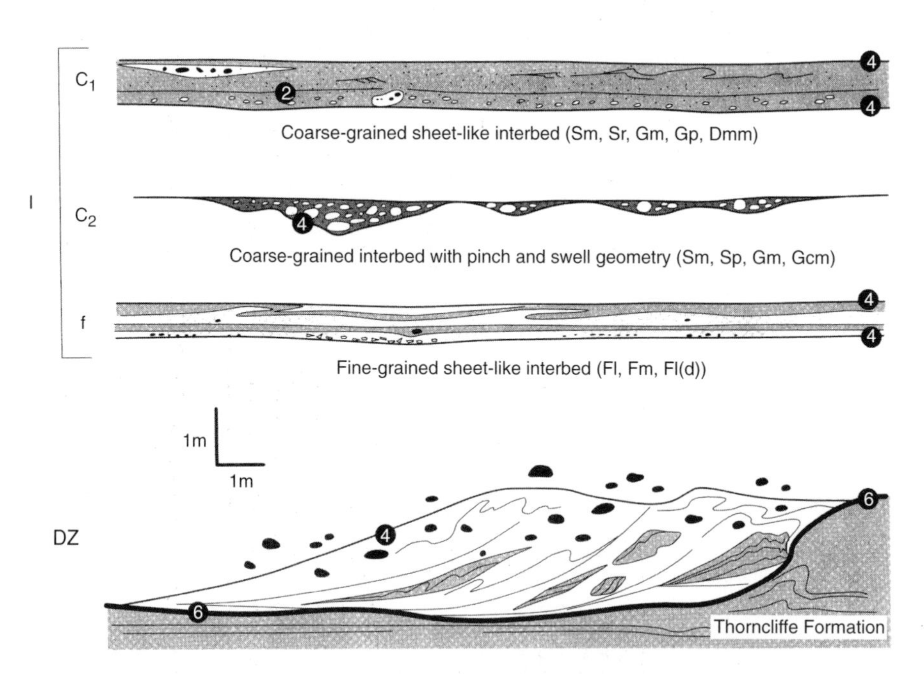

Deformed zone (Dmm, Dms, Sd)

FIG 2.22a(i)

for successive episodes of deformation till emplacement interrupted by subglacial sheetflow events (Fig. 2.22d). This study is particularly interesting because it employs both outcrop data and borehole stratigraphy to establish the three-dimensional form of sediment bodies beneath a drumlinised till surface.

ARCHITECTURAL ELEMENTS IN NORTHERN TILL		CODE	OUTCROP (2-D) GEOMETRY	APPROX. SCALE	L/T	LITHOFACIES ASSEMBLAGE	INFERRED PROCESSES
Diamict element		DE	Tabular diamict beds, planar to genrly undulating bounding contracts boulder pavement often marking upper surface	>100 (L) <10 m (T) $>10^3$ m^2 (A)	>10	Dmm, Dms, Dms, Sm, Sr	Subglacial aggradation of deformation till
Interbed	Coarse	I-c_1	Laterally continuous, sheet-like sands and gravels separating diamict elements	>100 m (L) <5 m (T) 10^3 m^2 (A)	>25	Sm, Sr, Sp, Gm, Gp, Dmm	Ice-bed separation; erosion and deposition by subglaciofluvial meltwater sheet-flow
		I-c_2	Laterally discontinuous sand gravel body, with pinch and swell geometry	<10 m (L) <1 m (T) 10^2 m^2 (A)	<10	Sm, Sp, Gm, Gcm	Ice-bed separation; localized incision by subglaciofluvial meltwater sheet-flow
	Fine	I-f	Laterally continuous tabular silt and mud units separating diamict elements	>10 m (L) <1 m (T) 10^2 m^2 (A)	>10	Fl, Fm, Fir, Fl	Ice-bed separation; low energy sedimentation in subglacial water body
Deformed zone		DZ	Undulatory zone of deformed till and thrusted sediments at base of till sheet; variable thickness and spatial extent	Variable (L) <10 m (T) 10^2 m^2 (A)		Dmm, Dms + included sub-till sediments	Subglacial deformation of pre-existing strata

FIG 2.22a(ii) Architectural analysis applied to multiple till sequences (Boyce and Eyles 2000) (a) (i and ii) Definition of architectural elements.

RANK	DEFINITION
0th order	Laminae (e.g. shear laminations in Northern Till).
1st order	Boundary separating like lithofacies within a lithosome.
2nd order	Boundary defining dissimilar lithofacies within a lithosome (e.g. minor sand/silt stringers in diamict element).
3rd order	Minor erosion surface; laterally discontinuous.
4th order	Laterally continuous surface defining boundary between individual architectural elements (e.g. diamict elements in Northern Till).
5th order	Laterally continuous erosion surface demarcating base of genetically-related architectural elements.
6th order	Surfaces marking boundaries of mappable stratigraphic units (e.g. top and base of Northern Till).
7th order	Surfaces bounding major depositional systems in a basin fill complex (e.g. base of Pleistocene succession).

FIG 2.22b Definition of bounding surfaces between and within elements.

STRUCTURE	OUTCROP (2-D) GEOMETRY	APPROX. SCALE	INFERRED PROCESSES
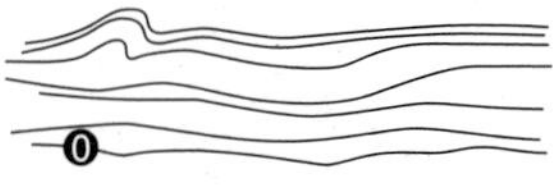 LAMINAE	Discontinuous horizontal to undulatory silt and clay partings in diamic matrix	<1m (L) <1mm (T)	Minor reworking of diamict by meltwater flows or subglacial shearing of included sediments within deforming till
 LENSES	Lenstate or planar beds of sorted sediment	<5m (L) <10cm (T)	Subglaciofluvial deposition at ice-bed interface
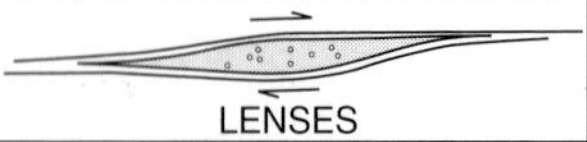	'Augen-like' shear lenses	<1m (L) <50cm (T)	Shearing of sediment inclusions within deforming till
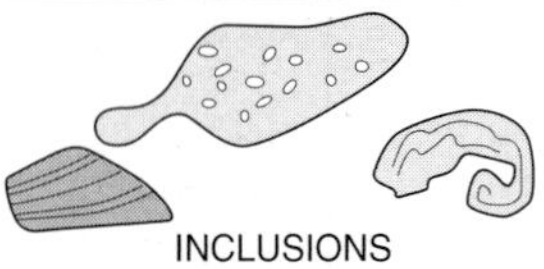 INCLUSIONS	Pods, blocks, and irregular-shaped rafts of sorted sediment	<5m² (A)	Incorporation and partial assimilation of sorted sediments within deforming till layer
 CLASTCLUSTERS	Irregular or pod-like concentrations of granule to pebble-sized clasts	<1m² (A)	Nucleation of clasts around lodged boulders in till
STONE LINES	Linear concentrations of granule to pebble-sized clasts	<1m (L) <10cm (T)	Localized erosion and reworking of upper surface of till
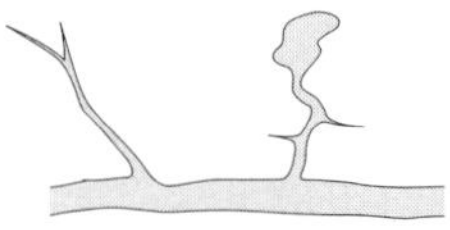 CLASTIC DYKES AND DIAPIRS	Vertical to subhorizontal clastic dikes and diapirs	<5m (L) <10cm (T)	Subglacial over-pressuring and liquefaction of sorted sediment interbeds

FIG 2.22c Internal structures and their interpretation.

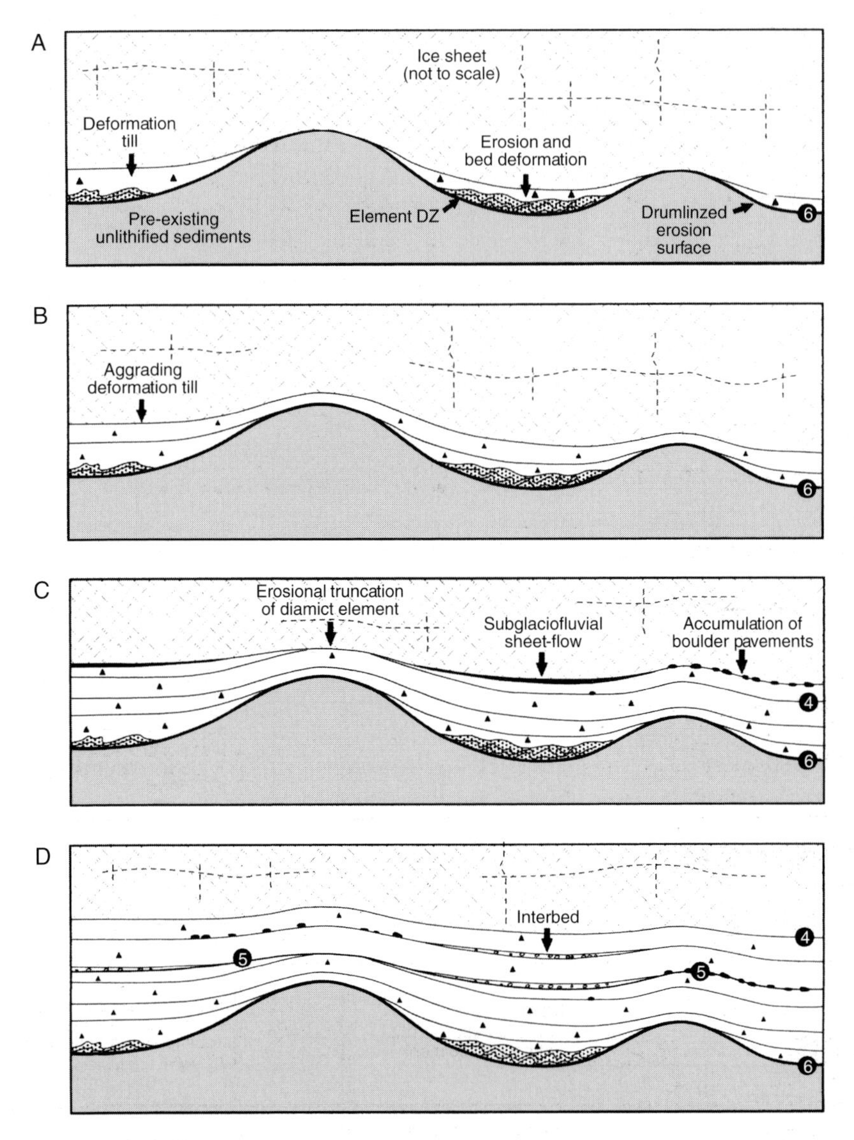

FIG 2.22d Reconstructed general sequence of events in the production of multiple subglacial tills reported by Boyce and Eyles (2000).

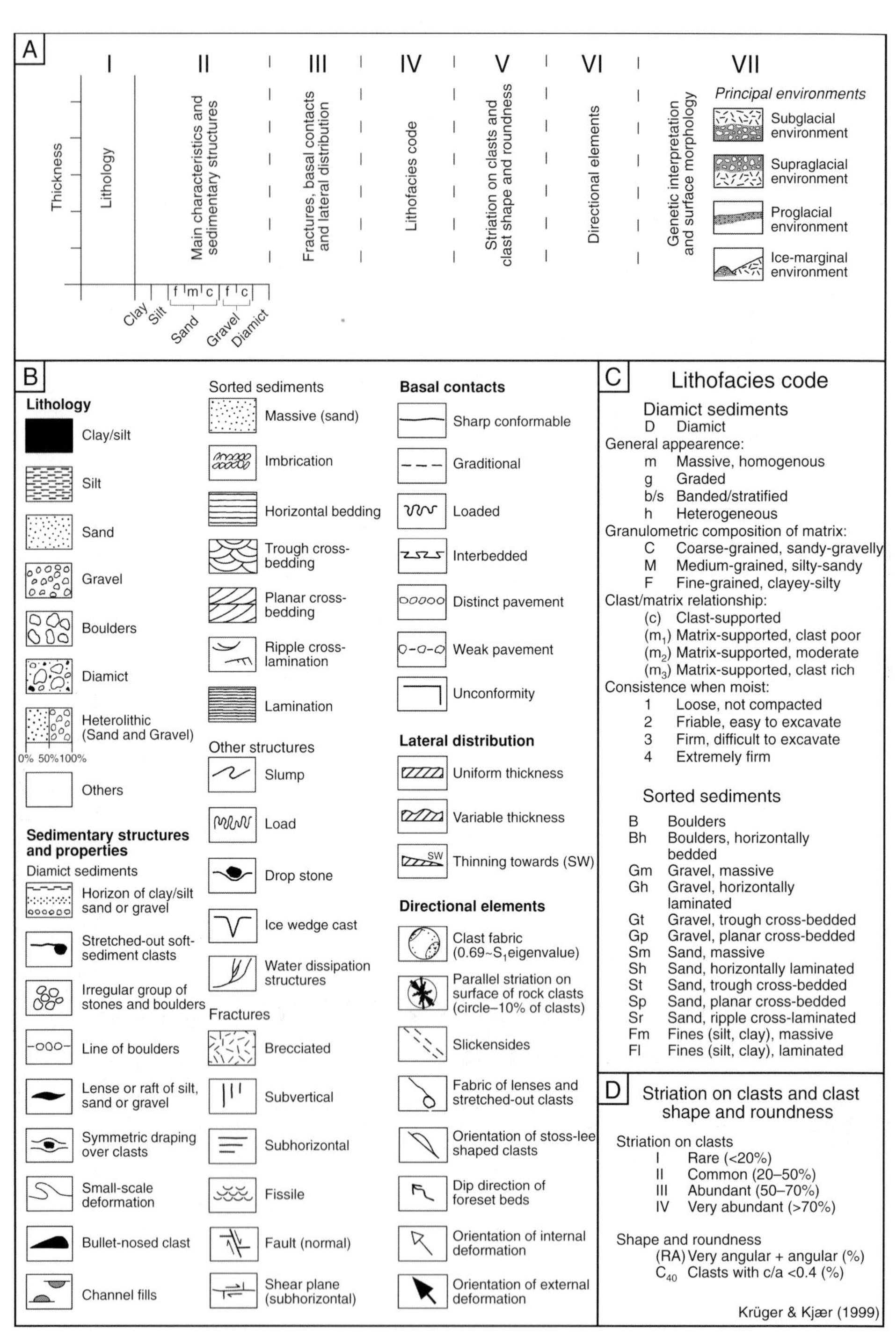

FIG 2.23 Data chart for recording field observations (Krüger and Kjaer 2000). A: Structure of data chart, showing information to be entered in each column; B: Symbols for recording lithological and structural data; C: Lithofacies codes; D: Definition of terms used to summarise clast form data.

E

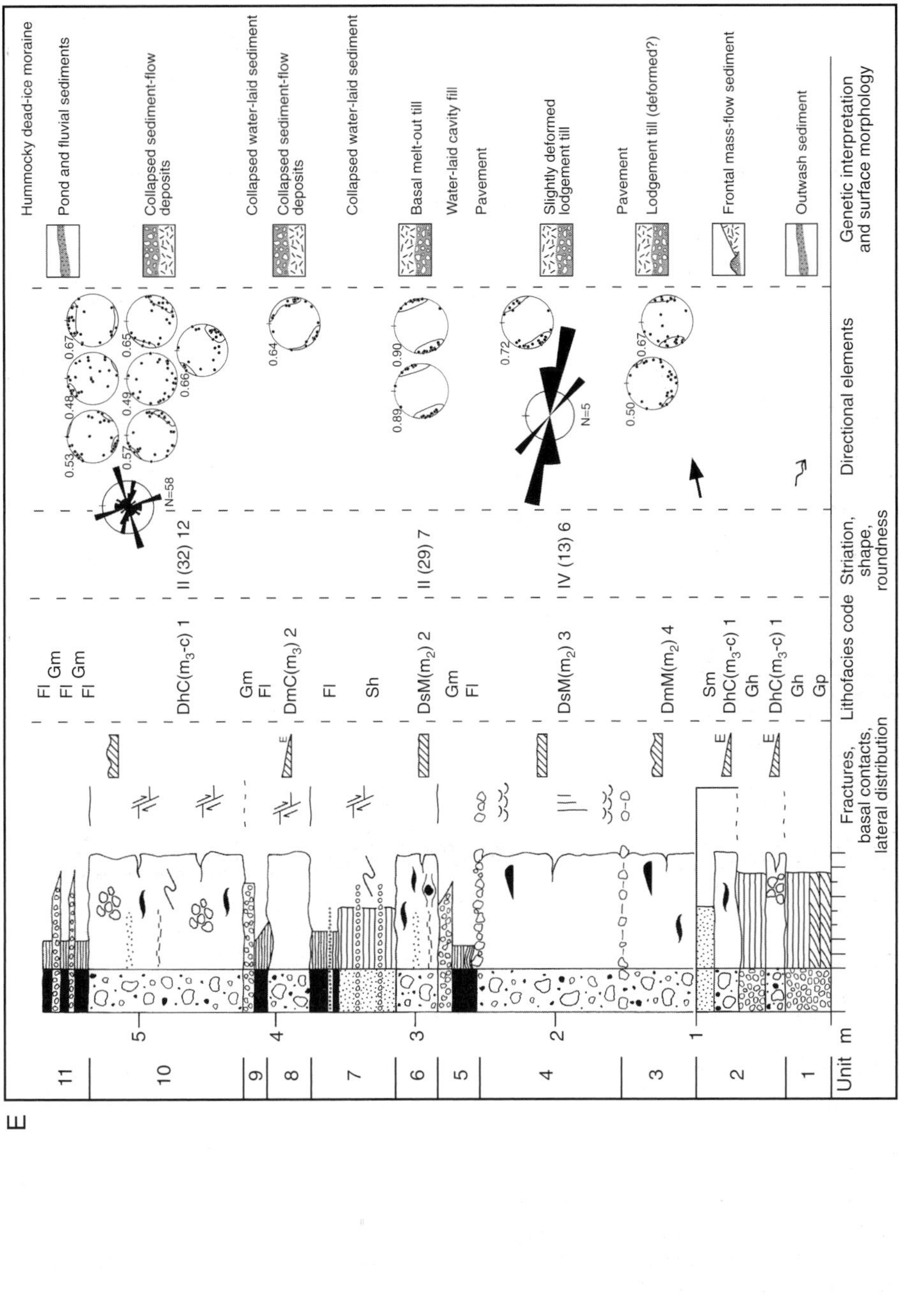

FIG 2.23 E: Example of completed data chart employing the scheme of Krüger and Kjaer (1999).

2.6.3 Lowland glacial diamicts and associated sediments

Krüger & Kjaer (1999) proposed a very useful method of recording field data on a standardised chart (Fig. 2.23), which incorporates many of the types of data discussed in this book. The chart format is particulary useful because it encourages a thorough, systematic approach and helps to ensure that important observations are not forgotten. Students may find it helpful to organise their field notebooks in this way. Such charts should be used within the context of careful research design and a flexible approach to fieldwork, as the work of Krüger and Kjaer clearly shows.

2.6.4 Two-dimensional logging of a complex till sequence and associated stratified sediments and inclusions

Where sediments exhibit significant lateral variability and/or complex deformation structures, vertical profiles such as those shown in Figures 2.21a and 2.23e cannot record all of the relevant information, and two-dimensional drawings are much more effective. Figure 2.24 shows a 2-D log of a coastal cliff where complex tills and associated stratified sediments were exposed (Evans et al 1995; Benn and Evans 1996). The figure illustrates the value of the technique for portraying lithofacies architecture, and the importance of detailed enlargements for the display of small-scale structures, particulary folded, faulted and contorted inclusions and interbeds. Bounding surfaces can be shown in their true position, and facies can be represented using different types of symbols or patterns. For example, in Fig. 2.24 sand facies are shown in a stipple pattern; the location of clast fabric samples are shown by circled numbers; lower case letters represent: (a) chalk stringers, (b) disaggregated chalk 'clouds', (c) deformed sand stringers and (d) sandy diamict/clayey sand laminae in a recumbent fold.

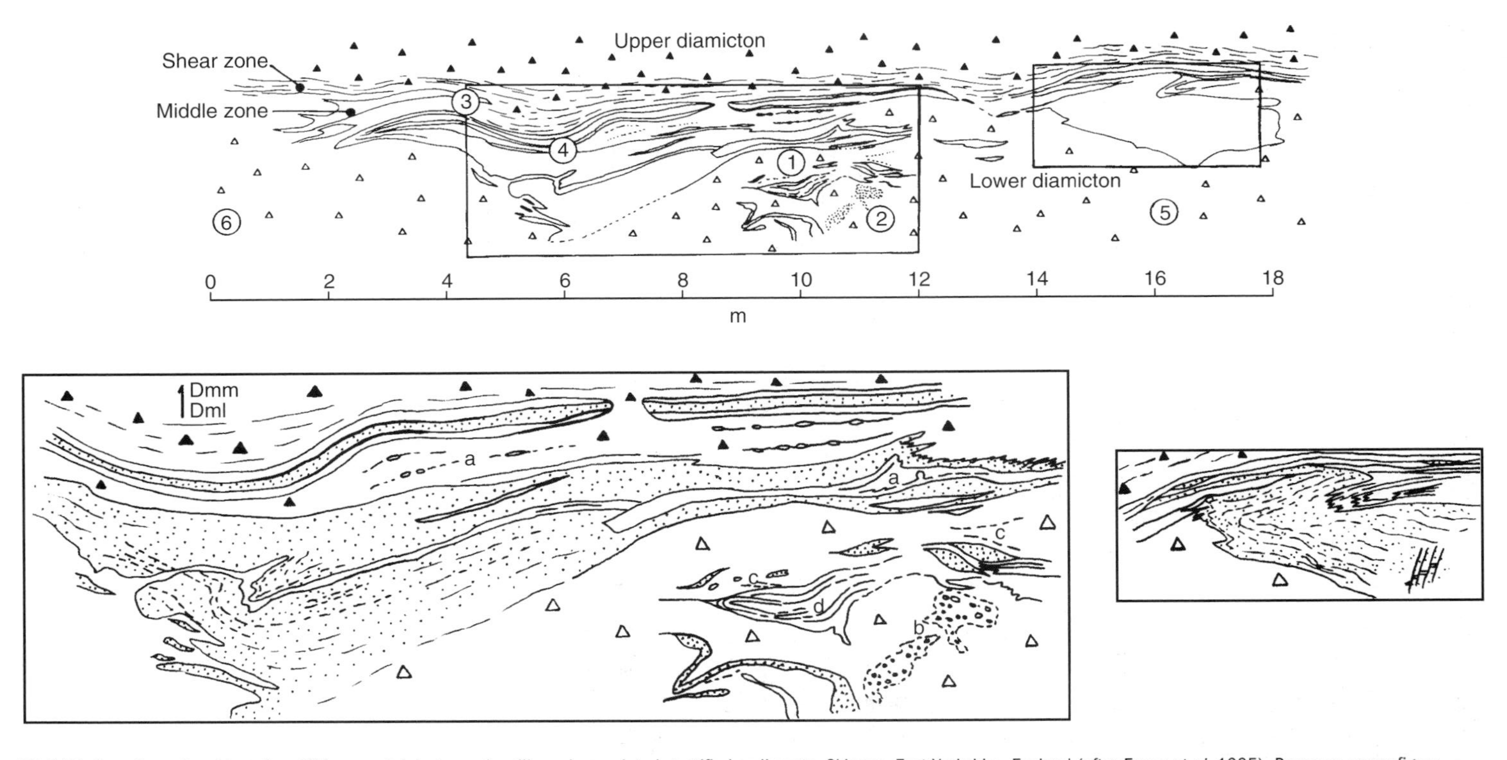

FIG 2.24 Two-dimensional log of a cliff face containing complex tills and associated stratified sediments, Skipsea, East Yorkshire, England (after Evans *et al.* 1995). Boxes on upper figure locate the enlarged areas depicted in lower figures.

3 The size of sedimentary particles

Trevor B. Hoey

3.1 INTRODUCTION

The sizes of particles found in a sedimentary deposit contain information about the source material, and about the processes of erosion, transport and deposition that have produced the deposit. It is now accepted that it is rarely possible to directly relate particle size to process because: (a) source materials produce variable grain sizes depending on lithology and prior processes such as weathering; and (b) previous phases of erosion, transport and deposition produce inherently biased grain-size populations from which the observed deposit is formed. However, principles of grain-size sorting can be developed that enable interpretation of sediments in combination with other forms of evidence, such as sediment sequences (Chapter 2), grain morphology (Chapter 4) and grain fabric (Chapter 5).

Grain size is characteristically described either as: (a) an entire *grain-size distribution* that shows the proportion (or percentage) of grains that fall within specified size classes; or (b) summary measures that describe the statistical moments of the distribution, including measures of central tendency (mean, median, mode), sorting (standard deviation, range), skewness and kurtosis. Whether the entire distribution or one or more statistical descriptors is appropriate in a particular case depends on the purpose of the investigation. Grain-size sampling and analysis are described further in sedimentological texts (e.g. Lindholm 1987; Tucker 1988; Gale and Hoare 1991; Syvitski 1991).

3.2 THE CHARACTER OF NATURAL GRAIN-SIZE DISTRIBUTIONS

Most natural sediments exhibit a characteristic skew (Figure 3.1a). Consequently, data are often shown on a logarithmic scale that produces a normal distribution in the ideal case of a log-normally distributed population (Figure 3.1b). To aid the description of grain-size distributions (referred to as *gsds* subsequently) the transformation to a logarithmic scale is often carried out at an early stage in the analysis. The most commonly used transformation is the *phi (ϕ) scale* (Krumbein 1938):

$$\phi = -\log_2 (D/D_0) \qquad (1)$$

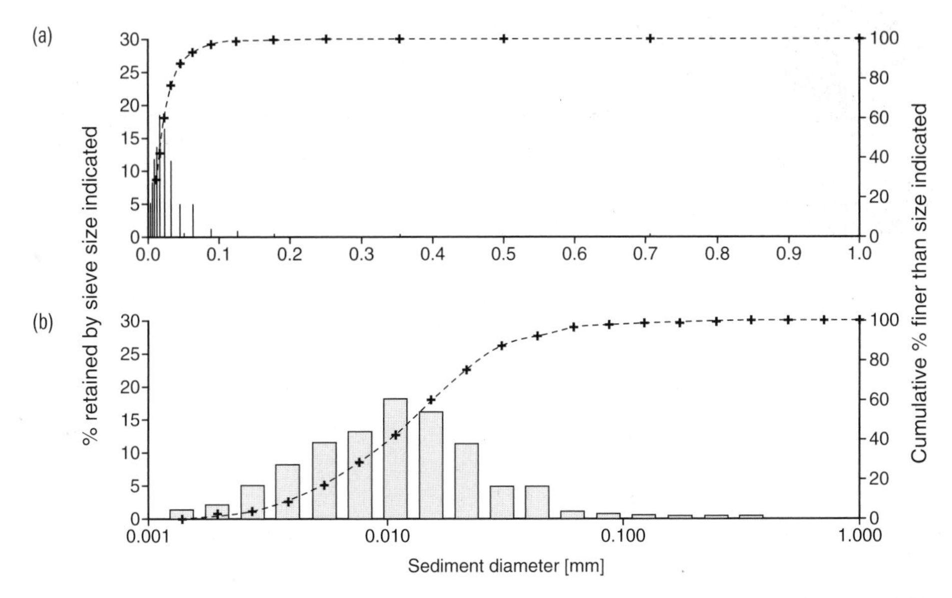

FIG 3.1 (a) Grain-size distribution for 'Sand subfacies 1' sampled by Glasser and Hambrey (2002) from Soler Glacier, Chile. Arithmetic scale used, with sizes from fine-to-coarse on the x-axis. (b) Data from Figure 3.1(a) plotted on a logarithmic x-axis.

where D is the grain size (mm), and D_0 is the grain size used as the origin of the scale. In most geological applications, D_0 is set to 1mm, such that equation (1) simplifies as follows:

$$\phi = -\log_2 (D) \qquad D = 2^{-\phi} \qquad \text{(2a,b)}$$

Equation (2b) shows how this transformation produces a convenient and easily memorable scale for which each *increase* of 1 in ϕ *halves* the grain size, and each *decrease* of 1 in ϕ *doubles* the grain size. Table 3.1 shows the phi scale along with verbal descriptors of grain size, after Wentworth (1922). One apparent anomaly in the phi scale is that fine grain sizes are associated with positive values of ϕ, whereas coarse sizes have negative ϕ values. This is acceptable when working mainly with fine grained (<1mm) sediments; workers who concentrate on coarser sediments have recently developed an alternative scale, the *psi scale* (Parker & Andrews 1985). This compensates for the negative value problem in coarse sediments by inverting phi as:

$$\psi = -\phi \qquad \psi = \log_2 (D) \qquad D = 2^{\psi} \qquad \text{(3a,b,c)}$$

Figure 3.2 shows a typical grain-size distribution from a fluviglacial terrace of the Kowai River, New Zealand (Blakely *et al.* 1981) plotted in mm, phi and psi forms.

Grain-size distributions can be measured using even or uneven phi/psi intervals. An interval of 1ϕ is often too coarse to reveal the details of the distribution and can lead to inaccurate moment statistics being determined. A half-phi interval is more common, although finer intervals may be required for some materials. There is no requirement for

Table 3.1 Particle size conversion table and nomenclature (after Wentworth,1922).

Millimetres (mm)	Phi unit ϕ	Psi unit ψ	Particle size category and sub-category		Micro-metres (µm)	Phi unit ϕ	Psi unit ψ	Particle size category and sub-category	
1024	-10.0	10.0			1000	0.0	0.0		
724	-9.5	9.5		boulder	707	0.5	-0.5		coarse sand
512	-9.0	9.0			500	1.0	-1.0		
362	-8.5	8.5			354	1.5	-1.5	Sand	medium sand
256	-8.0	8.0			250	2.0	-2.0		
181	-7.5	7.5			177	2.5	-2.5		fine sand
128	-7.0	7.0			125	3.0	-3.0		
90.5	-6.5	6.5		cobble	88.4	3.5	-3.5		very fine sand
64.0	-6.0	6.0			62.5	4.0	-4.0		
45.3	-5.5	5.5	Gravel		44.2	4.5	-4.5		coarse silt
32.0	-5.0	5.0			31.3	5.0	-5.0		
22.6	-4.5	4.5			22.1	5.5	-5.5		medium silt
16.0	-4.0	4.0		pebble	15.6	6.0	-6.0	Silt	
11.3	-3.5	3.5			11.0	6.5	-6.5		fine silt
8.00	-3.0	3.0			7.81	7.0	-7.0		
5.66	-2.5	2.5			5.52	7.5	-7.5		very fine silt
4.00	-2.0	2.0			3.91	8.0	-8.0		
2.83	-1.5	1.5		granule	2.76	8.5	-8.5		coarse clay
2.00	-1.0	1.0			1.95	9.0	-9.0		
1.41	-0.5	0.5		very coarse	1.38	9.5	-9.5	Clay	medium clay
1.00	0.0	0.0		sand	0.977	10.0	-10.0		
					0.691	10.5	-10.5		fine clay
					0.488	11.0	-11.0		

the same sampling interval to be applied across the whole of the distribution, although if the interval is varied then care must be taken with some of the calculations. A typical half-phi interval gives size class boundaries at 0.125mm, 0.177mm, 0.250mm, 0.354mm, 0.500mm, 0.707mm, 1.00mm, 1.41mm, 2.00mm, 2.82mm, 4.00mm etc.

In an ideal normally-distributed gsd (after transformation to the phi/psi scale) the mean, median and mode of the distribution are identical. Natural distributions are rarely symmetrical in detail, and central tendency is most commonly assessed by means of the distribution's *median* value (this being the size for which exactly 50% of the sample is finer, and 50% is coarser). The *mode* (grain-size class that contains the greatest proportion of the sample) and *mean* (true average) require accurate measurement of the entire distribution, and are less widely used. Figure 3.3 indicates the statistical properties of a typical gsd. The degree of *sorting* of a distribution can be described by its standard deviation. Because gsds have traditionally been analysed using hand-plotted graphs, approximations to the higher statistical moments of the distributions are conventionally used rather than full calculation using all available information. These graphic measures

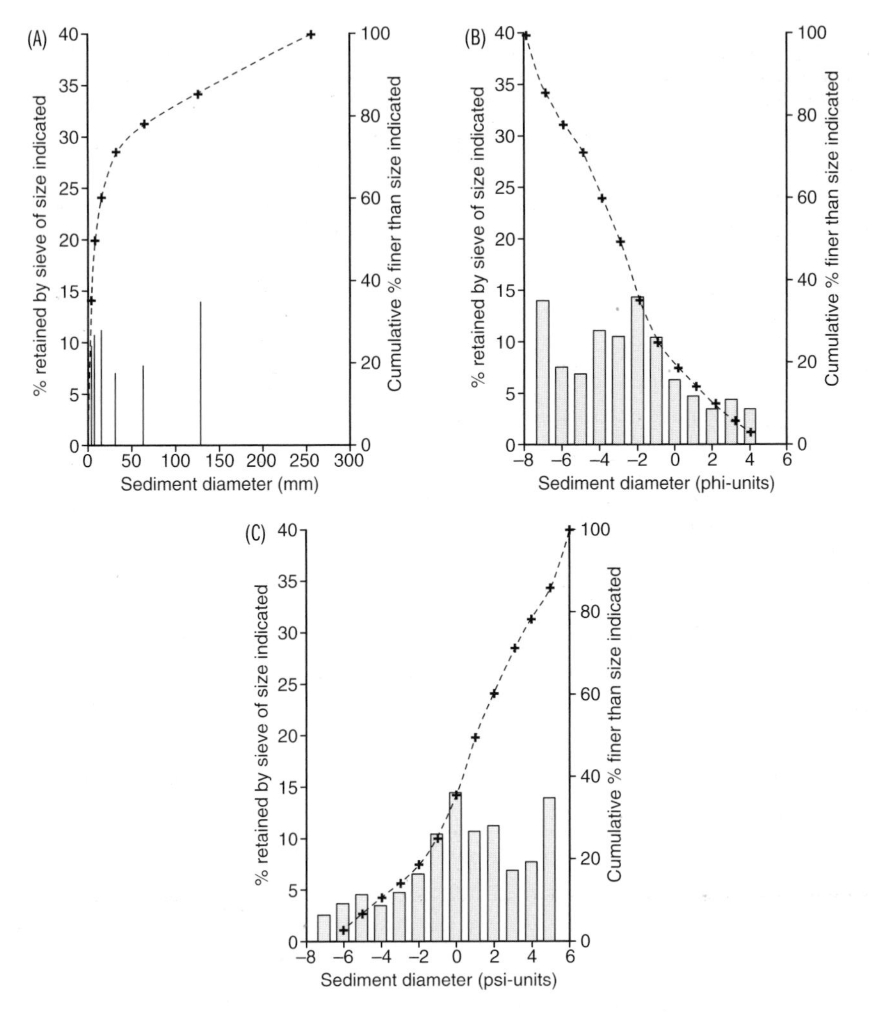

FIG 3.2 Grain-size distribution of a fluviglacial terrace (Woodstock formation, >250,000 years bp) in the Kowai River valley, New Zealand (Blakely *et al.* 1981), plotted in three different forms. (A) millimetre scale; (B) phi-unit scale; (C) psi-unit scale. Bars show the percentage of the bulk sample retained by the sieve of the size class shown. Dashed lines are cumulative percentages finer than the sieve size.

(Folk and Ward 1957) include the *graphic mean* (M_G equation 4) and the *graphic standard deviation* (σ equation 5; σ_I is the inclusive graphic standard deviation), defined as:

$$M_G = (\phi_{16} + \phi_{50} + \phi_{84})/3 \qquad (4)$$

$$\sigma = \frac{[\phi_{84} - \phi_{16}]}{2} \qquad \sigma_I = \frac{[\phi_{84} + \phi_{16}]}{4} + \frac{[\phi_{95} + \phi_5]}{6.6} \qquad (5a,b)$$

where ϕ_{nn} is the phi size for which nn% of the distribution is *coarser*. When the data are arranged as cumulative percentage *finer*, equation 5 still applies but provides results that are equal to $-\sigma$ and $-\sigma_I$. Equation 5 also applies to data in psi-unit, percentage-finer form.

The results of equations (4 and 5) are in phi/psi units as appropriate. Table 3.2 provides a guide to the interpretation of standard deviation results. A commonly used alternative measure of sorting is σ_g, defined as $\sigma_g = (D_{84} / D_{16})^{0.5}$ where D_{nn} is the millimetre size for which nn% of the distribution is finer. σ_g is equal to 2^{σ}.

Even after transformation using the phi scale many distributions remain skewed. The degree of skewness can be measured using the *graphic skewness* (as above, subscript I refers to the more comprehensive *inclusive graphic skewness*) described by equation (6a,b):

$$Sk = \frac{[\phi_{84} + \phi_{16} - 2\phi_{50}]}{[\phi_{84} - \phi_{16}]} \qquad Sk_I = \frac{[\phi_{84} + \phi_{16} - 2\phi_{50}]}{2[\phi_{84} - \phi_{16}]} + \frac{\phi_5 + \phi_{95} - 2\phi_{50}}{2[\phi_{95} + \phi_5]} \qquad (6a,b).$$

These measures of skewness are dimensionless (i.e. they have no units associated with them). When percentage *coarser* (phi-units) or percentage finer (psi-units) format is used a positive skewness implies a skew towards the coarse tail (small ϕ) end of the distribution (Figure 3.3). This is reversed for percentage *finer* (phi-units) data. Table 3.3 provides a guide to the interpretation of skewness results.

The fourth moment of a distribution is the *kurtosis*, which can be thought of as a measure of the 'peakedness' of the distribution. A normal distribution has neutral kurtosis. The *graphic kurtosis* is defined (Folk and Ward 1957) as:

$$K_G = \frac{[\phi_{95} - \phi_5]}{2.44[\phi_{95} - \phi_{25}]} \qquad (7)$$

Table 3.2 Verbal descriptions of standard deviation (sediment sorting) (after Folk and Ward 1957).

Descriptive category	Standard deviation (phi- or psi-units)	σ_g
Very well-sorted	< 0.35	< 1.27
Well-sorted	0.35 → 0.49	1.27 → 1.40
Moderately-sorted	0.50 → 0.99	1.41 → 1.99
Poorly-sorted	1.00 → 1.99	2.00 → 3.99
Very poorly-sorted	2.00 → 3.99	4.00 → 15.99
Extremely poorly-sorted	≥ 4.00	≥ 16.00

Table 3.3 Verbal descriptions of skewness. Note that when expressed in phi-units % finer form, a positive skew is towards the *fine* end of the size distribution and a negative skew is towards the *coarse* end. In psi-units positive and negative skews are towards the coarse and fine ends of the distributions, respectively.

Descriptive category	Skewness (data in phi- units, % finer form)	Skewness (data in psi- units, % finer form **or** phi-units, % coarser form)
Very coarse skewed	-1.00 → -0.31	1.00 → 0.31
Coarse skewed	-0.30 → -0.10	0.30 → 0.10
Approximately symmetrical	-0.10 → +0.09	0.10 → -0.09
Fine skewed	+0.10 → +0.29	-0.10 → -0.29
Very fine skewed	+0.30 → +0.99	-0.30 → -0.99

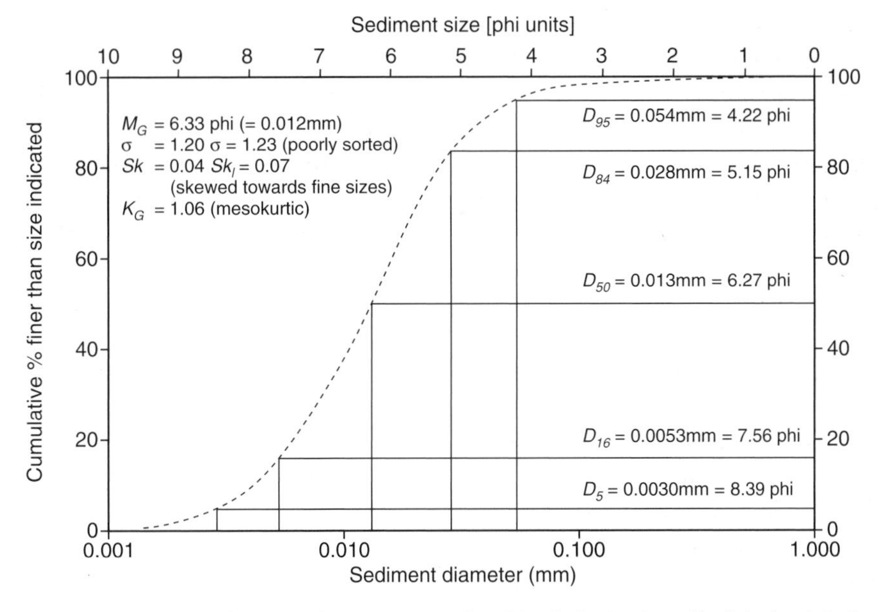

FIG 3.3 The grain-size distribution from Figure 3.1 with key percentiles of the distribution shown. The derived statistics have been calculated using equations 4-7, and their interpretation is based on Tables 3.2-3.4.

in which the scalar 2.44 is determined from the requirement that K_G for a normal distribution is 1.0. Table 3.4 provides a guide to the interpretation of kurtosis results.

Calculation of the parameters of grain-size distributions is most usually done using dedicated programs or spreadsheet calculations. A number of such programs have been developed and published (e.g. Blott and Pye 2001).

In many environments the largest grain size that is present provides an indicator of the intensity of the process involved. The interpretation of this indicator (often referred to as the *competence* of the transport process) is difficult, as it relies on the assumption that larger-sized sediment was available for transport and deposition at the time that the deposit formed. This may not always be a valid assumption to make.

Natural grain-size distributions can be more complex than those illustrated above, exhibiting more than one peak (mode), or occasionally having no apparent structure at all. Such distributions can arise due to the nature of the processes involved in forming a

Table 3.4 Verbal descriptions of kurtosis.

Descriptive category	Kurtosis
Very platykurtic	< 0.67
Platykurtic	0.67-0.90
Mesokurtic	0.90-1.11
Leptokurtic	1.11-1.50
Very leptokurtic	1.50-3.00
Extremely leptokurtic	> 3.00

deposit, or through inappropriate sampling procedures being applied. Distributions with two modes, that often reflect two distinct processes, are most commonly encountered. These *bi-modal distributions* can be formally described (Figure 3.4) using simple indices concerning the distance between the two modes and their relative sizes. Wilcock (1993) proposed the bimodality index *B* that Sambrook Smith *et al.* (1997) generalized as *B**:

$$B^* = |\phi_2 - \phi_1|\left(\frac{F_2}{F_1}\right) \qquad (8)$$

where subscript 1 refers to the primary (larger) mode and subscript 2 to the secondary (smaller) mode. F_1 and F_2 refer to the proportions of sediment in the two modes, where a mode is defined as the four contiguous ¼ phi units containing the largest proportion of sediment (where a coarser size interval is used, this definition can be amended appropriately). As the absolute value of the difference in grain-size is used, the psi-scale can be used interchangeably with the phi-scale shown in equation (8). Sambrook Smith *et al.* (1997) note that there are size distributions for which equation (8) will generate misleading results but that equation (8) is adequate for most applications. Where a distribution has a

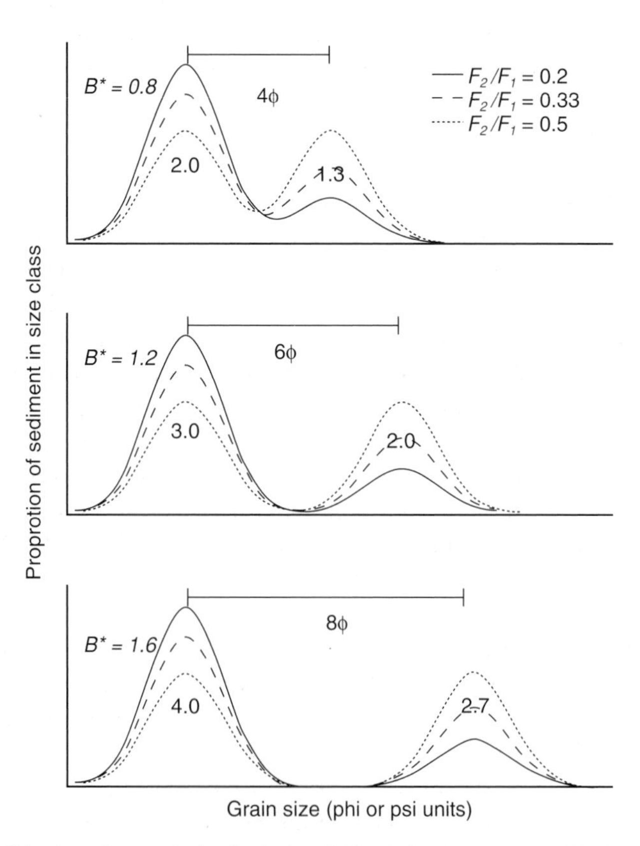

FIG 3.4 Definitions of bimodal sediment grain-size distributions. *B** is calculated using equation (Sambrook Smith *et al.* 1997), and distributions with values of *B** greater than 1.5–2.0 are considered to be bimodal.

value of B^* in excess of 1.5–2.0 then it can be described as being bimodal; values of <1.5 are unimodal, and values within the 1.5–2.0 range are transitional. The existence of a bimodal sediment implies modes of behaviour in entrainment and transport that are distinctive from unimodal sediments (e.g. Wilcock 1993). As a corollary, the existence of a bimodal deposit implies distinctive depositional processes from unimodal deposits.

3.3 SAMPLING ISSUES

3.3.1 Field sample collection

Before grain-size distributions can be measured, samples need to be collected in the field. Natural deposits are occasionally uniform, both laterally and vertically, and hence easy to sample. Most deposits, however, are *heterogeneous* both spatially and vertically. Where to take a sample thus becomes the single most important question to be answered in field sampling. Before asking this question, it is necessary to clearly define *why* a sample is to be taken, and what is it hoped that the sample will contribute to an understanding of the deposit being sampled. Once the 'why' question is clearly answered, the 'where' question can then be approached. These sampling issues are best illustrated in some of the examples below.

3.3.2 Bulk sampling

Ideally, samples should always be repeated so that sampling errors can be assessed. With fine-grained sediments this may be possible, but there will always be a trade-off between increasing the number of localities that are sampled and increasing the number of replicates that are taken of each sample. When sediment sizes increase to the gravel-size range (>4mm) the issue of *sample size* becomes important. In order to collect enough individual grains to have a meaningful, reliable sample the required sizes of sample increase rapidly for these coarser sediments. In addition, the required sample size increases more when the sediments are poorly sorted. Sampling large volumes of material is difficult and time-consuming, and inevitably requires that material from large areas is included in any one sample. There are several guidelines for sample-size determination in existence. Church *et al.* (1987), using the assumption that about 100 grains are required in each half-phi size class in order to produce stable and repeatable estimates of the properties of the distribution, showed numerically that this equates to the mass of the largest particle being equivalent to about 0.1% of the total sample mass. This rapidly produces very large sample sizes (Table 3.5), so Church *et al.* (*ibid*) recommended using a 1% criterion when the maximum particle size exceeds 32mm. Church *et al.* (*ibid*) analysed modern fluvial gravels that are characteristically moderately well-sorted and devoid of very fine material. Gale and Hoare (1991) applied the same approach to samples of beach, glacifluvial and till deposits and recommend significantly larger sample sizes for the latter two environments in the 2–30mm size range (Table 3.5; Figure 3.5). Because deposits containing large particles are usually very heterogeneous a case can be made for collecting multiple repeat samples from such deposits.

Table 3.5 Minimum mass of bulk sediment samples required to obtain reproducible estimates of entire grain-size distributions (and thus to enable accurate estimation of percentiles of those distributions) according to different criteria (River bed: Church *et al.* 1987; others: Gale and Hoare, 1991). Any suggested sample size of <0.5kg is shown as 0.5* in the table on the basis that a sample of this mass is the minimum that should be collected in any situation.

	Minimum total sample mass (kg)				
Maximum particle diameter (mm)	Till	Glacifluvial deposits	Beach	River-bed 0.1%	River-bed 1%
2	1	0.5*	0.5*	0.5*	0.5*
10	25	5	2	0.5*	1.3
50	625	580	200	17	170
100	2500	4600	1700	140	1400

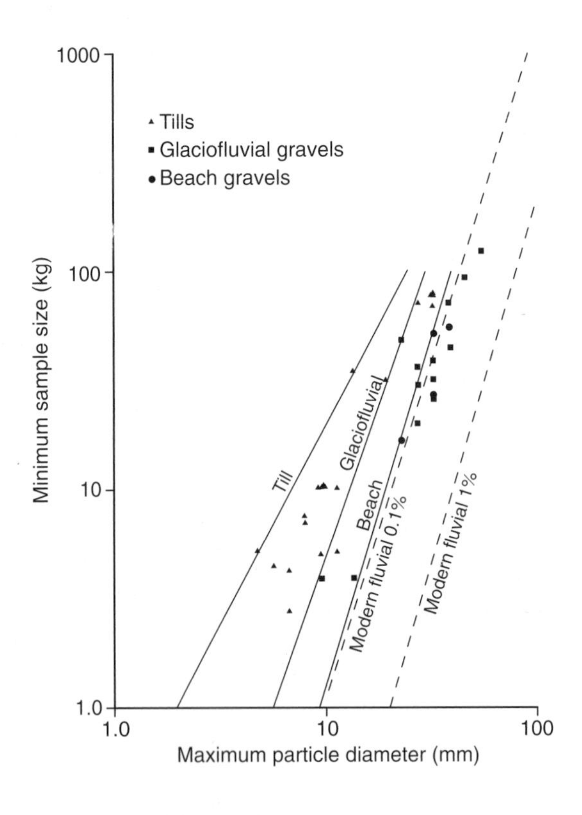

FIG 3.5 Minimum sample size required for reproducible measures of the entire grain-size distributions in different environments (Gale and Hoare, 1991). This graph was determined empirically and may not be generally applicable. It does, however, provide a guide as to the required sample sizes in different environments (see Table 3.5). The two suggested sampling criteria for modern fluvial gravels of Church *et al.* (1987) are also shown: maximum clast mass is 0.1% and 1% of sample mass, respectively.

This is unrealistic in most cases, on account of the enormous effort involved to collect and sieve tonnes of material. There are four possible approaches which can be applied when the deposit consists of identifiable sedimentary facies (Wolcott and Church 1990): (1) stratify samples with respect to the sedimentary facies present; (2) sample each facies independently and consistently; (3) sample only one facies, but sample this consistently at any different sites that are being compared; or (4) sample in a way that is random with respect to the facies. The appropriate approach depends on the aims of a particular study.

3.3.3 A procedure for coarse sediment sampling

An alternative approach is to separate the issue of total sample size from the reliability of the grain-size determination. Consider a deposit for which a sample of 1000kg is required according to the Church *et al.* (1987) guidelines. The recommended procedure for deposits in which the largest particle weighs 1kg or more is thus:

1| decide whether a single sample or aggregated small samples are required;

2| determine the total sample mass required (M_S) from M_S =1000*M_{Max} (or 100*M_{Max}), where M_{Max} is the mass of the largest particle in the sample. Weigh the entire sample in the field;

3| sieve and weigh the grain-size fractions for all grains coarser than a pre-determined threshold size. 90mm and 128mm are the most likely suitable thresholds (Table 3.6);

4| randomly sub-sample the sample that lies below the threshold size. Each sub-sample should be of at least the mass indicated by Table 3.6. Sieve this in the field down to a convenient limit (8mm or 16mm are often best);

5| bring two further sub-samples of the remaining material to the laboratory for drying, sieving and weighing. These samples should be re-weighed after drying, prior to sieving, and a wet-to-dry conversion worked out and applied to the material in this finer fraction (Table 3.7).

Table 3.6 Estimates of the sample size (Kg) required for maximum particle sizes shown. Both the 1% and 0.1% criteria of Church *et al.* (1987) are shown. The particle mass is calculated assuming a spherical particle of density 2650 kg.m^{-3}.

Maximum particle diameter (mm)	Approximate particle mass (kg)	Sample size for M_s = 100 M_{max} (1% criterion)	Sample size for M_s = 1000 M_{max} (0.1% criterion)
64	0.36	36	360
90	1.0	100	1000
128	2.9	290	2900
180	8.1	810	8100
256	23	2300	23 000
360	65	6500	65 000
512	186	19 000	190 000

Table 3.7 Example of a bulk sample obtained using the procedure outlined in the text.

Sample type: Sub-surface, fluviglacial riverbed

Mass of largest particle, M_{max}	2.06kg
Total bulk sample mass (in field)	209.2kg
Total sample mass after wet-dry conversion for <8mm fraction	204.8kg

Size class (mm)	Dry weight (in field) (kg)	Percentage of total size distribution (dry weight)
90–128	11.0	5.4
64–90	10.9	5.3

<64mm total mass	*187.3kg (A)*
<64mm sub-sample	*56.3kg (B)*
Sub-sample ratio	*3.33 (C= [A/B])*

	Sub-sample dry weight (in field) (kg)	Converted whole sample dry weight (kg)	
45-64	11.30 (D)	37.59 (=C*D)	18.3
32-45	7.68	25.55	12.5
23-32	6.78	22.56	11.0
16-23	4.60	15.30	7.5
11.2-16	5.88	19.56	9.5
8-11.2	3.47	11.54	5.6

<8mm sub-sample mass (wet)	*16.60kg (E)*	*<8mm total mass (wet)*	*55.22kg (H=C*E)*
<8mm 2nd sub-sample mass (wet)	*1.037kg (F)*	*<8mm total mass (dry)*	*50.85kg (I=H*G/F)*
<8mm 2nd sub-sample mass (dry)	*0.954kg (G)*		

	Sub-sample dry weight (in lab) (kg)	Converted whole sample dry weight (kg)	Percentage of total size distribution (dry weight)
5.6-8	0.164 (J)	8.74 (=J*I/G)	4.3
4-5.6	0.125	6.66	3.2
2.8-4	0.109	5.81	2.8
2-2.8	0.117	6.24	3.0
1.4-2	0.139	7.41	3.6
1.0-1.4	0.104	5.54	2.7
0.71-1.0	0.068	3.62	1.8
0.50-0.71	0.050	2.67	1.3
0.35-0.50	0.025	1.33	0.6
0.25-0.35	0.018	0.96	0.5
<0.25	0.036	1.92	0.9

For all samples of coarse sediment, it is possible to estimate the precision of the results obtained based upon sample characteristics. This procedure relies on empirical results that have, to date, been provided only for moderately well-sorted fluvial gravels (Ferguson and Paola 1997). Two separate issues arise: (1) small samples may provide biased estimates of distribution properties, due to the excessive influence of small numbers of grains in some size classes; and (2) small samples provide imprecise estimates of the same properties due to random variability between samples. The numerical experiments of Ferguson and Paola (*ibid*) provide empirical equations that can be used to estimate the bias of a specified percentile, the precision of this estimate and the minimum sample size required to estimate a percentile to a specified precision. Table 3.8 summarizes these equations, Table 3.9 provides sample calculations and Figure 3.6 shows the dependence of required sample volume (mass) on the sorting (standard deviation) of the sampled population.

3.3.4 Surface sampling

Where the surfaces of sedimentary deposits are exposed (e.g. on beaches, river terraces) samples can be taken of the surface material. These surface samples are fundamentally different to the samples of bulk material described previously. In terms of processes, the surface of a deposit is where direct interaction with the transporting medium occurs. Thus processes that, for example, selectively remove fine sediments operate on the surface and create a difference between the grain-size characteristics of the surface layer and those of the sub-surface (Figure 3.6). As well as these process differences, there are sampling differences between the surface and sub-surface which mean that samples of surface material are not always directly comparable to those of sub-surface material. However,

Table 3.8 Equations for calculating likely bias and precision (in psi units) of any percentile *p* of a bulk grain-size distribution (Ferguson and Paola 1997).

V = sample size (volume or mass); V_b = critical sample size for the elimination of bias; V_c = critical sample size for the onset of linear behaviour in the precision-sample size relationship; V_{50} = volume (or mass) of the median size grain in the sampled distribution; $\Delta\psi$ = bias in estimating a specified percentile, where $\Delta\psi$ is the mean difference between sample and population values of the *p*th percentile; σ = standard deviation of the grain-size distribution (in psi-units; use equation 5); z_p = the *p*th percentile point of the unit normal distribution; s_0 = standard deviation of percentile estimates for small sample sizes; s_p = standard deviation of percentile estimates for large sample sizes.

Bias of estimated percentile	
If $V < V_b$	$\Delta\psi / \sigma \approx 0.5 \log_{10}(V / V_{50}) - 0.65 - 0.5 \log_{10}(\sigma) - 0.45\sigma z_p$
	where V_b is estimated from
	$\log_{10}(V_b / V_{50}) \approx 1.3 + \log_{10}(\sigma) + 0.9\sigma z_p$
If $V > V_b$	zero
Precision of estimated percentile	
If $V < V_c$	$s_0 \approx 0.4\sigma^{0.6}$
	Where V_c is estimated from
	$\log_{10}(V_c / V_{50}) \approx 2.2 + 3 \log_{10}(\sigma) + 0.9\sigma z_p$
If $V > V_c$	$\log_{10}(s_p) = 0.7 + 2.1 \log_{10}(\sigma) + 0.45\sigma z_p - 0.5 \log_{10}(V / V_{50})$
Minimum sample size required to estimate percentile *p*	
Without bias	$\log_{10}(V_b / V_{50}) \approx 1.3 + \log_{10}(\sigma) + 0.9\sigma z_p$
To specified precision	$\log_{10}(V / V_{50}) = 1.4 + 4.2 \log_{10}(\sigma) + 0.9\sigma z_p - 2 \log_{10}(s_p)$

Table 3.9 Calculations of sample bias, precision and minimum size for the grain-size distribution given in Table 3.7.

Basic information on size distribution

D_{50} = 24.5mm (4.61 psi) $V_{50} = (\pi.D_{50}^3)/6 = 7.70 \times 10^{-6}\text{m}^3$

σ_I = 2.02 (eq. 5b)

Sample mass (M) = 204.8 kg $V = M/2650 = 0.077\text{m}^3$

Particle density = 2650 kg.m^{-3}

Bias of estimated percentile

For 50th percentile, $z_p = 0$.

$\log_{10}(V_b / V_{50}) \approx 1.3 + \log_{10}(\sigma) + 0.9\sigma z_p \approx 1.3 + \log_{10}(2.02) + 0.9 * 2.02 * 0 \approx 1.605$

$\therefore V_b = 0.00031\text{m}^3$ (for 50th percentile), equivalent to 0.82 kg.

For the 95th percentile, $z_p = 1.645$

$\log_{10}(V_b / V_{50}) \approx 1.3 + \log_{10}(2.02) + 0.9 * 2.02 * 1.645 \approx 4.60$ so $V_b = 0.304\text{m}^3$ (805 kg)

Precision of estimated percentile

For 50th percentile

$\log_{10}(V_c / V_{50}) \approx 2.2 + 3\log_{10}(\sigma) + 0.9\sigma z_p \approx 2.2 + 3\log_{10}(2.02) + 0.9 * 2.02 * 0 \approx 3.12$

$\therefore V_c = 0.010\text{m}^3$ (for 50th percentile), equivalent to 27 kg.

As $V > V_c$

$\log_{10}(s_p) = 0.7 + 2.1\log_{10}(\sigma) + 0.45\sigma z_p - 0.5\log_{10}(V / V_{50})$

$\log_{10}(s_p) = 0.7 + 2.1\log_{10}(2.02) + 0.45 * 2.02 * 0 - 0.5 * \log_{10}(0.077/7.7\text{x}10^{-6})$

$\therefore s_p \approx 0.22$

For the 95th percentile

$\log_{10}(V_c / V_{50}) \approx 2.2 + 3\log_{10}(\sigma) + 0.9\sigma z_p \approx 2.2 + 3\log_{10}(2.02) + 0.9 * 2.02 * 1.645 \approx 6.11$

$\therefore V_c = 9.84\text{m}^3$ (for 95th percentile), equivalent to 26 000 kg.

As $V < V_c$

$s_p \approx 0.4\sigma^{0.6} \approx 0.4 * (2.02)^{0.6} \approx 0.61$

Kellerhals and Bray (1971) showed that, in many cases, a random sample of surface clasts is directly equivalent to a sieved and weighed sample of the same material (this assumption is slightly over-simplified and correction procedures have also been devised; see Diplas and Sutherland (1988) and Fraccarollo and Marion (1995)). Thus, surface sampling (grid-by-number) is a commonly applied technique on sediment surfaces. When the method was originally proposed (Wolman 1954), use of a regular sampling grid was advocated. This has generally been replaced by a random technique, where operators collect the grains that lie beneath their foot when walking randomly around the sampling area. This method requires careful delimitation of the area being sampled, such that heterogeneity of the deposit is accounted for either by taking separate sub-samples of different units, or by stratifying the sample such that the number of particles taken from each unit is proportional to the size of that unit (Figure 3.7).

The surface count method is prone to operator variance (Wohl *et al.* 1996), and it is recommended that the same operator is responsible for collection of all samples in a given study, and/or that comparative tests are carried out whereby all operators sample the same

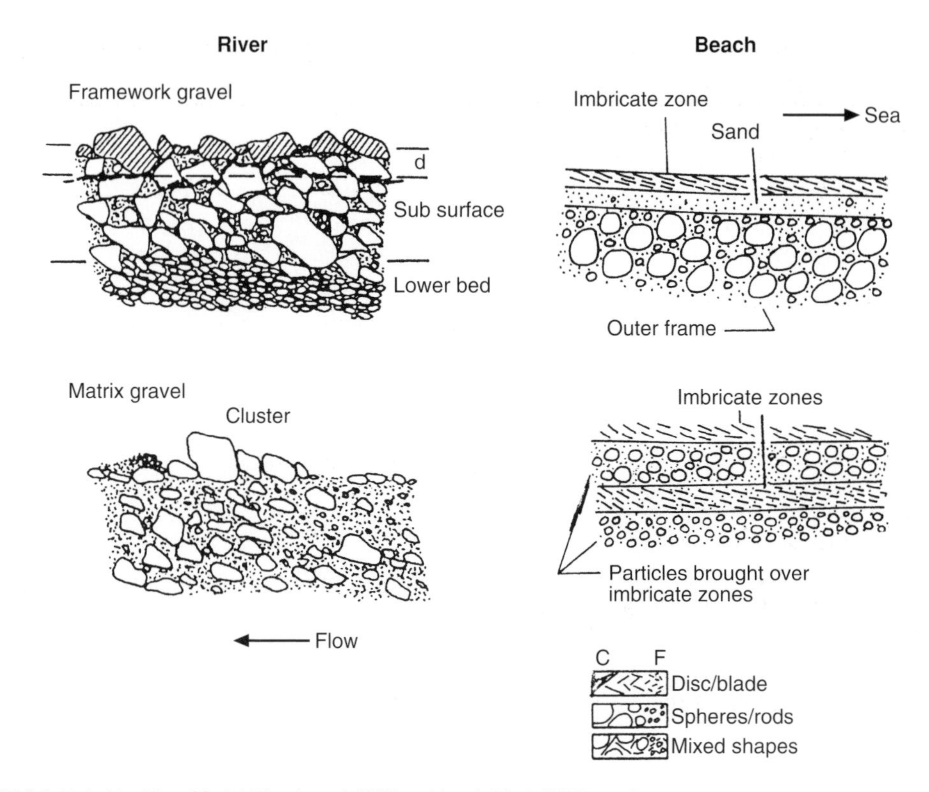

FIG 3.6 Typical bedding of fluvial (Church *et al.* 1987) and beach (Bluck 1999) gravels.

deposit using an agreed method and their results are compared. Surface sampling is a count-by-number method, so that the particles are not weighed but are classified into (usually) half-phi size classes. For ease of measurement, templates can be made that contain holes cut to half-phi sizes (Figure 3.8). Square sided openings replicate the shapes of laboratory sieve meshes, but circular openings are also used. The differences between these two shapes are only significant when the particles are generally blocky in shape. This technique is usually applied down to a lower size limit of 8mm, as identification of finer grains is not reliable. In coarse deposits, this lower limit may have to be increased.

The grid-by-number approach can be used when sampling vertical sections through deposits, but no assessment has been made of the bias that may be introduced in so-doing, so it is dangerous to compare results obtained in this way with those obtained using other methods such as bulk sampling.

Wolman (1954) originally suggested that a sample of 50 particles would provide a reliable estimate of the mean grain size in a sample. Many investigators subsequently have used samples of 100 particles, for no better reason than this is a round number. The actual number required to obtain a reliable estimate of the median size or any other percentile of a deposit depends upon the overall sorting of the deposit. Rice and Church (1996) recommend a sample size of 400 particles for fluvial samples. This provides 95% confidence limits of the order of 0.1-psi (or phi) unit for the 50th and coarser percentiles.

FIG 3.7 Possible sampling strategies for surface count sampling: (a) regular grid in which different facies are sampled according to their areal extent. Lines are marked with tapes and one particle collected from each intersection. Line spacing shown is indicative only - closer spacing would be required to take a sample size of *c*. 400 particles from the bar shown; (b) *a priori* determination of facies boundaries with separate samples then taken from each unit. Within each unit, tapes would be used as in (a) to determine sampling localities.

Precision of the fine percentiles (D_5, D_{16}) is significantly poorer than this in all cases that Rice and Church (*ibid*) examined.

3.4 ANALYTICAL METHODS

There are many published procedures for appropriate analytical methods for different ranges of grain size (see McManus 1988; Gale and Hoare 1991; Syvitski 1991; and references therein). The following are brief summaries of the main methods that are in use, and indications of the applicability of these methods are provided. Where sediments contain a very wide range of sizes, analysis may require the use of more than one method. As the different methods measure different properties of the sediment (b-axis length; submerged weight; light diffraction) they are only directly comparable in idealized cases. Thus, great care should be taken when presenting and interpreting results obtained using a combination of different methods.

FIG 3.8 A grain-size template for use in sampling gravel-size deposits. Numbers are lengths of the sides of the squares (mm). Photograph by Steve Rice.

3.4.1 Dry sieving

This is the standard method for analysing material in the size range *c.* 0.063mm to 4mm. Relatively small samples are required (50–250g) and replicates can easily be analysed. Samples must be carefully dried prior to analysis: if possible, and especially if organic material is to be retained, samples should be air-dried, otherwise oven-drying at 105°C is recommended. Samples may be prone to aggregation during drying.

3.4.2 Wet sieving

Wet sieving can be used for material finer than 0.063mm (63 microns) and sieves as fine as 5 microns are available. This method is now used infrequently due to being time-consuming in comparison to modern settling column and laser diffraction methods.

3.4.3 Settling column methods

Settling column methods rely on the settling of particles under gravity following Stokes' Law. The results are thus sensitive to particle density and shape as well as size. The standard traditional settling column method is that of pipette analysis. Samples in the size range 0.5–63 μm (0.00049–0.063mm) can be analysed using this technique, although analysis of the finer end of this range takes over 65 hours (McManus 1988). Similar to pipette analysis, the SediGraph starts with a homogeneous suspension. In this case, the concentration of particles at different depths in the column is measured from the attenuation of a finely collimated x-ray beam. Coakley and Syvitski (1991) present details of this technique and suggest that it is reliable within the range 1–70 μm. Alternative settling column methods that are based on settling of samples introduced at the top of the column utilize the same theoretical principles as the SediGraph. Errors in such apparatus can occur due to convection within the tube and interactions between grains and further details are given by McCave and Syvitski (1991).

3.4.4 Laser diffraction

Laser diffraction apparatus for particle size determination provides a rapid, accurate, precise and reproducible method for analysing small grain-size material (0.4–4000 μm; Buurman *et al.* 1997). The method uses the forward scattering of a laser beam by particles passing through it, and back-calculation of the sizes of these particles from the properties of the diffraction pattern that is produced. This is not directly equivalent to either wet sieving or settling column methods, as the behaviour of particles in a moving fluid is different to their settling behaviour in a static fluid and light diffraction is not perfectly correlated with any single measure of particle size (mass or diameter). Hence, several authors report systematic differences between results from laser diffraction methods and settling column analysis (e.g. the sieve-pipette method; Beuselinck *et al.* 1998; Loizeau *et al.* 1994). These differences may arise as particle density is correlated with size in many cases (Matthews 1991). As a consequence of these systematic biases, the shapes of grain-size distributions and the position of modes will be reliably determined using laser diffraction, but derived statistics may not always be reliable. Extreme care is required where results obtained by different methods are compared or combined.

3.4.5 Remote methods

An attractive way of obtaining grain-size information is to make measurements from digital images or photographs of the sediment, either by hand or automatically. This is a method best suited to coarse-grained deposits, although it can be used for fine-grained materials as well (especially if samples can be prepared and photographed in the laboratory). A high resolution photograph of the deposit is taken, with adequate calibration scales being included within the field of view. Conventional images can be scanned and then analysed automatically using either custom-developed or readily available image analysis software (Lane 2001; Sime and Ferguson 2003; van den Berg *et al.* 2003). Some care is required with this procedure for three reasons: (a) the method samples grains that are at the surface and, as previously noted, surface samples may give different results from volumetric samples; (b) due to illumination and image quality, some grains may be difficult to detect on the image: this especially affects fine grains and may lead to bias in the results; and (c) the grain axes that are projected on a two-dimensional image are likely not to be the a- and b-axes of the grains. The projected area will depend on grain shape and may be related to actual grain size in different ways for different grains within the same sample. Where grain boundaries are well-defined with respect to the rest of the image, these boundaries can be successfully identified. Because of grain-shape effects the reliable measurement of grain-size percentiles from images requires calibration. However, recent results show that such techniques can be used for the rapid mapping of spatial variation in grain sizes (e.g. Lane 2001; van den Berg *et al.* 2003).

3.4.6 Other methods

In addition to the methods discussed above, there are a range of other techniques that have been developed for grain-size analysis. These include resistance pulse counters (for example the Coulter counter; McManus 1988) and photon correlation spectroscopy (McCave and Syvitski 1991).

3.4.7 Issues for field sampling

With respect to analytical procedures there are two critical issues that need to be considered in the field:

1| the analytical method to be used affects the size of sample that is required and the number of samples that can be processed; there may be requirements for calibrations between different methods, especially where sediments with large size ranges are being analysed; this needs to be known at the outset and appropriate numbers of samples collected for this purpose;

2| comparing results obtained using different analytical methods is not always easy; there is a need for consistency in sampling within studies and between studies if comparison is desired.

These issues, and others raised previously, lead to the same general conclusions regarding field sampling. These are that: (1) it is important to plan sampling programmes carefully in advance so that the appropriate number of samples, of appropriate sizes, are collected in order to provide the characterization of the deposit that is required for later interpretation; (2) grain-size samples should always be collected with a specific purpose in mind, as different purposes lead to different sampling strategies; and (3) as the following examples illustrate, grain-size data alone are rarely able to answer the questions posed by Quaternary scientists; in all cases, multiple sources of evidence are required for reliable, robust interpretations to be made.

3.5 CASE STUDIES

Grain-size distributions and the parameters of size distributions are very widely used in the interpretation of Quaternary sediments. The following examples illustrate some of the possibilities and some of the limitations of grain-size analysis. They have been chosen to try to reflect the range of different sediment types, and thus sedimentary environments, that have been sampled and to illustrate the different ways in which the results can be used.

3.5.1 Process domains from grain-size characteristics

Climatic inference from grain-size properties (Prins and Weltje 1999)

The shapes of grain-size distributions have long been related to depositional processes. However, there is rarely a unique relationship between process and distribution shape, so supporting evidence is often required. Prins and Weltje (1999) present an example in which changing grain-size distribution shape allows river-derived sediments to be differentiated from aeolian dust. This differentiation is then used as a climate proxy (Figure 3.9). In this case, sufficient data were available to allow statistical analysis of the grain-size distributions and this leads to reliable palaeo-climatic interpretation.

Decomposition of multi-modal gsds to infer sub-glacial processes (Sharp et al. 1994)

In many cases, grain-size distributions alone provide less distinctive process signatures than in the preceding example. This is especially the case when the distributions are bi-modal or even multi-modal. Sharp *et al.* (1994) used Gaussian component analysis (Sheridan *et al.* 1987) to decompose grain-size distributions from basal ice layers into component sub-populations (Figure 3.10). Interpretation of these modes was made with reference to known sub-glacial processes, leading to inferences being made regarding ice-bed contact and thus sliding mechanisms in a surge-type glacier. As is the case in several of the examples quoted here, the authors relied on multiple sources of evidence to support their conclusions and did not rely on grain-size analysis alone.

Genesis of gsds during glacial transport (Fischer and Hubbard 1999)

The number of grains in individual size classes can be used to provide an indication of process. Grain number (N_i) is related to grain diameter (D_i) by a power-3 relationship, such that $N_i \propto D_i^{3}$. Departures from gradient –3 on a plot of N_i vs. D_i have been used to infer process domains (Hooke and Iverson 1995; Iverson *et al.* 1996; Fischer and Hubbard 1999; Benn and Gemmell 2002). Figure 3.11 shows the typical nature of these plots, in which the slope of the relationship is interpreted as the fractal dimension of the distribution. Fischer and Hubbard (1999) concluded that subglacial sediment deformation had caused a systematic spatial pattern of fractal dimension which changed from 2.47 for a glacier headwall sample to 2.64 at the glacier terminus. This increase in fractal dimension is due to an increase in fines in the more distal samples. Hooke and Iverson (1995) reported a similar finding that they also related to transport mode. Although these results are consistent, the technique is statistically weak since the slopes of N_i vs. D_i curves are strongly influenced by single, relatively high (or low) values toward the margins of the distribution. Further, Benn and Gemmell (2002) demonstrate that samples that are

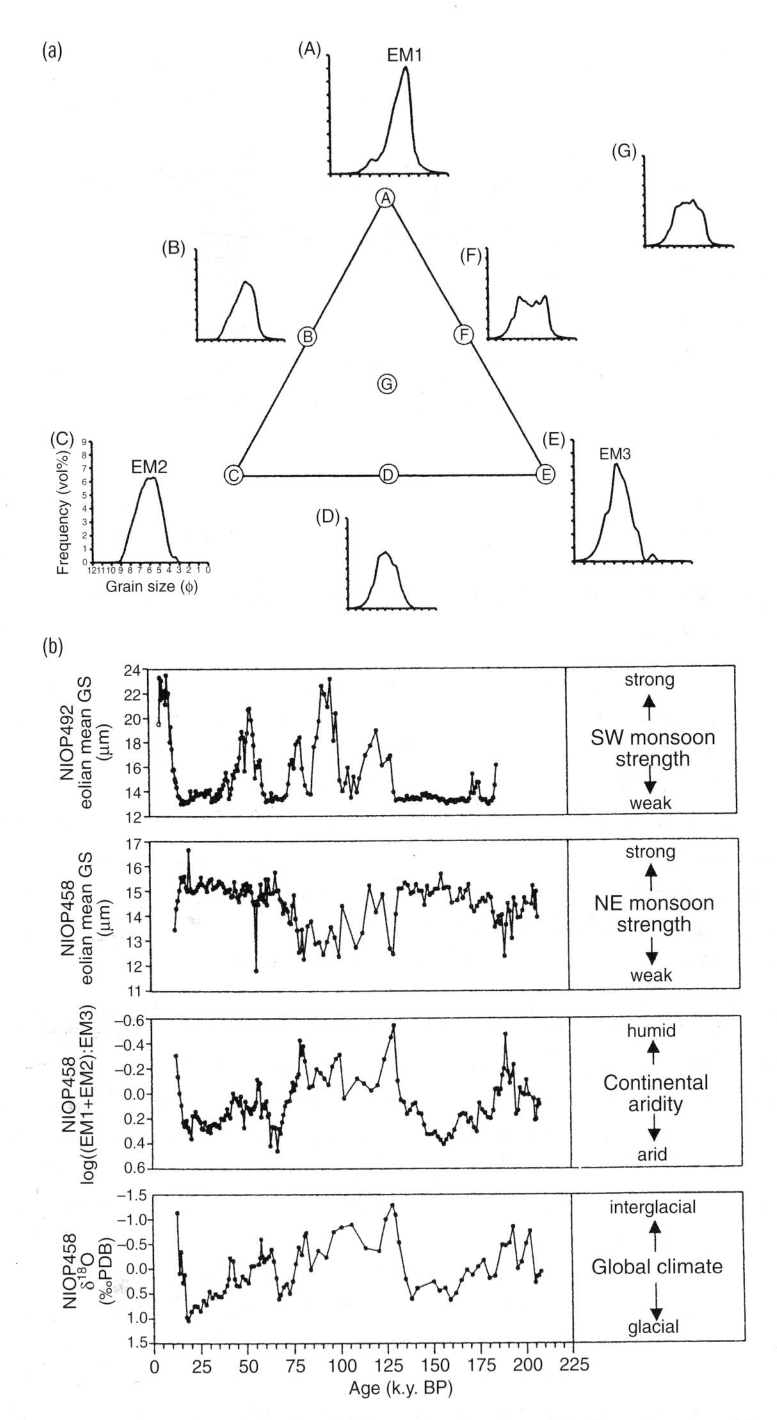

FIG 3.9 (a) The generation of grain-size distributions of different shape from three distinctive end-members (Prins and Weltje 1999); (b) Reconstructions of Late Quaternary variations in Arabian Sea monsoon climate based on end-member modelling of grain-size distributions (*ibid*). Different cores (NIOP492 and NIOP458) and different properties of the grain-size distribution are indicative of different aspects of palaeo-climate in this region.

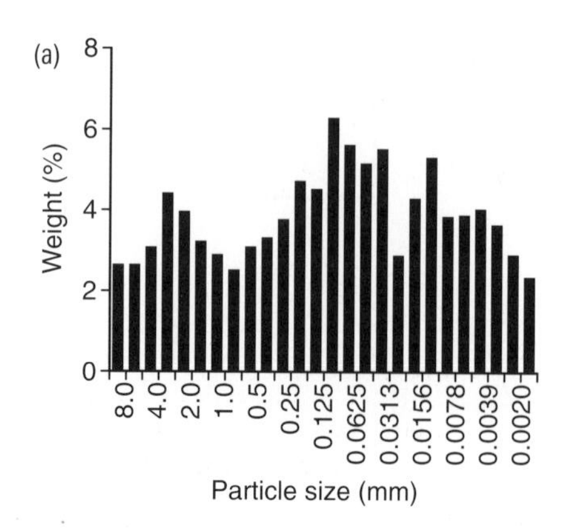

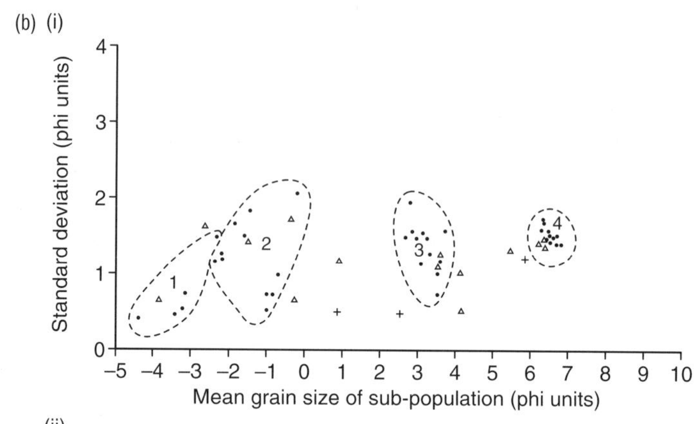

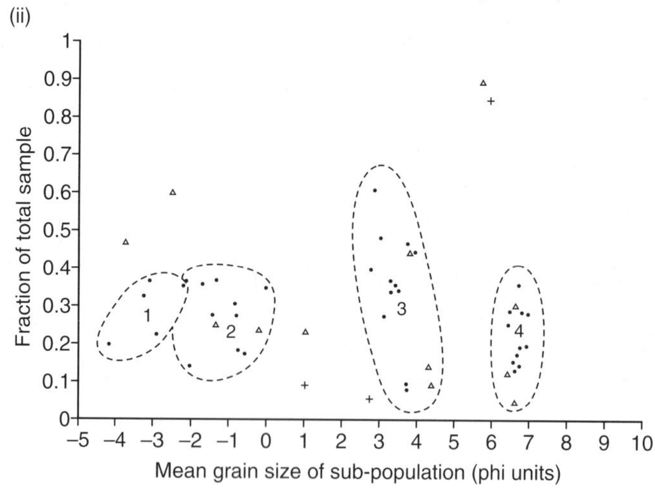

FIG 3.10 (a) A typical multi-modal grain-size distribution taken from a basal till sample (Benn and Gemmell 2002); (b) standard deviation and fraction of total sample in a given mode as a function of mean grain size of each sub-population in basal ice samples from Variegated Glacier, Alaska (Sharp *et al.* 1994). Individual modes were identified using Gaussian component analysis.

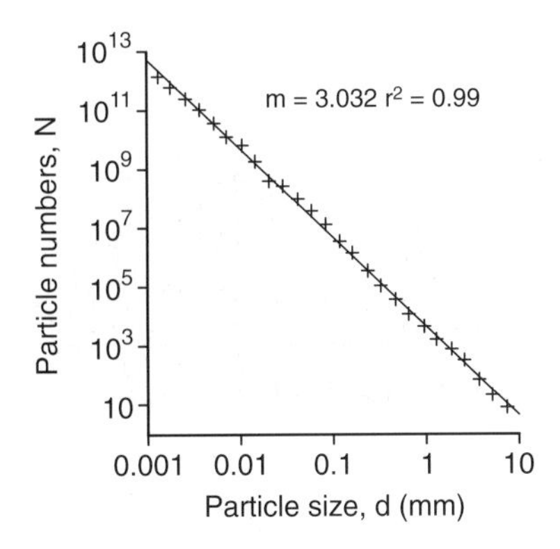

FIG 3.11 Particle number vs. size for the basal till sample shown in Figure 3.10(a) (Benn and Gemmell 2002).

composites of different source populations cannot be differentiated from those modified by weathering or transport processes on the basis of fractal dimension alone. Their results reinforce the recommendation made elsewhere in this manual that single measures of sediment character are rarely *uniquely* related to a particular genetic process and that multiple sources of evidence should be used as the basis for interpretation wherever possible.

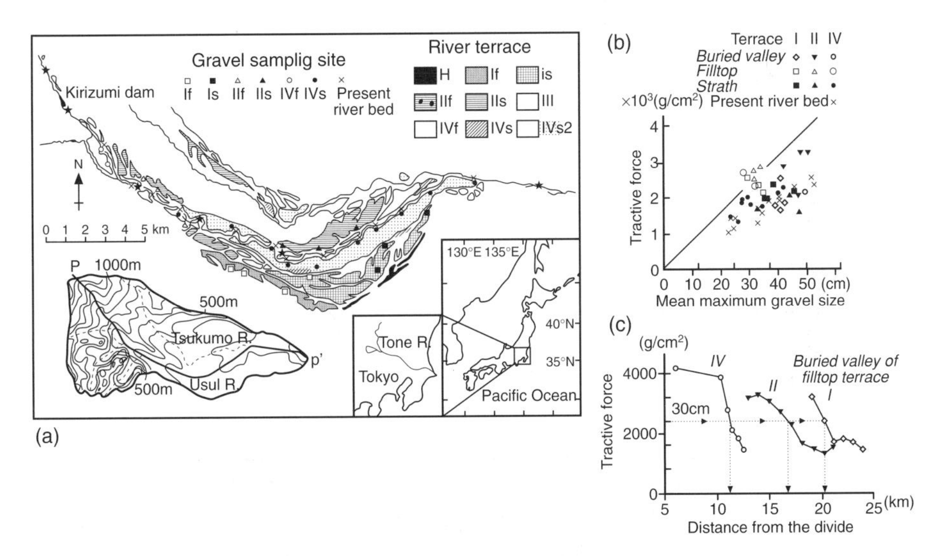

FIG 3.12 (a) The distribution of terraces along the Usui River, Japan; (b) Determination of critical tractive force corresponding to the largest material found on each terrace; (c) Downstream changes in tractive force on buried valley floor deposits. The valley fills correspond to glacials; fill I is dated as after 310ka, II as before 188ka, and IV as *c.* 20ka. Successive valley fills extend less down-valley as a consequence of river profile steepening due to degradation during inter-glacials. After Sugai (1993).

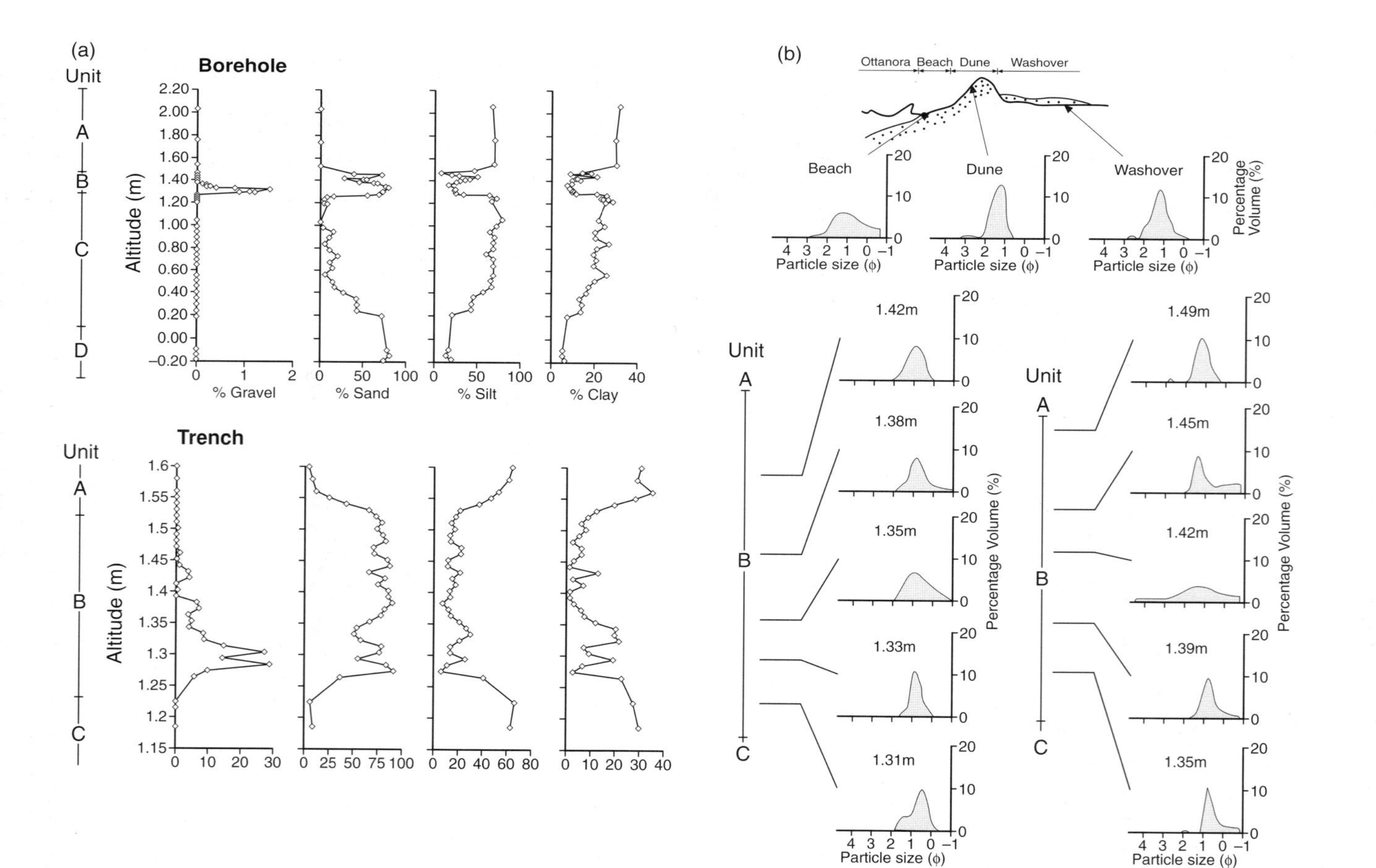

FIG 3.13 (a) Variation in sediment texture at Boca do Rio, Portugal in borehole and trench samples; (b) Comparison of grain-size distributions from modern sedimentary environments with samples from the inferred tsunami unit (unit B) in two sites at Boca do Rio. After Hindson and Andrade (1999).

Competence of flows depositing fluviglacial terraces (Sugai 1993)

The largest size of sediment present in a deposit is often assumed to reflect the ability of the depositional process to move sediment, i.e. the competence of the flow. This approach requires that: (a) larger sediment was available to be transported/deposited but was unable to be moved; and (b) the process that deposited the sediment has been correctly identified. In using competence as a measure of process intensity, it is essential that these issues are assessed in the field by searching for larger, unmoved materials in the sediment source area, and by collecting supporting evidence of the transport/depositional process. Sugai (1993) measured the a-axes of the 10 largest particles within a $5m^2$ area on terraces in the Usui River valley, Japan (Figure 3.12). The forces required to move these gravels were calculated and allow interpretation of the valley fills as being related to lower flood intensity (and hence typhoon frequency) during glacial periods than is the case at the present day.

3.5.2 Spatial patterns in grain size as process indicators

Dynamics of a Holocene tsunami (Hindson and Andrade 1999)

Vertical patterns of grain size may indicate progressive or episodic transitions in depositional process. Hindson and Andrade (1999) show abrupt changes in grain-size distribution that can be related to a specific depositional event (Figure 3.13a). Within the

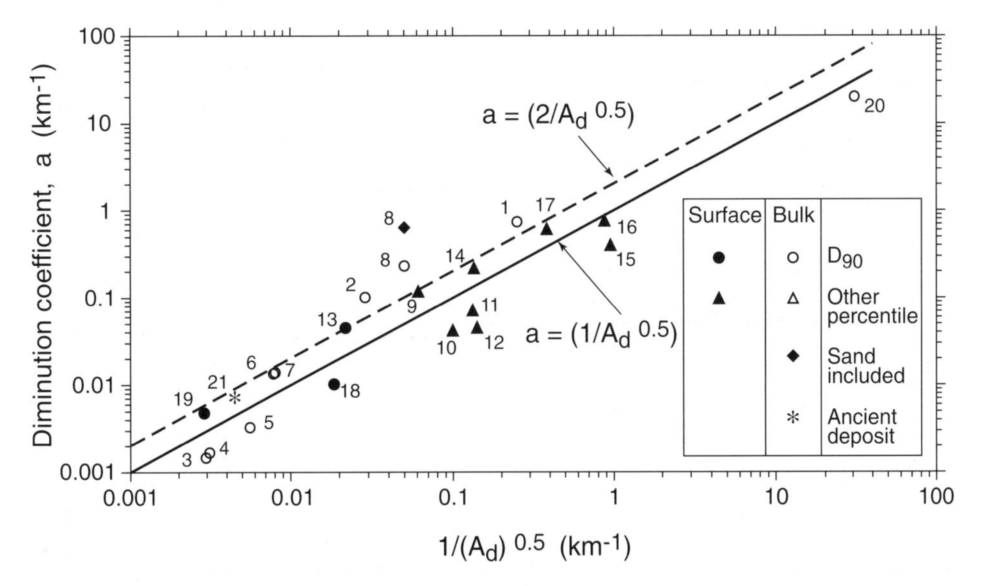

FIG 3.14 Relationship between the rate of downstream fining in river systems (expressed as the parameter *a* in an equation of the form $D = D_0.\exp(-ax)$, where D is grain size, D_0 is the grain size at the proximal end of the system, and x is distance downstream along the system [km]) and catchment area, A [km^2] expressed as $1/(A)^{0.5}$ for consistency of units between the two axes of the graph (Hoey and Bluck 1999).

main sedimentary unit deposited during this tsunami event (their unit B) changes in the shape of the grain-size distributions can be related to modern depositional environments (Figure 3.13b). These changes are thought to indicate changing hydrodynamic conditions during and after the tsunami as deposition of finer sizes may have taken several days to complete.

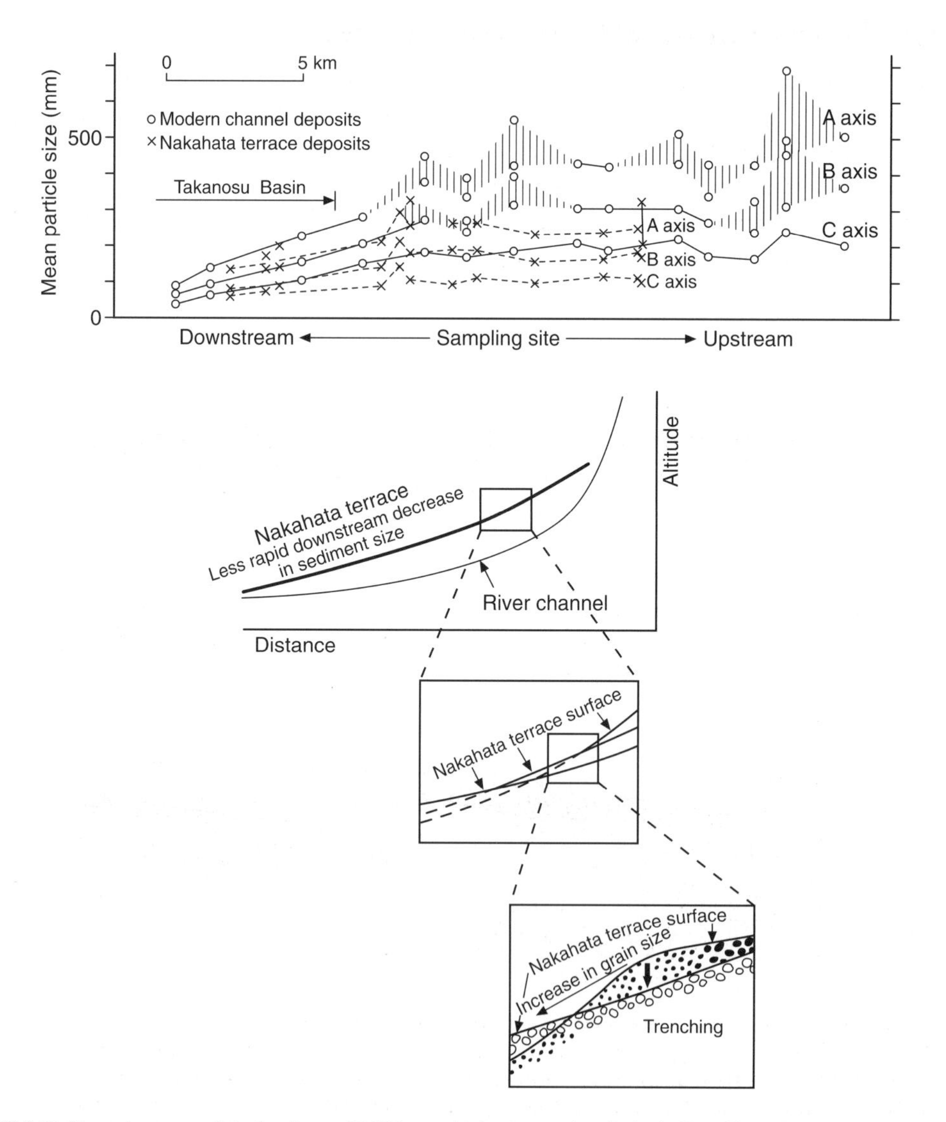

FIG 3.15 Change in mean particle size along *c.* 20 000 year old river terrace deposits in the Dewa Mountains, Japan, and their genetic interpretation (Toyoshima 1987).

Downstream fining in river systems as a palaeo-environmental indicator (Toyoshima 1987; Hoey and Bluck 1999)

Proximal-to-distal changes in sediment characteristics provide indications of sedimentary process. Figure 3.14 shows how the rate of downstream fining in river sediments is related to the length of the river system. This property of river systems allows palaeo-environmental reconstruction from preserved river deposits (Figure 3.15). Such inference is especially strong when combined with reliable slope and age estimates for the deposits (Plumley 1948).

Local sorting on gravel beaches (Bluck 1999)

Small-scale patterns in sediment size and sorting provide indications of process environment type and energy regime. Characteristic patterns of sorting in beach deposits (Bluck 1999; Figure 3.16) allow depositional processes to be inferred. Similarly characteristic patterns in fluvial deposits allow fluvial style to be determined (e.g. the differentiation between meandering and braided river deposits; Brierley 1991) that can be directly related to depositional environment.

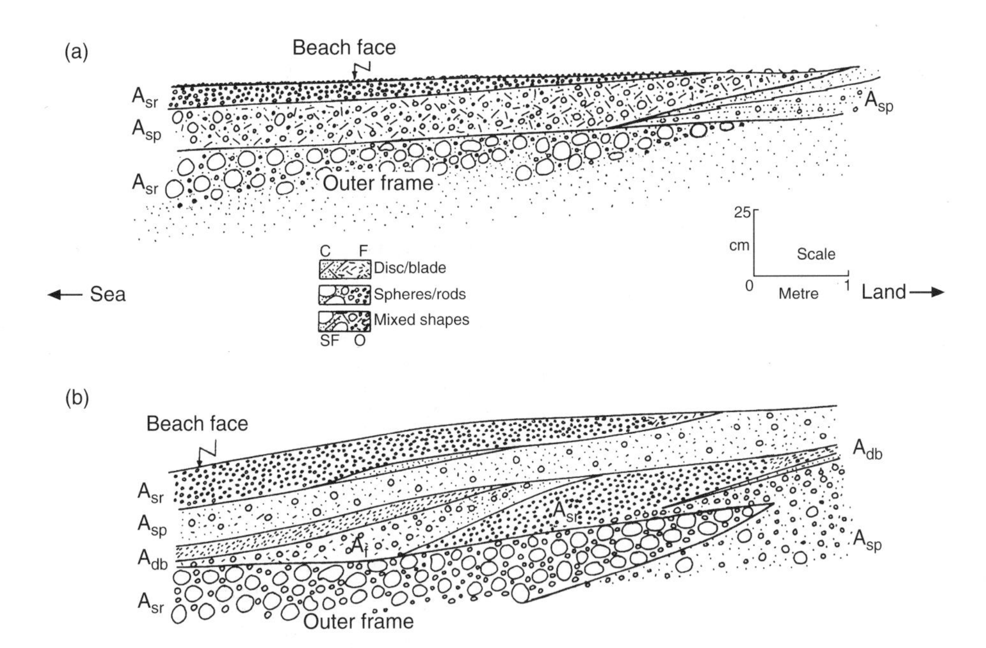

FIG 3.16 Sections through the seaward edges of gravel beaches, showing both seaward progradation and vertical accretion. (a) Predominantly vertically accreting sheets, interrupted by a mass flow deposit (Asp) derived from a steep, coarse part of the beach to the landward of this section; (b) Predominantly laterally accreting gravel sheets formed at low elevations where sediments are mobilized frequently due to changing wave and tidal conditions. Notation: A indicates a clast-supported sedimentary unit ; s,r,d,b indicate the dominant clast shapes in each unit (spheres, rods, disks, blades); f indicates a unit with poor size and shape sorting; SF is a sand-filled fabric and O is an open fabric. (After Bluck 1999).

4 Clast morphology

Douglas I. Benn

4.1 INTRODUCTION

Clasts in different depositional environments commonly exhibit contrasting morphologies as a result of their erosional, transport and depositional histories. Quantitative analysis of clast morphology therefore has the potential to provide information on aspects of debris history prior to and during deposition. In studies of glacigenic sediment, clast morphological analysis is particularly useful for differentiating subglacial debris from englacially, supraglacially or fluvially transported sediment, and in conjunction with macrofabric analysis (Chapter 5) can be used to infer processes of subglacial transport and deposition. Clast morphological analysis has generated a very large literature, and a wide and often confusing range of methods has been proposed (see Barrett, 1980; Illenberger, 1991; Gale and Hoare, 1991 for good reviews). This chapter focuses on methods of measurement and data interpretation that have proved robust and versatile in field studies in a variety of glacial environments.

4.2 MEASUREMENT AND DESCRIPTION OF CLAST MORPHOLOGY

Casual examination of any stone shows that its surface, in all its minute detail, has a very complex form. A full description of clast form is clearly impractical, so instead we focus on particular characteristics of clasts which are both (a) measurable and (b) useful as indicators of clast history. It is useful to identify three types of form characteristics, which form a hierarchical sequence (Barrett 1980; Benn and Ballantyne 1993):

1| *shape*, or the relative dimensions of the clast;

2| *roundness*, or the overall smoothness of the clast outline;

3| *texture*, or small-scale surface features.

Shape is the first-order property; roundness is the second-order property and is superimposed on shape; and texture is the third-order property and is superimposed on roundness. (Note that some authors use the term *form* to refer to shape as defined here, and use *shape* to refer to the overall morphology.)

4.2.1 Shape

The *shape* of a clast, in the sense used here, can be visualized as the shape of the smallest box that the clast will fit into. The dimensions of the box are determined by the length of the three orthogonal axes of the clast: these are termed the Long (L), Intermediate (I) and Short (S) axes. Some authors use the conventions a, b and c to refer to the long, intermediate and short axes, respectively. Shape depends simply on the relative dimensions of the three axes and forms a continuum between three possible end members:

1| a cube, with L = I = S;

2| a rod, with L >> I = S;

3| a disc, with L = I >> S.

All possible shapes can be represented on a triangular diagram in which each apex represents one of the end members (Sneed and Folk 1958; Fig. 4.1a). Shapes near the 'cube' apex are described as 'blocky' or 'equant', those near the disc apex are 'slabby' or 'oblate', and those near the rod apex are 'elongate' or 'prolate'. Sneed and Folk (*ibid*) also defined several categories of intermediate shapes, but these are rarely used (Fig. 4.1b).

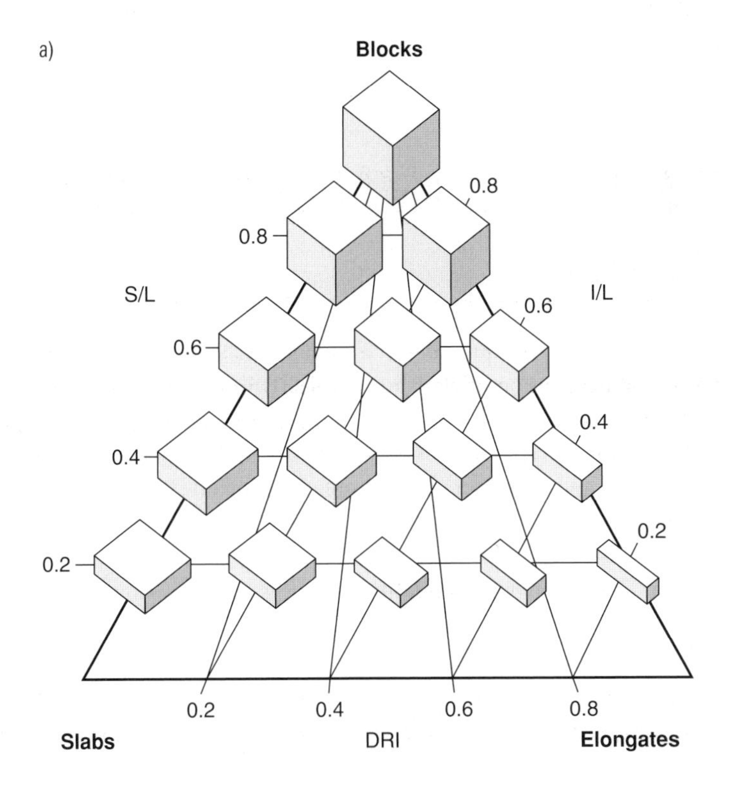

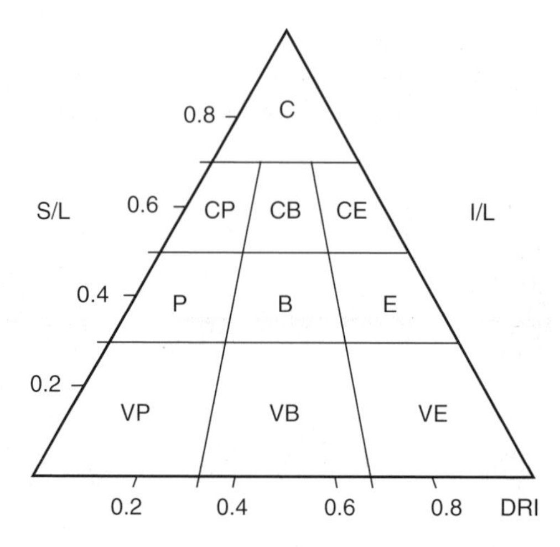

FIG 4.1 a) The continuum of clast shapes depicted on an equilateral triangular diagram (general shape triangle). The diagram is scaled using the c:a (S/L) and b:a (I/L) axial ratios, and the disc-rod index (DRI = (a-b)/(a-c) *or* (L-I)/(L-S)). B) Clast shape categories defined by Sneed and Folk (1958). C: compact; P: platy; B: blade; E: elongate; V: very.

4.2.2 Roundness

Roundness refers to the degree of wear exhibited by a clast, and is more difficult to quantify than shape. Several methods of quantifying roundness have been proposed based on the radii of curvature of particle edges. For example, the *Wentworth–Cailleux index* (2000r/L), compares the radius of curvature of the sharpest edge of the clast (r) with its longest dimension (L). Values of r can be determined from templates, although in practice r is very difficult to determine for angular or very angular clasts. Furthermore, the Wentworth–Cailleux index may be misleading, as the roundness of the sharpest edge may not be representative of the clast as a whole. As a result, the Wentworth–Cailleux and similar indices are no longer employed in many field studies. Several attempts have been made to apply digital imaging and processing techniques to the problem of quantifying clast form (e.g. Orford and Whalley 1991). Although rigorously objective, such techniques are of limited usefulness in field studies of clast-sized particles, because they require large numbers of particles or images to be collected for later analysis. They may, however, be valuable in studies of sand, silt and smaller grains. In field studies, clast roundness is now usually quantified by assigning a clast to one of a number of roundness categories (e.g. Krumbein 1941; Powers 1953; Pettijohn 1957; Olsen 1983). This can be achieved by comparing a clast with a set of standard images (e.g. Krumbein 1941; Fig. 4.2; inside cover), or a set of descriptive criteria (e.g. Benn and Ballantyne 1994; Table 4.1).

In studies of glacigenic sediments, the six roundness categories defined in Table 4.1 are normally sufficient to capture the important differences between samples. For some purposes, however, it may be necessary to add additional, intermediate categories (e.g. between angular and sub-angular clasts). The nine classes shown in Figure 4.2 can be very useful in this regard. The important point is to record the genetically significant variations between populations, while maintaining the objectivity and reproducibility of results.

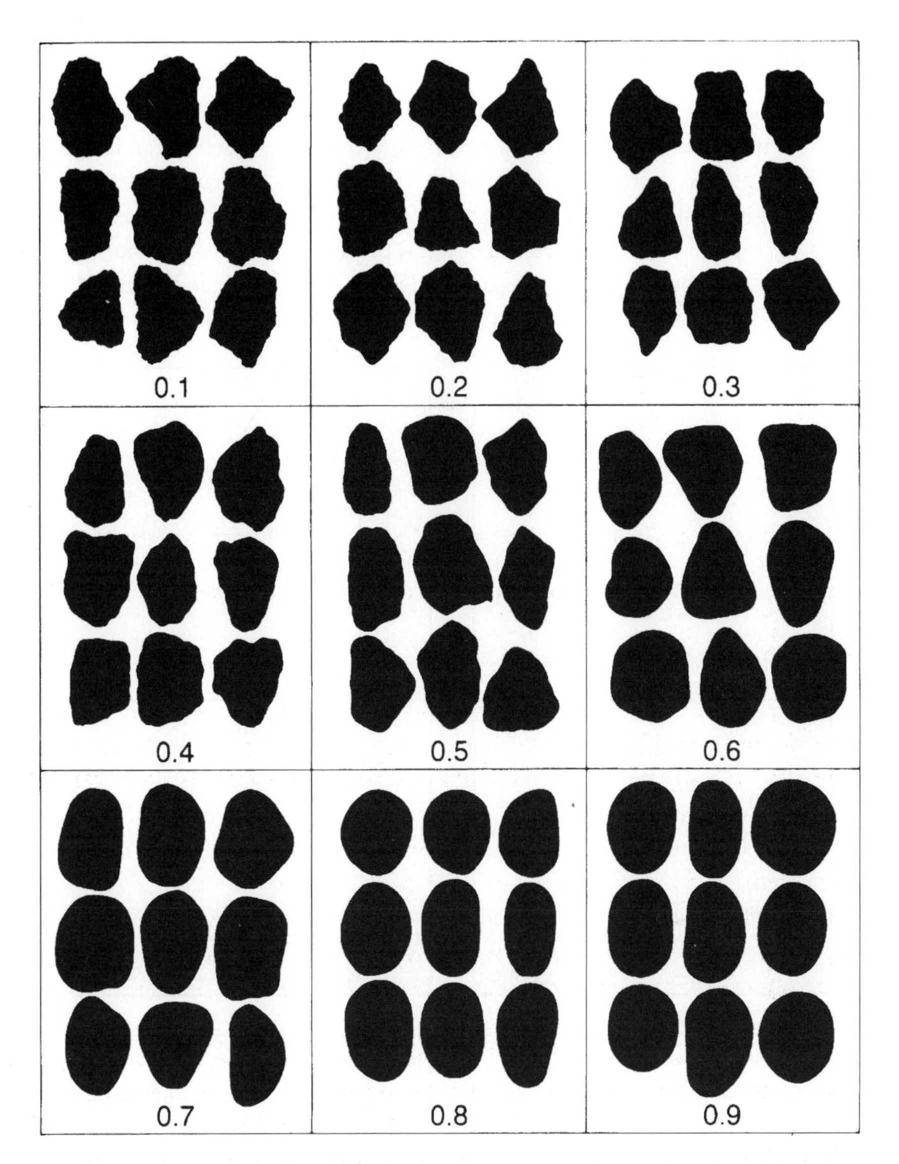

FIG 4.2 Comparison chart for clast roundness categories (Krumbein 1941).

Table 4.1 Descriptive criteria for clast roundness categories.

Class	Description
Very Angular (VA)	Edges and faces unworn; sharp, delicate protuberances
Angular (A)	Edges and faces unworn
Sub Angular (SA)	Faces unworn, edges worn
Sub Rounded (SR)	Edges and faces worn but clearly distinguishable
Rounded (R)	Edges and faces worn and barely distinguishable
Well Rounded (WR)	No edges or faces distinguishable

4.2.3 Clast asymmetry

Most measures of clast morphology provide a value or set of values describing the clast as a whole. Such measures will overlook *asymmetric wear patterns*, which are an important feature of some process domains, most notably subglacial environments. Debris-rich ice over-riding lodged clasts focuses abrasion on the upglacier (stoss) faces of clasts, and encourages fracture on the downglacier (lee) faces, producing distinctive asymmetric stoss-lee clasts (Fig. 4.3a; Boulton 1978; Krüger 1979; Sharp 1982; Benn and Evans 1998). Such forms are also known as 'bullet-shaped' or 'flat-iron' clasts. Where clasts are dragged

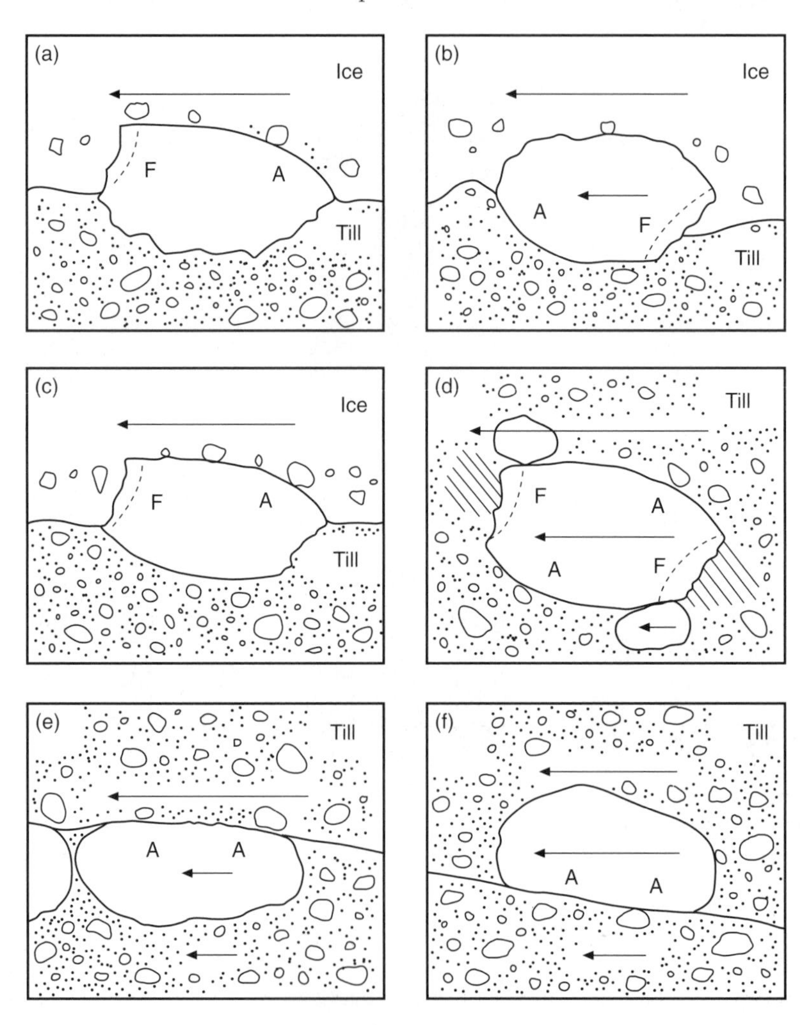

FIG 4.3 Hypothetical development of asymmetric clast wear patterns below sliding ice and within deforming till, showing principal locations of abrasion (A) and fracture (F). Arrow lengths show relative velocities. (a) Lodged clast with stoss-lee form owing to abrasion on the upglacier (stoss) side and fracture on the low-pressure downglacier (lee) side. (b, c) Double stoss-lee form resulting from a two-stage process of ploughing (b) and lodgement (c). (d) Double stoss-lee form resulting from a one-stage process within a deforming layer. Low-pressure zones are shaded. (e, f) Flat, polished facets eroded on the upper and lower surfaces of clasts adjacent to shear planes. (Modified from Krüger 1984, and Benn and Evans 1996.)

along at the base of a glacier prior to deposition, or undergo wear within shearing subglacial till, double stoss-lee forms can develop (Krüger 1984; Benn and Evans 1996, 1998; Figure 4.3b–d, G33). In contrast, clasts subjected to transport in fluvial or coastal environments tend to be worn more evenly, with no marked asymmetry. Thus, asymmetric wear patterns are an important source of information that are well worth recording. In the absence of easily applicable quantitative criteria, this can be done by assigning clasts to nominal categories.

4.2.4 Texture

The surface texture of clasts can yield useful information on transport mechanisms and depositional environments, although relatively little quantitative work has been done. A simple but effective approach is to record the presence/absence of particular surface textures or features. Potentially useful information includes:

- fresh, unworn surfaces indicative of no wear since fracture;
- rough, weathered surfaces;
- smooth, polished surfaces;
- striations;
- crescentic fractures on polished pebbles, which may indicate high impact forces in shoreface environments;
- conchoidal fractures;
- ventifacts, recording polishing by wind-blown sand.

The identification of textures is necessarily subjective, but provides a useful means of recording evidence often noticed in the field.

4.3 FIELD MEASUREMENT OF CLAST MORPHOLOGY

4.3.1 Sample design

Sampling design should be tailored to particular research questions. As a general principle, samples should be representative of the whole population of clasts in a deposit, and we should therefore aim to take a number of samples from different locations or facies, each sample being of a sufficient size to minimize random sampling effects. Very little work has been done on the statistical variance of clast morphological data, so there is little theoretical basis for determining the optimum sampling strategy or sample size. Many workers use sample sizes of 50, which seems to be a reasonable compromise between maximizing data quality and minimizing time spent sampling. Care should be taken to sample randomly, avoiding bias in the choice of clasts.

Clast shape can be influenced by clast size, due to scale-dependent variations in the distribution of joints or other weaknesses in the parent rock, and other factors (Benn and Ballantyne 1994). Thus, clast samples should lie within relatively narrow size ranges. There are no standard recommended size ranges, although Benn and Ballantyne (1994) have found that sampling within the range 35–125mm yields reproducible results. This is unlikely to be the case for all rock types and process domains. It is important to confine samples to single rock types, because different lithologies can respond differently to similar wear processes. For example, sedimentary rocks with well-developed bedding planes commonly yield slabby clasts, whereas clasts derived from massive igneous rocks tend to be more equant.

Before setting out on a sampling program, it is useful to think carefully about how the data will be analysed. Large datasets will be of little use if there is no clear and unambiguous method of interpreting them. Too often, students (and some professionals) collect data, then attempt to make sense of the results with *post hoc* 'explanations', which tend to be unscientific and of dubious value. Data interpretation should be guided by clear principles and be as unambiguous as possible. In clast morphological studies, the best method of achieving objectivity is through the use of *control samples*. That is, data collected from deposits of unknown origin should be compared with samples that have a known history. For example, Benn and Ballantyne (1994) used samples of scree, supraglacial debris and subglacial tills to guide the interpretation of samples from lateral moraines. The use of such control samples provides known reference points against which other data can be compared, either informally or statistically. The choice of control populations should be guided by theory, and should incorporate all of the process domains thought to be important in a given environment (see Section 4.6.1).

4.3.2 Data collection

Shape

The long, short and intermediate axes of clasts can be measured quickly and easily using a tape measure or ruler. It is usually sufficient, for clasts >5cm, to measure to the nearest 0.5cm, probably the best that can be achieved by eye with a ruler or tape. More accurate data can be obtained using calipers, although this is much more time consuming. If one is doing a lot of shape measurements, it is worth constructing a *shape box* (Shakesby 1979). This consists of three sides of a box with scales marked on the inner faces. A clast is placed against the inside corner of the box, and the dimensions read off from the three scales, allowing the rapid collection of accurate data.

Roundness

The assessment of particle roundness is more subjective than measurement of shape, and care must be taken to avoid bias. The definitions of the roundness categories in Table 4.1 are unambiguous, but the boundaries between classes are less so and may be interpreted in different ways by different workers. Data collection by a single operator will help to maintain internal consistency in a dataset, but this is not always possible and does not address the problem of comparability with other studies. Perhaps the best solution is to cross check for consistency between operators and to ensure that inexperienced operators are properly trained.

Ambiguities in roundness categories often arise when clasts display varying degrees of roundness across their surfaces. For example, a rounded pebble with broken edges may be classified as rounded by one worker (emphasizing the rounding), angular by another (emphasizing the breakage) and sub-rounded by yet another (attempting to compromise). The best approach in such cases is to record additional information (e.g. 'rounded (broken)'). Detailed data can always be simplified later, but simplified data can never be made more detailed.

4.4 DATA PRESENTATION AND ANALYSIS

The graphical representation of data is very important. Few people can gain an appreciation of the meaning of a dataset by scanning a column of numbers or statistics, but most can do so by viewing a clear diagram or graph. Shape data are best represented on triangular diagrams, which reflect the three-way variation of all possible orthorhombic particle shapes. Several forms of triangular shape diagram have been proposed (e.g. Illenberger 1991; Hofmann 1994), but the most versatile is the linearly-scaled *general shape triangle* (Sneed and Folk 1958; Hockey 1970; Ballantyne 1982; Benn and Ballantyne 1993; Fig. 4.1). This diagram is scaled using the S/L (c:a) ratio, which measures 'blockiness', and the I/L (b:a) ratio, which measures 'elongation'. These two indices plot as isolines parallel to the basal and left edges of the triangle, and can be used together to uniquely define any clast shape. The third 'edge-parallel' index is (L–I)/L (Benn and Ballantyne 1993, 1995). The general shape triangle can also be scaled using isolines that plot radially from the three apices. Sneed and Folk (1958) used the Disc-Rod Index, (L–I)/(L–S), which plots radially from the 'cube' apex, to distinguish disc- and rod-shaped clasts. The other members of the family of 'radial' indices are: S/I, which plots as radial lines from the 'elongate' apex, and S/(L–I), which plots radially from the 'slabby' apex. These appear not to have been used in data interpretation and are only mentioned here for completeness. A useful Excel spreadsheet for plotting clast shape data on the general shape triangle has been produced by Graham and Midgely (2001).

Clast-shape data are sometimes summarized using measures of *sphericity*. For example, Sneed and Folk (1958) defined Maximum Projection Sphericity (MPS) as $\sqrt[3]{(S^2/LI)}$. This index is derived from the ratio of the maximum projection area of a clast to that of a sphere with the same volume, and reflects the balance of drag and gravitational forces acting on a particle immersed in a fluid. As such, MPS is a good index of the behaviour of clasts transported by or settling in water, but is of limited usefulness outside this hydraulic context. The same remarks apply to other indices of sphericity, such as Krumbein intercept sphericity (Krumbein 1941). For most purposes, the simple blockiness and elongation indices, S/L and I/L, are sufficient to summarize shape.

Clast-shape data may be plotted on biaxial diagrams such as the Zingg plot, which graphs S/I against I/L. Such diagrams have been shown to be less useful and versatile than the general shape triangle (Benn and Ballantyne 1993), although they still appear in the literature from time to time. Clast-roundness data are most simply displayed in histograms. Bars can be given different shadings or colours to simultaneously display textural data.

It is useful to employ summary statistics to characterize samples of clasts and discriminate between pairs of samples. The aggregate form characteristics of samples may be summarized using either: (1) index means and standard deviations; (2) medians and interquartile ranges or (3) statistics based on the percentage of a sample with index values above or below some reference value (Benn and Ballantyne 1993). Parametric statistics are only applicable where the indices for a sample are normally distributed (Dobkins and Folk 1970), and in many cases non-parametric methods must be used (Ballantyne 1982). No single statistic or pair of statistics will serve to characterize all samples for all purposes, and the choice of analytical method must be guided by the nature of the data and the purpose of the research. If chosen arbitrarily, summary statistics may be insensitive to the actual variability in the dataset and of limited usefulness. Therefore, when choosing summary statistics it is important first to view the data graphically (e.g. on the general shape triangle). Viewing the data allows the relative distributions of samples to be identified, thus pointing to the most effective statistic to distinguish between them. For example, Ballantyne (1982) found that the main difference between frost-shattered and subglacially-transported gneiss clasts in Jotunheimen, Norway, was the degree of 'blockiness', which is measured by the S/L ratio (Fig. 4.4). He found that samples from these two populations could be most readily differentiated using the percentage of clasts in a sample with S/L ratios ≤0.4. This statistic, subsequently termed the C_{40} index (Benn and Ballantyne 1993), has proved to be of particular value in distinguishing actively-transported (subglacially modified) and passively-transported (i.e. englacially, supraglacially or subaerially transported) clasts of a wide range of crystalline lithologies (Ballantyne 1986; Vere and Benn 1989; Benn 1992; Benn and Ballantyne 1994; Krüger and Kjaer 1999). It must be emphasized, however, that this statistic will not be effective in all situations, and that other statistics (perhaps of the same general type) may need to be used in other contexts. For example, where it is important to distinguish subglacially- and fluvially-transported debris, the C_{40} index is of little or no use, because neither type of deposit is likely to contain many angular or very angular clasts. In this case, the percentage of rounded and well-rounded clasts will provide a much more powerful discriminant statistic, because rounded clasts tend to be abundant in fluvial sediments but rare in glacially-transported debris (see Section 9.4.1).

Mean roundness statistics have been calculated by some workers (e.g. Matthews and Petch 1982; Olsen 1983) by converting roundness categories into an arithmetic scale. This procedure is questionable, however, because clast-roundness data are commonly not normally distributed, and there is a logical difficulty in converting ordinal data into rational form. This problem does not arise for summary indices that express the percentage of clasts in particular categories or groups of categories.

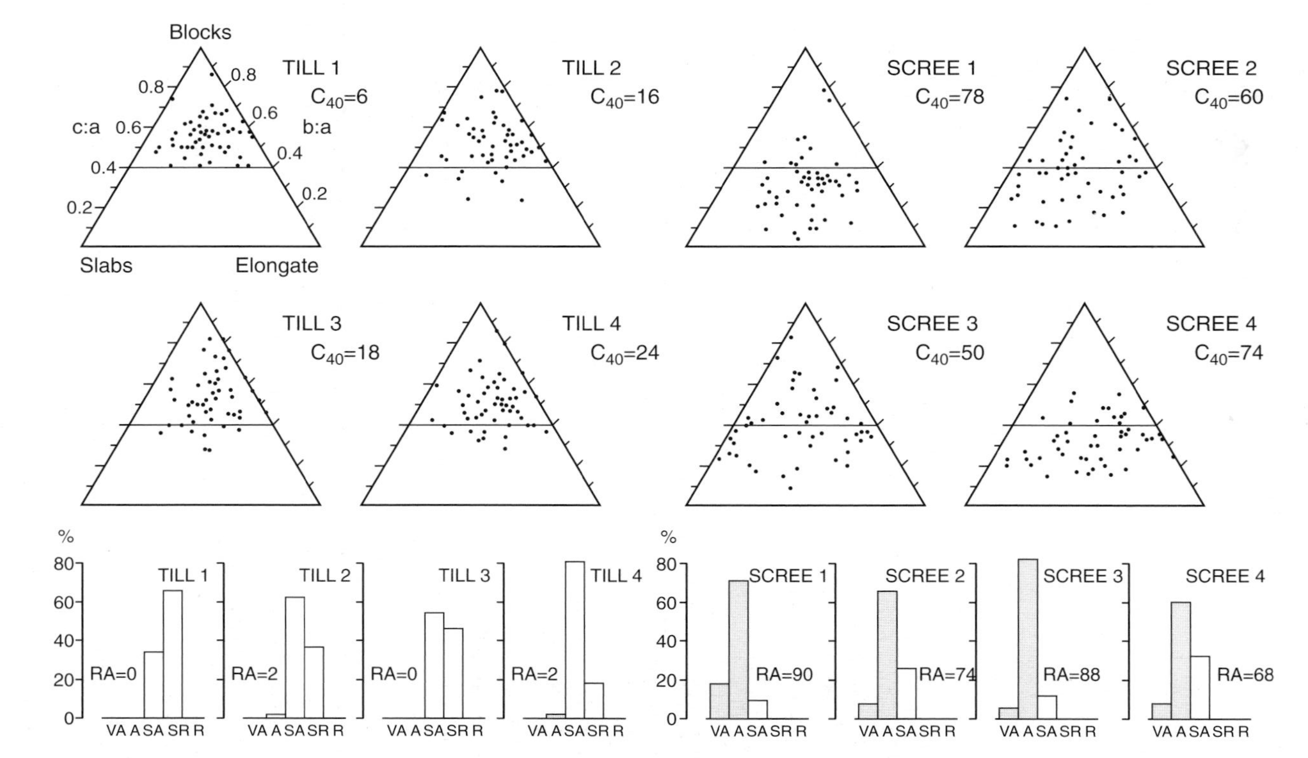

FIG 4.4 Clast-shape and roundness data for subglacially-deposited debris (basal till) and rockfall debris (scree) near the margins of Storbreen, a cirque glacier in Norway. (From Ballantyne and Benn 1994).

4.5 CO-VARIANCE ANALYSIS

Analysis of the way in which two aspects of clast form (e.g. shape and roundness) vary, can yield much more information than the study of each characteristic in isolation (Pettijohn 1957). However, the usefulness of this approach depends on the choice of shape and roundness indices. For example, Boulton (1978), Domack *et al.* (1980) and Dowdeswell *et al.* (1985) plotted Krumbein's intercept sphericity against roundness indices for samples of clasts from different glacial environments. Their plots show clear between-sample differences in roundness, but the sphericity (shape) values show little or no between-sample variation. Thus, in these studies co-variance analysis yielded no more information than the study of roundness alone, because Krumbein's sphericity is an insensitive measure of the variation found in the shape of glacially-transported clasts.

In contrast, Benn (1992) and Ballantyne and Benn (1994) showed that co-plots of RA (roundness) and C_{40} (shape) clearly differentiated groups of samples, allowing detailed insights to be gained into the provenance and transport history of glacial deposits. Figure 4.5 shows summary clast-form data from Storbreen, a cirque glacier in eastern Jotunheimen, Norway. The plots show samples taken from the crest of the Little Ice Age maximum moraine, together with two groups of control samples: (1) scree and passively-transported supraglacial debris and (2) clasts from basal debris layers and subglacial tills from the glacier foreland. The two sets of control samples show clearly distinct characteristics; the scree/supraglacial samples have high RA and C_{40} values (i.e. high percentages of angular and slabby/elongate clasts), whereas the till/subglacial samples have low RA and C_{40} values. The moraine samples plot in a band between the two control groups, indicating that they represent mixtures of debris from supraglacial and subglacial

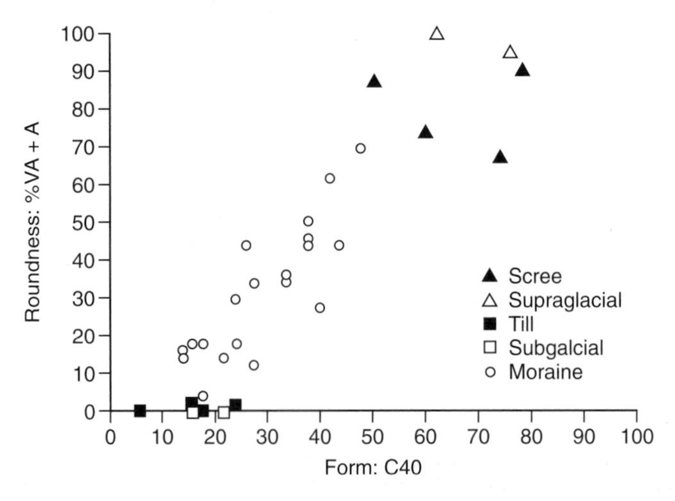

FIG 4.5 RA-C40 diagram showing clast-shape and roundness data from Storbreen. Control samples are from scree, passively-transported supraglacial debris, basal till and debris from basal debris-rich ice. The moraine samples plot as a band between the two sets of control samples, suggesting that they comprise mixtures of debris from passive and active transport paths.

sources. The position of particular moraine samples on the diagram allows the relative proportions of the two sources to be estimated.

As noted above, the RA and C_{40} indices will not be suitable measures in all cases, depending on the nature of the datasets in question. The important point is that co-variance analysis is a powerful technique only if *both* form indices have been shown to be useful for distinguishing between the debris types of interest. Further examples of co-variance analysis are presented in Section 4.6.

4.6 CASE STUDIES

4.6.1 Co-variance of aggregate characteristics: Glen Arroch, Skye

Figure 4.5 shows how clasts from moraines on the foreland of Storbreen have shape and roundness characteristics intermediate between those of two control groups, suggesting that the moraine clasts represent mixtures of the two. However, these two control populations are not the only two possible debris sources in all cases. Figure 4.6 shows co-plots of shape and roundness data for Younger Dryas moraines in Glen Arroch, Skye, together with control samples (Benn 1992). In contrast with Figure 4.5, the moraine samples do not form a band between the scree and till controls. Instead, the moraine

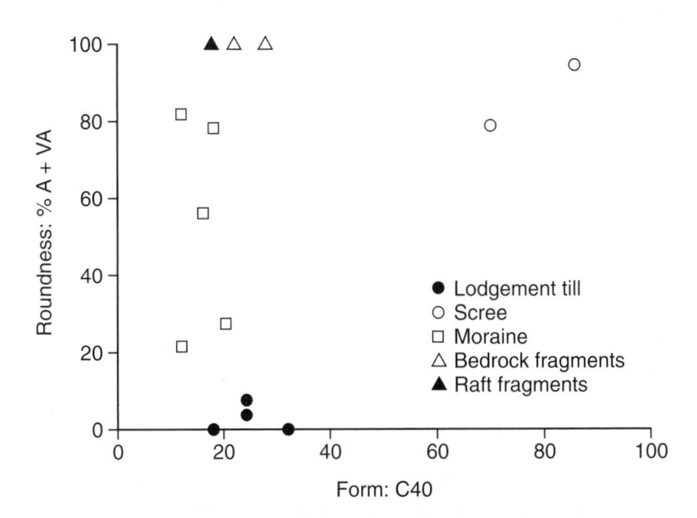

FIG 4.6 RA-C40 diagram for data from Glen Arroch, Skye. Control samples are from scree (representing passively-transported rockfall debris), basal till (representing subglacially-transported debris) and fragments prised from roadside bedrock exposures and rock rafts contained in the moraines. Clast samples from the moraines plot as a vertical band between the rock fragment controls and the basal till controls, suggesting that they consist of rock fragments quarried from the glacier bed, which have undergone variable amounts of edge-rounding but little or no change in shape. (From Benn 1992.)

samples have a narrow range of C_{40} values, similar to those of the till controls, despite the fact that the RA values display a broad range. This pattern caused major confusion in the early stages of the investigation, when it was believed that passively-transported scree and actively-transported subglacial debris represented the only two possible inputs to the moraines. A solution was found when it was realized that, in this case, there was a third possible source: rock fragments quarried from the bed relatively close to the point of deposition. A further set of control samples was taken to test this idea. Joint-bounded rock fragments were prised from bedrock outcrops and from a large 'raft ' in one of the moraines, and samples were found to have low C_{40} values (predominantly 'blocky') and high RA values (fresh and unworn, as would be expected). The position of the rock fragment controls on the RA/C_{40} diagram indicates that the moraine samples most likely represent quarried rock fragments which have undergone variable degrees of edge-rounding (but no overall change in shape) prior to deposition in the moraines.

This example illustrates the importance of maintaining an open mind during clast-form studies, and not allowing a single ruling hypothesis to dominate data interpretation. Recent work has shown that fluvial transport (subglacial, englacial and supraglacial) can form a significant component of the debris cascade on many glaciers (Kirkbride and Spedding 1996; Spedding 2000), so future glacial clast-form studies should also include fluvial controls where possible.

4.6.2 Co-variance of sub-samples: Storbreen, Norway

In Section 4.5 we concluded that the moraine samples from Storbreen probably consist of mixtures of actively- and passively-transported debris. Critical consideration of Figure 4.5, however, suggests that there are in fact two possible interpretations of the data: (1) the moraine samples are mixtures of two distinct clast populations and (2) the moraine samples represent different stages of the evolution of clast populations from an initial state (scree-like) towards a final state (till-like). Benn and Ballantyne (1994) devised a strategy to test these competing hypotheses. They reasoned that, if (1) is true, then the angular clasts in the moraines should have shape characteristics similar to those of the scree/supraglacial controls, whereas the sub-angular and sub-rounded clasts should have shape characteristics similar to those of the till/subglacial controls. Conversely, if (2) is true, the shape characteristics of the moraine clasts should be intermediate between those in the control groups, for all roundness classes. To determine which is the case, additional sub-samples were taken at each site, consisting of 25 clasts in each roundness category. The results (Fig. 4.7) clearly show that the angular and very angular clasts in the moraines are similar to those in the scree controls, whereas the sub-angular and sub-rounded clasts in the moraines closely resemble those in the same roundness classes in the till control group. This pattern, which holds whether either C_{40} or median values are used to summarize aggregate clast shape, indicates that hypothesis (1) is true: that the clasts in the moraines are mixtures of actively- and passively-transported material.

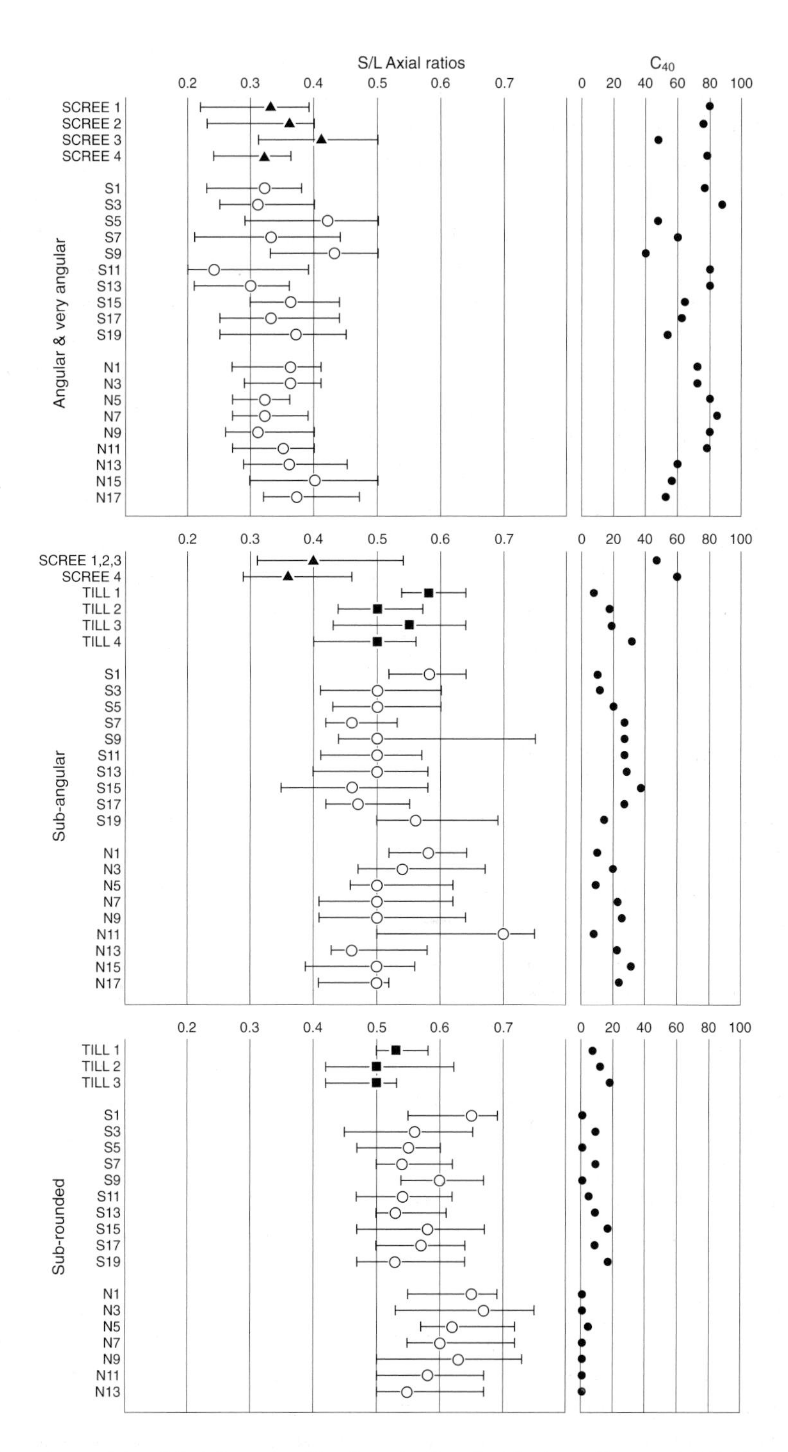

FIG 4.7 Shape characteristics of clasts from scree, till and moraine deposits at Storbreen, Norway, subdivided by roundness category. The data show mean and interquartile ranges of S/L axial ratios (left panel) and C40 values (right panel). (From Benn and Ballantyne 1993.)

Interestingly, Figure 4.7 also shows that the shape characteristics of sub-angular clasts in the scree are similar to those of angular/very angular scree clasts. This probably means that the sub-angular scree clasts have undergone some edge rounding by weathering processes, rather than abrasion, with no corresponding change in clast shape.

These case studies show how co-variance analysis is a powerful tool for exploring debris transport processes in glacial environments. With modification, these and similar methods could also be applied to other sedimentary systems, such as facies deposited by shoreface processes or gravel-bed rivers. The examples also demonstrate how important it is not to be 'locked in' to preconceived ideas about the system under study, and the need for a flexible, critical approach. If applied uncritically, clast-shape studies may yield little of value. With thoughtful research design, they can provide valuable windows on the dynamics of sediment transport.

5 Macrofabric

Douglas I. Benn

5.1 INTRODUCTION

'Fabric' refers to all of the directional properties of a sediment, including the orientation of particles, bedding planes, folds, faults, erosion surfaces and others (*cf.* Derbyshire *et al.* 1976). 'Macrofabric' is applicable to those properties visible to the unaided eye. Many authors have used 'fabric' in a more restrictive sense, to refer to the orientation of particle long axes alone (e.g. Bennett *et al.* 1999), but this usage is not followed here because particle long axes are only one of several useful fabric elements.

It has long been recognized that sediment fabric can yield valuable information about former depositional processes and environments, such as the flow directions of glacier ice, rivers and winds (e.g. Miller 1884; Holmes 1941). More recently, fabric data have been used to infer depositional processes and deformational mechanisms (e.g. Dowdeswell and Sharp 1986; Hart 1994; Benn 1994a, b). As a result, fabric analysis has become a standard technique in Quaternary sedimentology (Gale and Hoare 1991; Krüger and Kjaer 1999). Recently, the value of fabric analysis has been questioned (Bennett *et al.* 1999). While it is true that poorly thought-out fabric studies will yield poor results, we will show in this chapter that well-designed fabric analyses can yield valuable insights into former processes and environments. We review the main types of fabric data that are collected in Quaternary studies, then examine some of the most useful methods of data analysis and presentation, and conclude by illustrating the potential of fabric analysis in case studies.

5.2 BEDDING, FOLDS AND FAULTS

5.2.1 Bedding

Measurement of the tilt of bedding surfaces is a standard technique in geology, used as a means of determining the tectonic deformation of sedimentary sequences. For Quaternary sediments, such measurements are useful where tectonic or glacitectonic deformation has occurred. It is also useful to measure the orientation and gradient of bedding planes in sedimentary systems that typically have sloping depositional surfaces, such as delta fronts, alluvial or subaqueous fans, or lateral-frontal dump moraines (Owen 1991; Ballantyne and Benn 1994; Benn and Owen 2002). In geology, measurements are made of the *strike*, or the orientation of a horizontal line across the tilted plane, and the *dip*, or gradient at which the plane slopes (i.e. the gradient in the direction normal to the strike). It is also necessary to record the direction towards which the plane slopes. Alternatively, one may measure the *dip* and *azimuth*, or compass bearing, of the slope (Fig. 5.1).

5.2.2 Cross-bedding

Many fluvial and aeolian lithofacies exhibit *cross-bedding*, that is, internal bedding which is inclined relative to the upper and lower bounding surface of the unit (Chapter 2; Figs. G2, G4 and G6). Cross-bedding records the lateral or downstream migration of bedforms such as ripples, bars or dunes, by the build-up of successive layers of sediment (Reinick and Singh 1980; Allen 1982). As such, they are widely used as palaeoflow indicators, on the reasoning that the dip of cross-beds records the downstream (or downwind) direction. This reasoning, however, must be applied with caution. Some types of fluvial bedform (such as point bars) migrate laterally across-channel, in which case cross-bedding represents *lateral accretion surfaces* which dip approximately at right angles to the flow direction. Many other fluvial and aeolian bedforms (e.g. trough cross-bedding; Fig. 2.3) have similarly complex relationships with flow direction. Consequently, to obtain reliable palaeoflow directions from cross-bed orientation, it is necessary to (a) determine exactly what types of cross-beds are being examined, and/or (b) take a large number of measurements and obtain a general palaeoflow from their mean. The best way to do (a) is to dissect the sediments to obtain a three-dimensional view. This may be very difficult to do from limited 2-D exposures and (b) may be the best option.

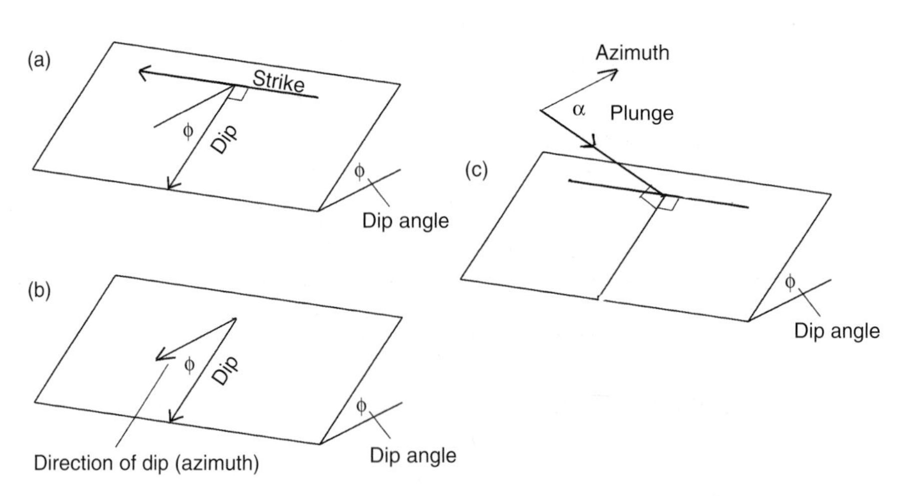

FIG 5.1 Three methods of recording the orientation and gradient of a sloping surface. (a) Strike and dip: measurement is made of the bearing of the horizontal line (strike), the angle of dip (phi), and the downslope direction. (b) Azimuth and dip: the same information is obtained by measuring the dip angle and the direction of dip (azimuth). (c) Pole-to-plane: the plane is represented by the line drawn at right angles to the plane (the 'normal' or the 'pole'), and the angle of plunge and the azimuth of the line are measured. The angle of plunge alpha = 90° – phi, and the azimuth in (c) is 180° from the azimuth in (b).

5.2.3 Faults

Faults are fractures along which displacement has occurred. The definition of basic terms and a general classification of fault types is shown in Figure 5.2. In the case of inclined fault planes, *normal faults* are those in which the upper (hanging wall) block is displaced downward relative to the lower (foot wall) block. *Thrust faults* (or *reverse faults*) occur where the hanging wall block is displaced upward relative to the foot wall block, while *strike-slip faults* record lateral, but no vertical, displacement. *Oblique-slip* and *rotational*

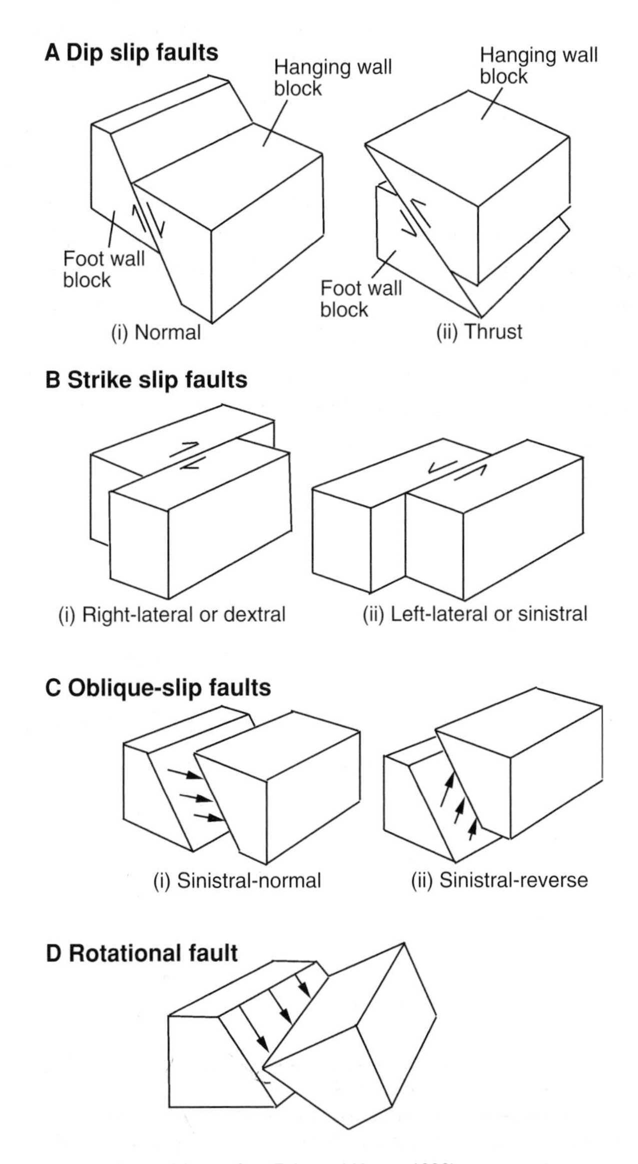

FIG 5.2 Terminology for describing faults. (Adapted from Twiss and Moores 1992)

faults record some combination of vertical and lateral displacements. In Quaternary settings, normal faults generally occur where sediments have undergone subsidence or collapse, such as slope failures or around the margins of melted ice blocks (Fig. G13). Thrust faults commonly occur in glacitectonized sediments and the toe of large slope failures, where the sediment pile has been shortened by compressional deformation. Faults therefore represent valuable records of the deformational history of sediments, and their orientation and directions of displacement should form part of a section log.

The orientation of fault planes is measured in the same way as bedding surfaces, i.e. the strike and dip (or azimuth and dip) should be recorded. In addition, the direction and amount of displacement should be measured where possible. Direction and amount of displacement can usually be established with reference to recognizable markers such as distinctive beds on either side of the fault plane.

5.2.4 Folds

Folds are wavelike undulations caused by deformation of rock or sediment layers. In Quaternary sediments, folds may be due to subglacial or proglacial glacitectonic deformation, slope failures or, in seismically active regions, tectonic deformation (e.g. Owen and Derbyshire 1988; Hart *et al.* 1990; Benn and Evans 1998). Deformation structures are very useful palaeo-environmental indicators, and can provide detailed information on post- and syn-depositional processes (Chapter 2). Analysis of sequences of deformation structures (sometimes known as kineto-stratigraphy) is among the most important tools used in reconstructing glacier dynamics, especially in situations where sediment successions have been dislocated and the law of superposition does not apply (Berthelsen 1978, 1979; Rose and Menzies 1996; van der Wateren 1999; van der Wateren *et al.* 2000).

Some basic terminology used to describe folds is illustrated in Figure 5.3. The *hinge line* or *hinge* is the line in a folded surface where the fold curvature is greatest, and the less tightly curved parts of the fold on either side are known as the *fold limbs*. The surface joining all the hinge lines in a particular fold is called the *hinge surface* or, more commonly, the *axial plane*. The outcrop of the axial plane on a surface is known as the *axial trace*. The orientations of folds are defined with reference to the hinge and axial plane. Folds with a vertical axial plane are referred to as *upright folds*, whereas those with tilted axial planes are *inclined*. *Recumbent folds* have horizontal axial planes. Hinge lines may be *horizontal*, *plunging* or *vertical*. Further details of the typology of folds can be found in any good structural geology textbook (e.g. Ramsay and Huber 1987; Twiss and Moores 1992).

In the field, measurements should be made of the orientation and plunge of the axial plane. It should be remembered that the apparent geometry and orientation of folds seen in section may be misleading. Exposures are very unlikely to lie at right angles to the true axial planes of folds, and it is therefore necessary to make careful excavations to determine where axial planes actually lie. This is generally easiest where distinctive and cohesive beds have been affected by the fold. Having identified the axial plane, the strike and dip should be measured with a compass and clinometer, in the same way as for bedding surfaces.

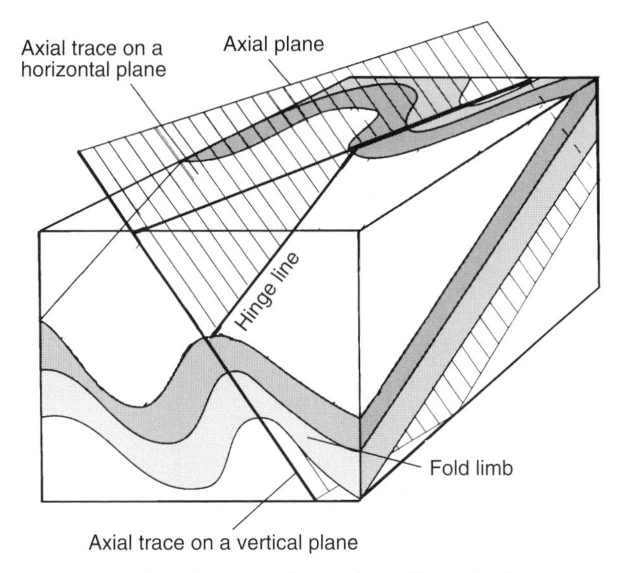

FIG 5.3 Some terminology used for describing folds. Adapted from Twiss and Moores (1992).

Measurements are most easily made if a portion of the axial plane is excavated, creating a flat surface. The plunge (azimuth and gradient) of the fold hinge should then be noted.

Folds are not, of course, infinite in horizontal extent. All die out at their edges, often curving backward giving the whole fold a 'sock-like' form. It should therefore be borne in mind that the part of a fold seen in section may not be representative of the fold as a whole, and may not bear a simple relationship with the large-scale stress field that formed it. Therefore, it is good practice to take as many measurements as possible, to provide a representative sample.

5.3 PARTICLE FABRIC

There is a very long tradition of interest in particle fabrics, especially in glacial studies where a-axis fabric data have been used to reconstruct ice flow directions for over a century (Miller 1884; Krumbein 1938; Holmes 1941). More recently, the potential for fabric signatures to yield information on genetic processes has been widely recognized (e.g. Glen *et al.* 1957; Lawson 1979; Domack and Lawson 1985; Dowdeswell *et al.* 1985; Dowdeswell and Sharp 1986; Hicock 1991; Hart 1994; Benn 1995; Larsen and Piotrowski 2003). Much of this work has focused on a-axis fabrics, because a-axes are known to rotate under applied stresses and their orientations thus provide information on former depositional and deformational processes (Benn 1994b; Hooyer and Iverson 2000a). However, a-axes represent only one of many possible directional properties of sedimentary particles, including the orientation of particle a-b planes, polished facets and plucked lee faces (Fig. 5.4). Particle a-b planes are commonly oriented so as to minimize particle resistance to water or ice flow, and are typically *imbricated* or tilted up-flow. In

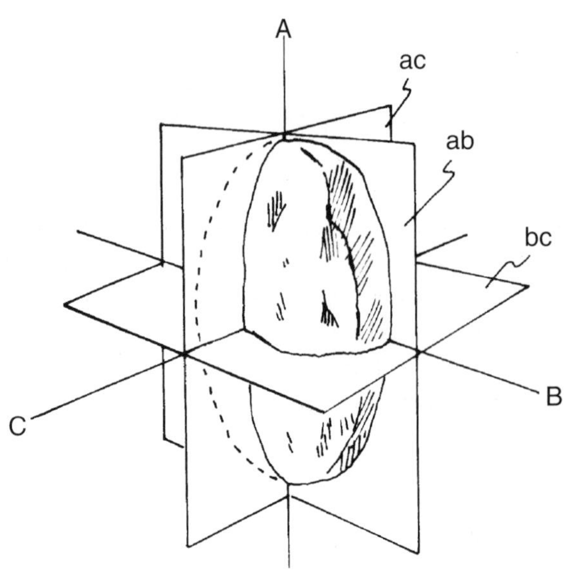

FIG 5.4 Definition of clast axes and planes. Note that each axis is at right angles (normal) to a plane: the c-axis, for example, is normal to the a-b plane. This means that the orientation of a plane can be specified by measuring the azimuth and dip of its normal.

fluvial gravel deposits, a-axes tend to lie transverse to flow and b-axes imbricated upstream (a(t)b(i) fabrics); whereas in mass-flow deposits a-axes tend to be flow-parallel, and a-axes imbricated upstream (a(p)a(i) fabrics; Boothroyd and Ashley 1975; Hein 1984).

Clasts in tills commonly exhibit *polished facets* on their upper and lower surfaces, parallel to the glacier bed or planes of shearing within the till, recording clast modification at or within the glacier bed (Hicock 1991; Benn 1994a; Fig. 4.3). Additionally, lodged clasts typically have asymmetric stoss-lee forms, like miniature *roches moutonnées*, as a result of abrasion and plucking by overpassing ice. Krüger (1984) described *double stoss-lee forms*, which he attributed to sequential ploughing and lodgement of clasts at the glacier sole, although it is likely that such forms can also develop within deforming till (Benn 1995; Benn and Evans 1998; Fig. 4.3, G33). Thus, many fabric elements are potential sources of information, and for detailed process studies (e.g. of subglacial till) it is often worth the considerable effort to measure all or most of them (Hicock 1991; Benn 1995).

5.3.1 Sampling

Choice of clasts

Usually, some subset of the total population of clasts in a lithofacies is chosen for sampling. First, for studies of a-axis fabrics, sampling is commonly restricted to clasts in which the a-axis is significantly longer than the b-axis, on the grounds that clasts with similar a- and b-axes will not be sensitive indicators of orienting forces and processes. A common choice of limit is L/I (a:b) ratios >1.5:1, but some authors have used a limit of 2:1 (Andrews 1971). The important point is that, if the results are to be compared with other studies, comparable methods must be used. In studies of fabric elements other than a-axes, it may be useful to sample from the whole population of clasts, regardless of axial ratios. Second, particular size limits may be chosen. Fabric characteristics may vary with clast size for a variety of reasons (Krüger 1970; Benn 1994a; Kjaer and Krüger 1998), so it is important to avoid introducing unintentional bias. Third, particles should be sampled from as small an area of the sediment as possible, to avoid variability associated with subtle facies changes. Fourth, within the constraints set by choice of particle shape and size, clasts for sampling should be chosen at random. Special care should be taken to avoid selecting clasts that appear to support a preferred hypothesis. Finally, one should have a very clear idea *why* data are being collected. Collection of fabric data can be a tedious and time-consuming business, and it is clearly desirable to avoid wasted effort. Data collection should always be guided by a clear sense of what one wants to know (see Chapter 9).

Field measurements

The equipment needed to measure pebble fabric consists of a compass-clinometer or equivalent instrument, and a knitting needle or long pencil. To make a measurement, randomly select a clast and carefully remove it from the surrounding matrix. Identify the fabric element you wish to measure (e.g. a-axis, a-b plane, or erosional facet) then replace the clast, removing additional matrix if necessary to retain a clear view of the fabric element. For a-axis measurements, place the knitting needle or pencil parallel to the axis and, with this as a guide, use the compass-clinometer to measure the gradient of the fabric element relative to the horizontal (*dip* or *plunge*) and the direction the dip is towards (*orientation* or *azimuth*; Figs. 5.4, 5.5). If a knitting needle is used, ensure that it does not attract the compass needle! The orientation of planes can be determined by measuring the azimuth and dip of a line *normal* (at right angles to) the plane. In the case of the a-b plane, the normal is the particle c-axis (Fig. 5.4). Record each pair of measurements (or multiple pairs if you are measuring more than one fabric element), then remove the clast and discard it to avoid duplication of measurements. Some researchers have suggested that fabric measurements are most accurate if made on a horizontal surface excavated into the deposit (e.g. Andrews 1971). This is straightforward for tills exposed on modern glacier forelands, but in quarries and coastal and riverbank exposures this is often not possible, and measurements are commonly made on vertical or near-vertical exposures.

Lengthy sampling programs can prove very rough on the hands. Fine rock flour (till matrix) has a powerful drying effect on the skin and can leave fingers chapped and painful. It is therefore worth using some form of barrier cream or surgical gloves to protect the hands. Studies of clast fabrics in glacier ice can be particularly trying, as chipping clasts out with an ice-axe will leave hands wet, cold and often numb. A good trick is to employ a blow-torch to melt clasts out of the ice. This has the added advantage of drying and warming the clasts, thus making them more pleasant to handle.

Sample sizes

There is at present no quantitative basis for the choice of sample sizes in fabric studies. The question of sample size is, however, an important one. Because the actual sample

FIG 5.5 Measurement of the azimuth (a) and dip (b) of clast a-axes using a compass-clinometer.

collected in the field is only one of many possible samples that could have been chosen, sample characteristics (e.g. preferred orientations, fabric shape, etc.) will deviate from the 'true' population values. Such statistical variance is routinely accounted for in univariate statistics, but the equivalent variance in 3-D data has received very little attention. It is not at present possible to determine the variance of 3-D data analytically, although Ringrose and Benn (1997) and Benn and Ringrose (2001) found a numerical solution using 'bootstrapping' techniques. This involves taking a sample, then constructing additional datasets by randomly resampling from the original data. Any one data point may appear more than once in the resamples (this is called sampling 'with replacement'). The group of resamples can be thought of as a set of samples taken from a population, and can be used to estimate the statistical variability associated with sampling effects. The technique is known as 'bootstrapping' because, by appearing to generate more data than we began with, it is the statistical equivalent of pulling yourself off the ground by your bootstraps. Although it may appear to be a statistical conjuring trick, bootstrapping has been shown to work well in comparative studies. Ringrose and Benn (1997) showed that, as would be expected, the amount of statistical variance (and hence potential 'error' in sample characteristics) is inversely proportional to sample size: the smaller the sample, the larger the variance around the 'true' population values. In general, statistical variance is unacceptably large for samples of 25. The situation is improved for sample sizes of 50, and improved further for samples of 100. Increasing the sample size, however, proportionately increases the amount of time and effort required for sample collection in the field. In most situations, sample sizes of 50 appear to represent a reasonable compromise. Larsen and Piotrowski (2003) have shown that statistical variance is small for sample sizes of 30, provided there is a strong preferred orientation in the data. Of course, one must conduct some sampling and analysis to determine whether this is the case. Thus, it is recommended to conduct a pilot study to establish whether the data have strong preferred orientations: if so, sample sizes of 30 may be adequate, if not, larger samples must be taken. This is a much more respectable basis on which to choose sample sizes than the boredom threshold of the investigator.

5.4 GRAPHICAL REPRESENTATION AND SUMMARY STATISTICS

Fabric data can be presented in several ways. The simplest is the rose diagram, or 2-D histogram of directional data (Fig. 5.6a). The number of observations falling into particular ranges (e.g. 1°–15°) can be represented as 'wedges' or as points which are joined by straight lines. In either case, the diagrams can be symmetrical (if a measurement of, say, 90° is regarded as equivalent to one of 270°) or asymmetrical. Rose diagrams can be used to present either orientation (azimuth) or dip data, and thus represent only part of a 3-D dataset. Nevertheless, they provide an effective means of conveying overall patterns in a dataset, and have a high visual impact.

5.4.1 Stereonets

Stereonets are powerful and widely-used diagrams for presenting 3-D orientation data, in which azimuth and dip data are represented by points on a circular net (Fig. 5.6b). A good way to visualize this is to imagine the orientation data (i.e. the azimuth and dip of each item in the sample) radiating outward from the centre of a sphere. Because dips are

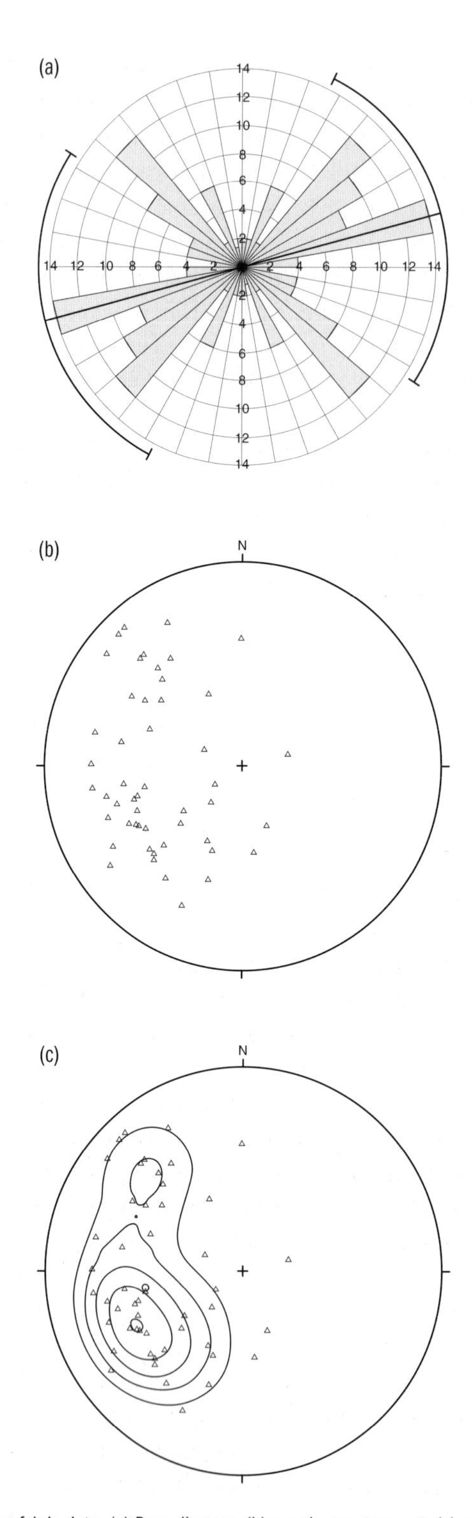

FIG 5.6 Three ways of presenting fabric data. (a) Rose diagram, (b) equal-area stereonet, (c) contoured stereonet.

measured downward from the horizontal, the data are actually confined to the lower hemisphere. Each line (data item) can be represented by a point where the line intersects the surface of the hemisphere. The polar coordinates of the points on the hemisphere are simply the azimuth and dip of each data item. Stereoplots and similar diagrams are 2-D projections of this 3-D data distribution. As in cartography, there are many ways of representing a spherical surface on a 2-D plane. The most widely-used method for orientation data is the *equal-area projection*, which projects data points from the hemispherical surface on to a horizontal plane in such a way that the areas of particular regions on the hemisphere are proportional to those of the equivalent regions on the diagram. Many commercial programs are available which allow the rapid plotting of data on stereonets.

Fabric data on stereonets may be contoured to provide a better visual impression of variations in point density over the diagram (Fig. 5.6c). Isolines are most usually labelled in standard deviation units, which give an indication of the statistical significance of point clusters. Most commercial computer fabric programs have contouring functions, and the better ones have a choice of contouring methods.

5.4.2 Eigenvalues

The comparison of fabric samples on stereonets provides a good visual impression, but it is useful to have some quantitative means of comparing fabric data. Several methods have been developed, but the most widely used in recent years is the *orientation tensor* or *eigenvalue method* (e.g. Mark 1974; Woodcock 1977; Benn 1994a). Eigenvalues reduce large datasets to simple descriptive statistics describing the strength and orientation of directional properties of a sediment, thus allowing the ready comparison of fabric data from many localities. Furthermore, facies with contrasting depositional and deformational histories may have distinctive ranges of eigenvalues, so that eigenvalues determined for deposits of known origin can be used to guide the interpretation of sedimentary facies, and to infer depositional processes and strain histories (Hart 1994; Benn 1994a, b, 1995; Evans *et al.* 1995; Benn and Evans 1996).

The eigenvalue method resolves sets of observations into three orthogonal eigenvectors, V_1, V_2 and V_3, where V_1 (the principal eigenvector) parallels the axis of maximum clustering in the data, and V_3 is normal to the preferred plane of the fabric (Scheidegger 1965; Watson 1966; Mark 1973). The degree of clustering of the data about the respective eigenvectors is given by the normalized vector magnitudes (eigenvalues) S_1, S_2 and S_3, where $S_1 > S_2 > S_3$ and $S_1 + S_2 + S_3 = 1$. The mathematics of the method is well explained by Woodcock (1977), and several commercially available computer programs do the job quickly and easily.

The three eigenvalues can be visualized as the axes of an ellipsoid that approximates the 'shape' of the data distribution in three dimensions (Fig. 5.7; Watson 1966; Woodcock 1977; Benn 1994b). Isotropic fabrics (with data points evenly distributed over a sphere) have $S_1 \approx S_2 \approx S_3$; planar girdles (with points evenly distributed around a great circle) have $S_1 \approx S_2 >> S_3$; and linear clusters (with all observations approximately parallel) have

$S_1 >> S_2 \approx S_3$. In Figure 5.7, the continuum of fabric shapes is represented on the *equilateral* or *general shape triangle* introduced by Benn (1994a, b), which is scaled using an *isotropy index* $I = S_3/S_1$, and an *elongation index* $E = 1-(S_2/S_1)$. This diagram is analogous to that introduced by Sneed and Folk (1958) for plotting particle shape data and represents the continuum of all possible ellipsoids (Benn and Ballantyne 1993; see Chapter 4). Indices based on eigenvalues can thus be thought of as representing *fabric shape*. Other plotting methods are also in use (e.g. Rappol 1985; Dowdeswell and Sharp 1986; Fig. 5.8), and their relative merits are discussed by Benn (1994b) and Ringrose and Benn (1997).

Different types of sediment tend to be associated with characteristic ranges of fabric eigenvalues (Dowdeswell and Sharp 1986). The fabric characteristics of sediments of known origin can, in some instances, be used to aid the interpretation of facies whose origin is not known (Dowdeswell *et al.* 1985; Benn 1994b; Fig. 5.9). Bennett *et al.* (1999) questioned the utility of this approach, and argued that fabrics from different types of sediment may be indistinguishable in practice. While it is undoubtedly true that there are overlaps between the fabric-shape fields for some types of sediment, the method is very useful for constraining depositional and deformational processes in some important cases. A quantitative basis for testing whether fabrics are likely to have come

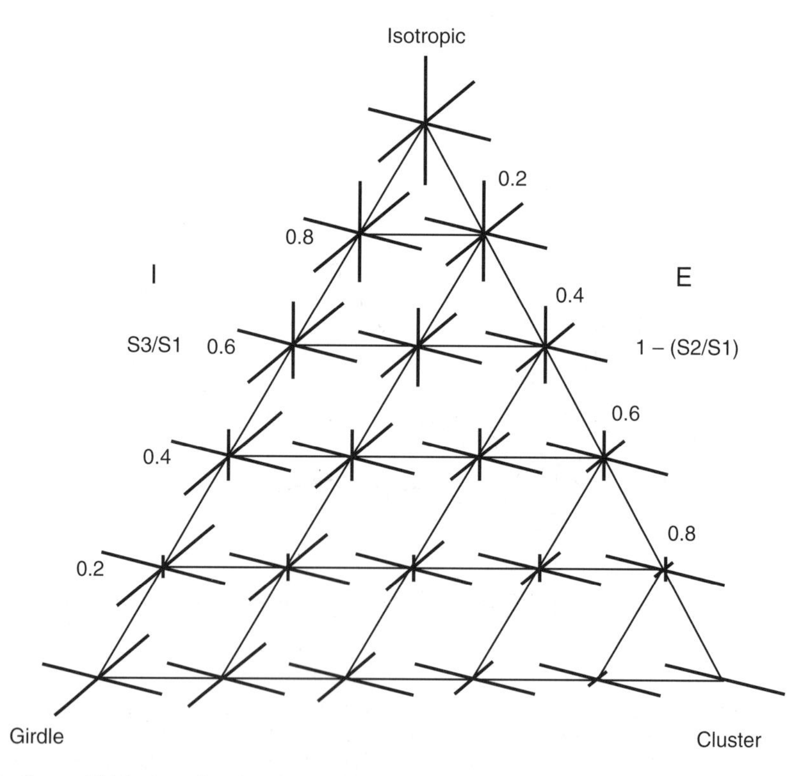

FIG 5.7 Continuum of fabric shape plotted on the general shape triangle. The magnitude of the normalized eigenvalues S_1, S_2 and S_3 is represented by the length of the three orthogonal lines at each node on the diagram.

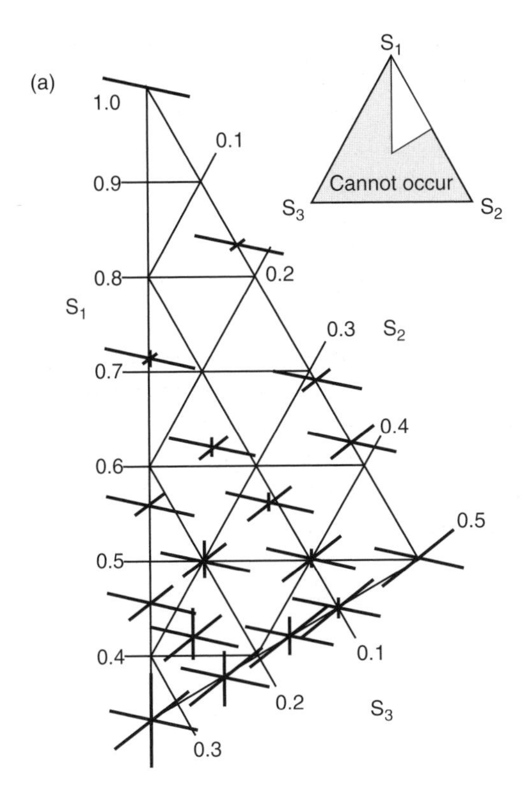

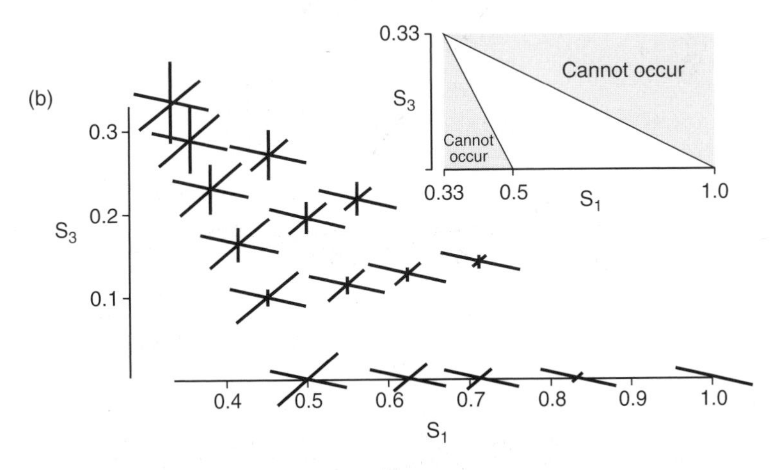

FIG 5.8 Alternative methods of plotting eigenvalue data. a) Rappol diagram. This diagram plots S_1, S_2 and S_3 on a standard ternary diagram. Because $S_1 > S_2 > S_3$, only one-sixth of the diagram is useable. b) Biaxial plot of S_1 and S_3, as used by Dowdeswell *et al.* (1985). Because 1.0 > S1 > 0.3, and 0.3 > S3 > 0, data points cannot occur in some parts of the diagram. Both diagrams are topologically equivalent to Figure 5.7, but perhaps provide a less intuitively obvious impression of fabric shape.

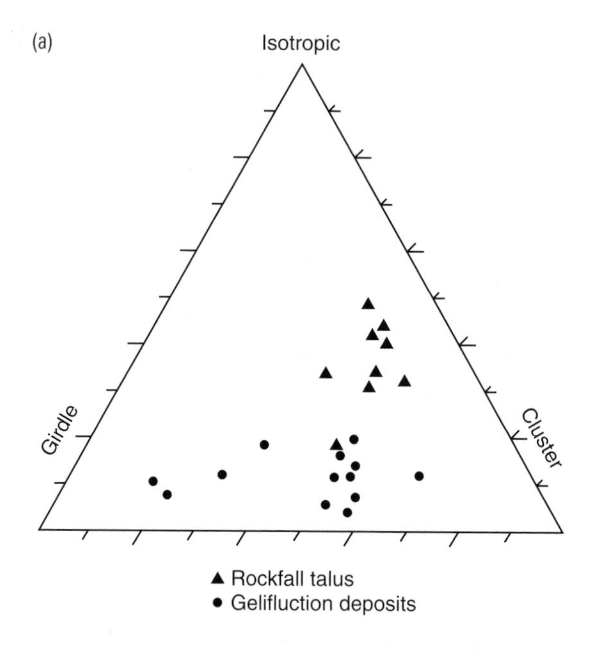

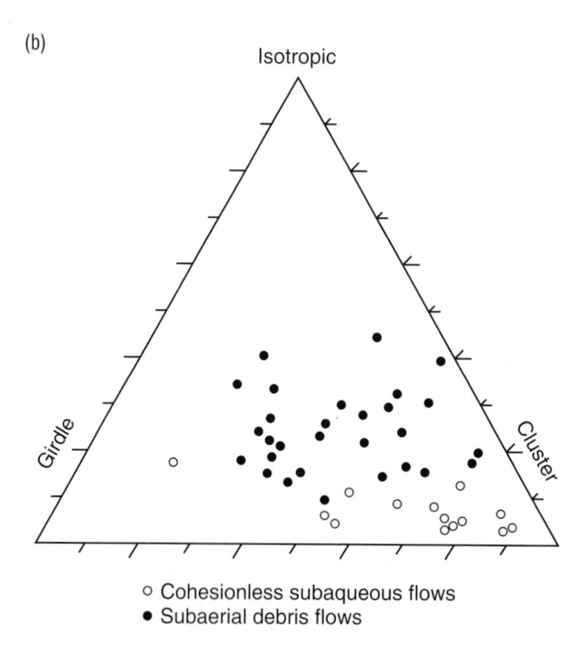

FIG 5.9 Examples of fabric shape fields (key as for Fig. 5.7). a) Fields for gelifluction deposits and rockfall talus. (b) Fields for subaerial debris flows and cohesionless subaqueous mass flows. Note that although the fields overlap with those for basal tills (Fig. 5.13), they do not overlap with each other. Thus, if other sedimentological data can be used to infer a mass flow origin, fabric data can be used to constrain the type of flow.

from different populations has been developed by Ringrose and Benn (1997) and Benn and Ringrose (2001), discussed in Section 5.5.3. It must be emphasized, however, that fabric data cannot be used as the sole criterion for determining the origin of sediment facies.

The eigenvalue method is only valid where the shape of the fabric is unimodal and approximates an ellipsoid. In cases where the distribution is bi- or multimodal, calculated eigenvalues may be spurious, lying between actual fabric modes. Consequently, fabric distributions should be checked visually on stereonets to ensure that the data are approximately unimodal before the eigenvalue method is applied. To the author's knowledge, rigorous statistical methods have not been used to test whether fabric data are unimodal. Development of such methods should be regarded as an important goal.

5.5 CASE STUDIES

5.5.1 A-axis fabrics and patterns of deformation in till

A-axis fabrics in till have been used in many studies to infer former ice-flow directions, on the assumption that clasts are oriented parallel to ice flow during lodgement below the glacier sole. Rose (1989b) adopted this approach to reconstruct patterns of ice flow in a detailed investigation of the origin of fluted moraines on the foreland of Austre Okstindbreen, Norway. Rose found that fabric maxima converged at a small angle towards the flute axis in the ice-proximal part of the flute, then diverged away from the flute axis at the distal end (Fig. 5.10). This pattern was interpreted as evidence for a convergent-divergent pattern of flow in the basal ice, apparently in response to large lodged boulders on the bed. However, Benn (1994b) argued that 'herringbone' fabric patterns in fluted moraines may reflect patterns of strain in the *till*, rather than in the overpassing ice. According to this interpretation, till will deform towards areas of low pressure in the lee of large boulders, and the resulting fabric patterns may be oblique to the local ice flow direction. This hypothesis is consistent with data from Slettmarkbreen, Norway, where fluted moraines exhibit 'herringbone' fabric patterns, but striae on embedded boulders in the flute show consistent orientations parallel to the flute axis (Benn 1994b). No boulder striation data are available for the flutes studied by Rose (1989b), but striations on boulders in other flutes on the foreland record parallel ice flow over the features (Rose 1992).

This case study is important not only for understanding the origin of fluted moraines, but also because it shows that long-standing assumptions about the meaning of data (i.e. a-axis fabrics faithfully record ice-flow direction) may prove to be incorrect in some circumstances. It also shows how such assumptions can be tested using independent evidence for ice flow (e.g. striae). Where such tests are not possible, caution should be exercised when interpreting a-axis fabric data.

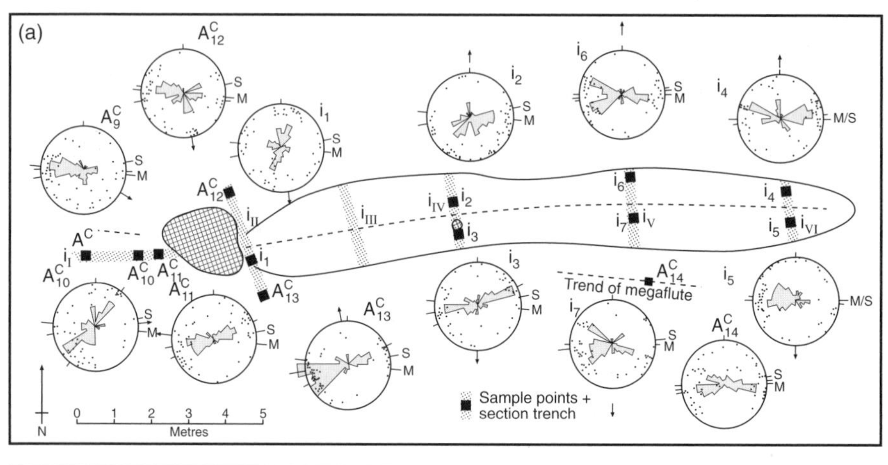

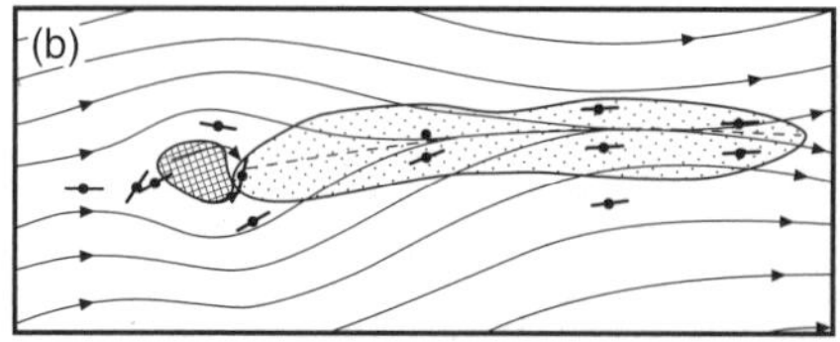

FIG 5.10 (a) A-axis fabrics in a flute on the foreland of Austre Okstindbreen, Norway. (b) Reconstructed flow patterns from preferred a-axis orientations. These have been interpreted as ice flow directions, or patterns of strain within the till. (From Rose 1989b.)

5.5.2 Basal till facies in Iceland

The role that deforming subglacial till plays in the motion of glaciers has been the focus of much recent research. Till deformation beneath modern glaciers has been measured using geophysical methods (e.g. Iverson *et al.* 1995; Truffer *et al.* 2000), but for former glaciers, mechanisms of basal motion must be inferred from geological evidence, such as the characteristics of basal till. However, the origin of tills deposited by past glaciers has been the subject of heated debate (e.g. Dreimanis 1989), largely because of the scarcity of well-known modern examples to guide interpretation. To obtain data on modern deformation tills, Benn (1995) conducted detailed fabric studies on recently exposed areas of the foreland of Breidamerkurjøkull, Iceland, where till deformation beneath the glacier margin was reported by Boulton and Hindmarsh (1987). The till consists of two horizons: an upper (A) horizon of porous, weak till, and a lower (B) horizon of more compact till with higher shear strength. According to Boulton and Hindmarsh, the A horizon underwent ductile deformation when it was beneath the glacier, encouraged by high porewater pressures, whereas the stiffer B horizon experienced much more limited deformation along discrete shear planes. Benn (1995) found that the A and B horizons also have contrasting fabric characteristics (Figs. 5.11, 5.12 and 5.13). In comparison with the upper till, clasts in the B horizon have stronger a-axis fabrics, more consistent up- and down-glacier orientations of plucked faces (double stoss-lee forms; Fig. 4.3), and stronger clustering of poles to facets on their upper and lower surfaces. The fabric characteristics of the lower horizon are thus

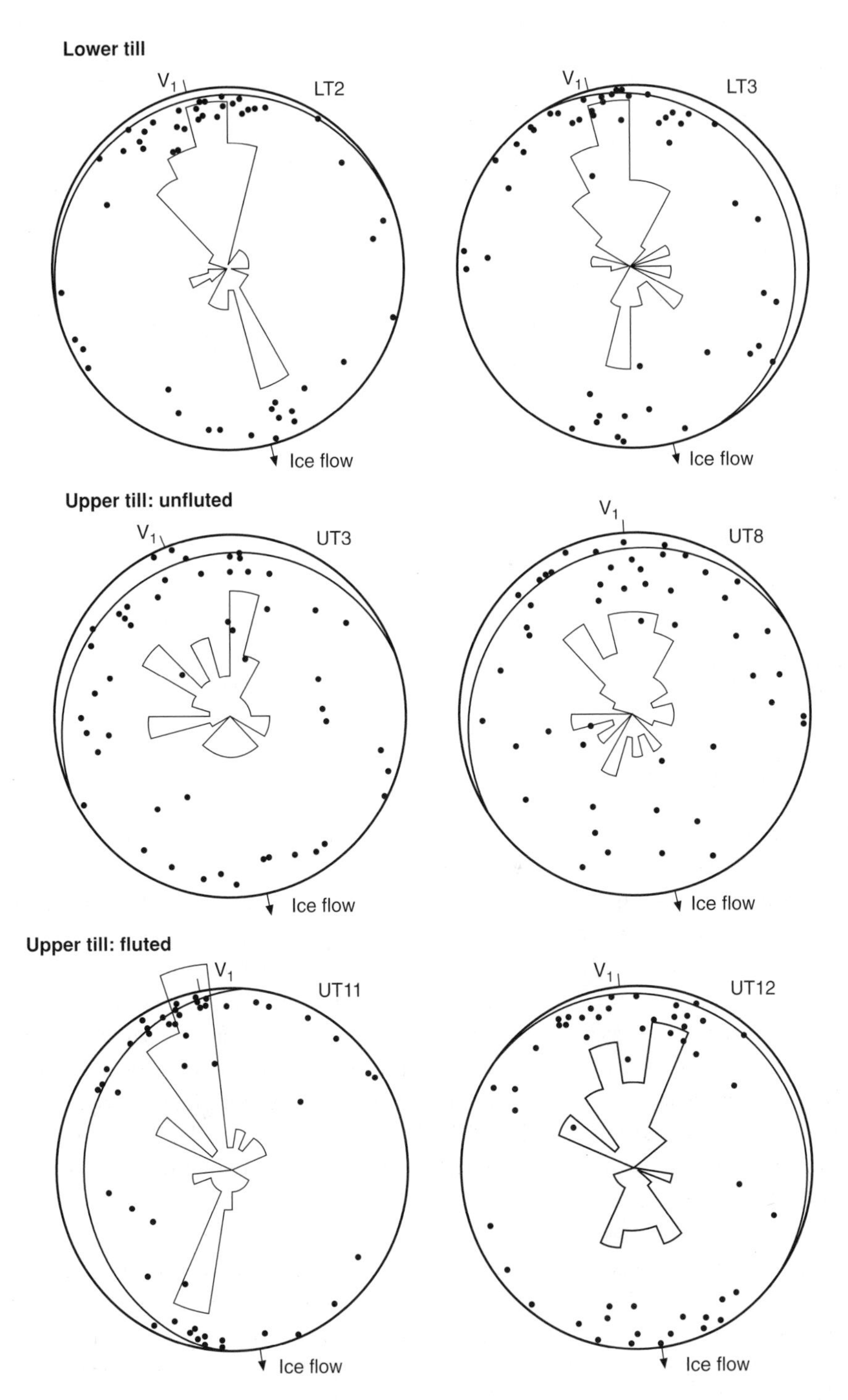

FIG 5.11 A-axis fabric data from the lower till and fluted and unfluted parts of the upper till at Breidamerkurjøkull, Iceland, plotted as lower hemisphere stereonets and rose diagrams. (From Benn 1995.)

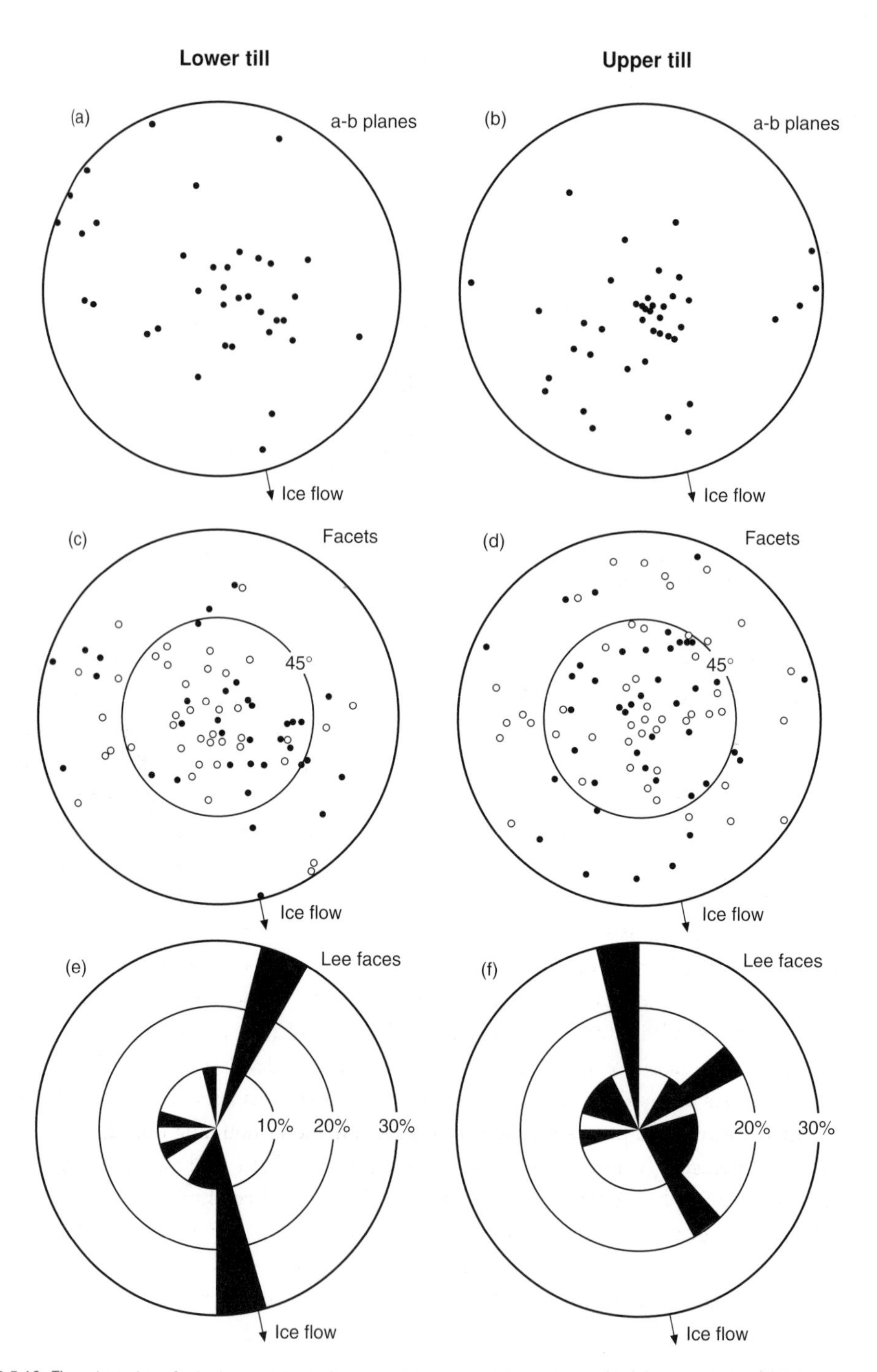

FIG 5.12 The orientation of a-b planes, poles-to-facets, and fractured lee faces for the lower and unfluted upper tills at Breidamerkurjøkull. For the pole-to-facet data (c, d), facets on the upper surfaces of clasts (open circles) are plotted on upper hemisphere nets, and facets on the lower surfaces (solid circles) are plotted on standard lower hemisphere nets. (From Benn 1995.)

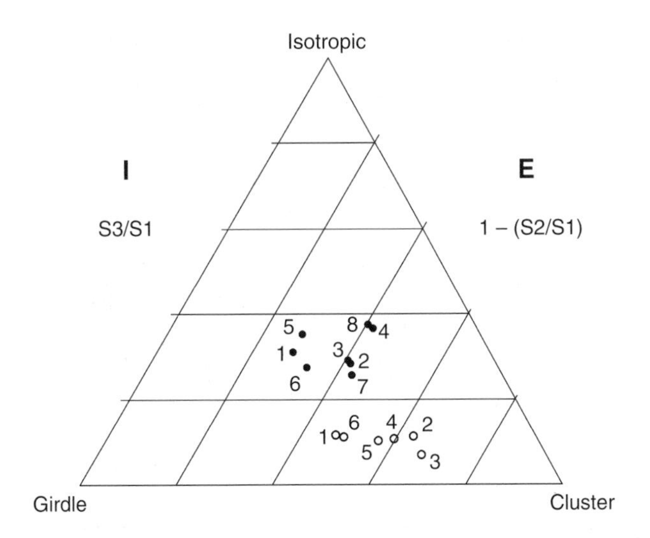

FIG 5.13 A-axis fabric data from the upper and lower tills at Breidamerkurjøkull plotted on the general shape triangle. Note that some of the sample points were plotted incorrectly in Benn (1995) and Benn and Evans (1996).

consistent with clast modification by abrasion and fracture within a shearing till. In contrast, the lee-face and pole-to-facet data from the A horizon are not consistent with *in situ* clast modification, indicating that the clasts rotated into their observed positions after their wear patterns were established. Benn (1995) and Benn and Evans (1996) argued that this reflects pervasive deformation of the upper horizon, during which clasts were rapidly rotated under transient stresses (e.g. in the vicinity of other clasts). The fabric data from the A-horizon apparently confirm the idea that highly-strained deformation tills can have 'weak' a-axis fabrics (e.g. Hicock and Dreimanis 1992; Hart 1994). However, on the basis of laboratory experiments, Hooyer and Iverson (2000a) argue that clast a-axes rapidly rotate into parallelism with the direction of shear, and that highly-strained deformation tills should always have strong fabrics. Although support for the Hooyer and Iverson model has been provided by detailed fabric studies by Larsen and Pietrowski (2003), this issue remains unresolved, and it is not known whether deformation tills can develop weak fabrics in response to various clast-orientation processes or transient stresses, or whether weak fabrics reflect post-deformational processes, such as liquefaction and remoulding during deglaciation. Interestingly, Benn (1995) found strong a-axis fabrics in fluted moraines at Breidamerkurjøkull, where cumulative strains are likely to have been higher than in surrounding areas of the till (Fig. 5.12). The meaning of fabric data have important implications for the interpretation of the Earth's glacial geologic record, and unresolved problems should be addressed in carefully designed field and laboratory investigations.

5.5.3 Statistical significance testing

In the foregoing discussion, we have identified and interpreted fabric shape 'fields' (Figs. 5.9 and 5.13) in a rather descriptive, non-quantitative way. This is usually sufficient where there is no overlap between the shape fields of interest, but in other cases it is useful to have some quantitative basis for testing whether a particular fabric belongs to one field or another. Before addressing this issue, it is useful to consider some wider problems regarding significance testing. Several statistical tests have been proposed to assess the 'significance' of fabric data. One widely-used method uses a form of the isotropy index to test whether a fabric significantly differs from 'random' (Woodcock and Naylor 1983; Fig. 5.14). If fabric isotropy is higher than some threshold value, any apparent preferred orientation is likely to be the result of chance sampling effects, whereas if the isotropy is lower, the preferred orientation is considered to be 'significant'.

This approach is unsatisfactory for two reasons (Benn and Ringrose 2001). First, the test is based on the probability of obtaining a given S_3/S_1 value by unbiased sampling from an isotropic ('random') population, but does not allow any significance to be attached to the orientation or strength of the principal eigenvector V_1 (which is what we usually mean by 'preferred orientation'). A girdle fabric (Point G, Fig. 5.14) may be statistically 'non-random', but only in the sense that more observations lie close to the V_1-V_2 plane than would be expected by chance. Nothing can be said about the significance of any apparent preferred orientations within the V_1-V_2 plane. Thus, it is incorrect to use the Woodcock–Naylor test to assess the 'significance' of apparent preferred orientations. Second, 'random' (or isotropic) fabrics are not a good standard against which to compare fabric data. Because clast orientations in sediments are usually constrained in some way (either by depositional surfaces or applied forces), isotropic fabrics are rare and arguably never 'meaningless'. Thus, it is more useful to ask: are the observed fabric characteristics close to those of the parent population, or could they have been significantly influenced by chance sampling effects? In univariate statistics, this question is addressed by specifying

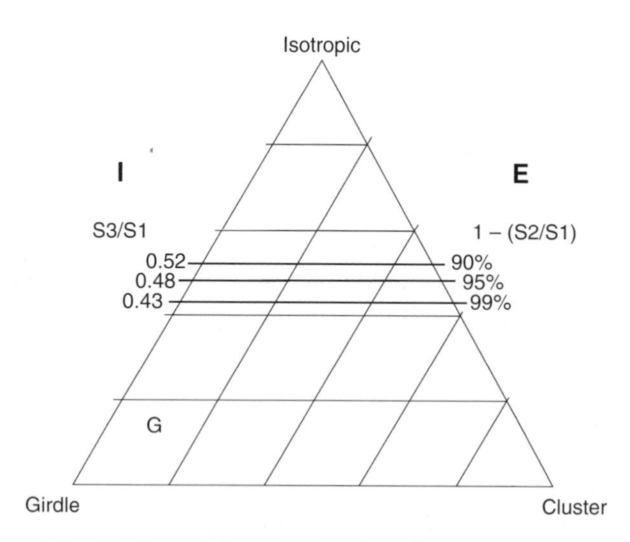

FIG 5.14 Graphic representation of the Woodcock-Naylor (1983) test for randomness on the general shape triangle. The bold lines indicate the minimum isotropy values that could be obtained by unbiased sampling from a random population at specified confidence levels.

confidence limits around sample means. For fabrics, the equivalent procedure is to establish confidence *regions* around points on the general shape triangle, which will encompass the 'true' fabric shape at some specified confidence level.

Ringrose and Benn (1997) and Benn and Ringrose (2001) approached this problem by using the bootstrapping technique to generate 'clouds' of resample points from original samples. A line connecting all of the outermost points (a *convex hull*) defines the limit of the entire distribution, but the size and shape of this region may be unduly influenced by anomalous outlying points, giving an inflated impression of the statistical variance. This problem can be overcome by 'peeling off' successive convex hulls and drawing new ones around the remaining points. Benn and Ringrose (2001) found that the 10th convex hull enclosed the parent population point in 90–95% of cases, based on 10 000 resample points generated from an original fabric sample. That is, the 10th hull can be regarded as an approximate 90–95% confidence region around a fabric shape, providing a quantitative basis for testing whether any two fabrics are likely to belong to different populations. Specifically, if a sample point lies outside the 10th hull of another, the two samples can be said to be different with a high level of confidence. It should be emphasized that this is not equivalent to a formal statistical test for difference, but in the absence of analytical methods it provides the best available approximation.

There is a further interesting complication that arises when plotting 'confidence regions' around points on the general shape triangle. Because of sampling effects, the preferred orientation of a resample, $V_1(R)$ may differ substantially from the population value, $V_1(P)$. Indeed, it may actually lie closer to $V_2(P)$ or $V_3(P)$, particularly for points plotting close to the left or right edges of the triangle, respectively. This 'swap-over' effect can seriously distort the distribution of resample points, producing skewed and misleading 'confidence regions'. Ringrose and Benn (1997) and Benn and Ringrose (2001) addressed this problem by using a six-way version of the general shape triangle to represent the six possible orderings of the resample eigenvalues, where S_1, S_2 and S_3 are defined by their *orientations* relative to the respective population eigenvalues rather than their relative magnitude in the resample (Fig. 5.15). Depending on the ordering of the resample eigenvalues, they can plot in any of the six panels, producing point distributions that tend to be more symmetrically arranged around the original population point.

Tenth convex hulls for the till samples from Breidamerkurjøkull discussed in Section 5.5.2, are shown in Figure 5.16. The sample points for the A Horizon (upper till) plot outside the 10th hulls for all of the B Horizon (lower till) samples, and the points for the B Horizon plot outside all but one of the 10th hulls for the A Horizon samples. With this possible exception, therefore, we have good evidence to conclude that the fabric characteristics of the two sets of samples are significantly different. Larsen and Piotrowski (2003) used the bootstrapping approach in their detailed study of Pleistocene deformation tills in Poland. The fabrics they measured had exceptionally strong preferred orientations, and conventional methods could not distinguish any between-sample differences in fabric characteristics. However, they were able to identify distinct groups of samples on the basis of 10th hulls generated by the bootstrapping method, which were used as the basis for inferring subtle differences in strain history.

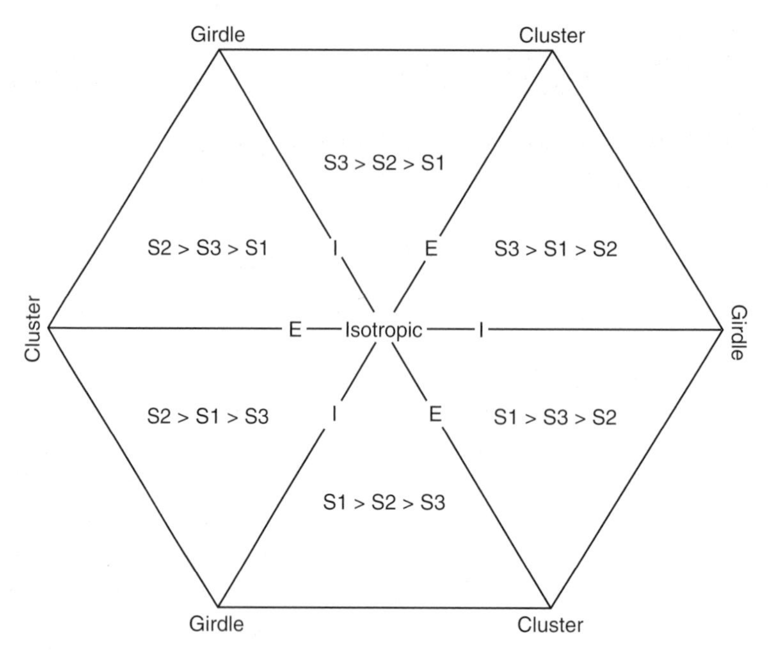

FIG 5.15 The 6-way extension of the general shape triangle (from Benn and Ringrose, 2001).

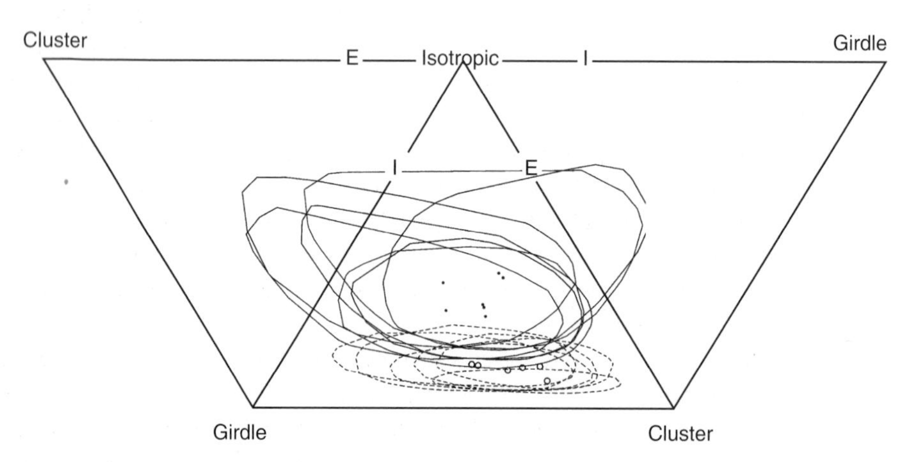

FIG 5.16 Tenth hulls for the Breidamerkurjøkull upper and lower tills (see Fig. 5.13), plotted on the lower half of a six-way diagram.

The rigorous statistical analysis of fabric data is still in its infancy, and many interesting and fundamental problems remain. Bootstrapping has many potential uses, although it represents only one of many potential numerical tools. Copies of a Matlab version of the bootstrapping program, written by Trevor Ringrose, may be obtained from D. Benn.

6 Micro-scale features and structures

Simon J. Carr

6.1 INTRODUCTION

As noted in previous chapters, a specific lithofacies will normally preserve textural and structural characteristics that will enable an interpretation of the process and environment of sediment deposition. Increasingly, however, it is becoming apparent within Quaternary and glacial sedimentology that macro-scale analysis does not always yield sufficient information to reliably demonstrate the depositional or deformation history of many lithofacies (van der Meer 1993; Carr *et al.* 2000). The complex interaction of a range of sediment erosion, transport and deposition mechanisms within the glacial context is such that different processes may result in what seem to be identical sediment facies when viewed at a macro-scale.

This situation is exacerbated if the spatial extent of visible exposures is limited, or if the sediments under investigation have been recovered from onshore or offshore locations using coring equipment (van der Meer and Laban 1990; Carr *et al.* 2000; Hiemstra 2001). This is highly relevant within the context of Quaternary glacial sedimentology, because many of the major Quaternary ice sheets terminated in regions currently located below sea level. This means that many of the key lithofacies formed by these ice sheets are inaccessible, except through coring. However, the limited sample volume recovered from cores and boreholes places a major constraint on the range of macro-scale techniques, including many of those outlined in this book, which can be applied to characterizing and interpreting lithofacies.

One of the most significant approaches taken to address these issues in recent years has been the analysis of micro-scale textures and structures of undisturbed sediment samples, more commonly termed *micromorphology*. Typically, micromorphological analysis of sediments is undertaken either at low magnification (10× to 100×), using thin sections and petrographic microscopes, or at higher magnifications (100× and greater) using a sediment stub and scanning electron microscopy (SEM). Micromorphology has long been applied to the investigation of lithified sediments, and forms a significant component of Quaternary soil science, such that it has been considered to be the only method capable of revealing the nature and complexity of soil formation and palaeosol history (Kemp 1998). However, the approach has only been applied to the study of Quaternary glacigenic sediments relatively recently, with the contribution of micromorphology becoming increasingly significant to our understanding of such sedimentary environments.

Whilst both SEM and thin section approaches have yielded significant new data enabling classification and interpretation of many Quaternary sediments, it should be stressed that neither should be viewed as a replacement for field description of lithofacies and the various textural, structural and geotechnical approaches outlined in this book. As clearly demonstrated in Chapter 9, micromorphology forms just one component of the multidisciplinary approach required to accurately characterize and interpret complex Quaternary glacial sediment sequences.

This chapter introduces the sampling, preparation and description of the micromorphology of sediments, focusing primarily on glacigenic sediments. Both SEM and thin section analysis will be discussed, and a range of examples of the application of these techniques presented. It is not within the scope of this chapter to discuss the formation of specific structures in great detail, and the reader should consult some of the excellent reviews on specific environments for further information (e.g. Bull 1981; Kemp 1985; 1999; Menzies and Maltman 1992; van der Meer 1993; 1996; Mahaney 1995; Whalley 1996; Menzies 2000).

6.2 FIELD SAMPLING AND LABORATORY PREPARATION

The intensive nature of sample preparation for micromorphological analysis is such that careful thought is required regarding appropriate field sampling strategies. As with many other techniques, the relatively small size of samples for either SEM or thin section production must be considered in terms of how representative samples are of individual lithofacies. Many micromorphological studies also focus on the contacts between lithofacies to evaluate the detailed nature of bounding structures, requiring careful sample selection. Furthermore, for both SEM and thin sections it is vital to obtain an oriented, undisturbed sample for micromorphological analysis. Oriented samples are often taken using double-hinged 'Kubiena' tins (Fig. 6.1) of different sizes (8×6×4cm or 15×8×5cm), which are transported, where possible, back to the laboratory in sealed bags, to maintain field moisture content and prevent disturbance.

The preparation of thin sections of unconsolidated sediments involves three main stages; the drying of samples, impregnation of dry samples with resin and finally cutting and grinding of resin impregnated blocks to produce the thin section (Fitzpatrick 1984; Murphy 1986; Lee and Kemp 1992; Carr and Lee 1998). The overview of the entire process is best illustrated as a flow diagram (Fig. 6.2). Dependent upon their clay content, samples may be air-dried or undergo water removal in an acetone bath. A number of different types of resin have been used in sample impregnation, including synolite, crystic and epoxy resins. Whilst all are adequate for thin section production, the harder the cured resin block, the better the completed thin section is likely to be (Tippkötter and Ritz 1996). Glacigenic sediments often pose further problems due to the overconsolidated

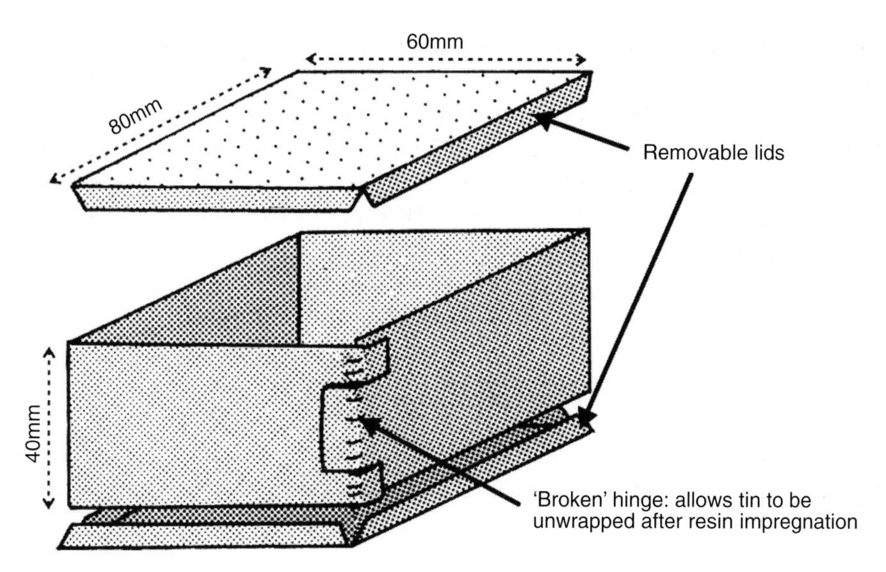

FIG 6.1 Schematic diagram of a Kubiena tin. The double-lidded and hinged construction allows the transport of an undisturbed sample back to the laboratory, and removal of the tin subsequent to resin impregnation.

nature of recovered samples, leading to incomplete impregnation of samples, and in many cases surface impregnation of cut blocks is required to produce an adequate sample (Carr and Lee 1998).

The sampling and preparation of sediments for SEM analysis is less complex, with samples usually air-dried prior to temporary or permanent mounting on aluminium stubs (Smart and Tovey 1982; Murphy 1982). As Whalley (1996) notes, however, it is important to consider the type of SEM analysis that will be undertaken on samples. In most cases a 'sputter' coating of gold is used, to provide a path for the incident electron beam to leak away, to maintain image quality and prevent image degradation through 'charging'. However, if elemental investigation of a sample is desired using an analytical EDS (energy dispersive spectrometer) probe, the sample has to be polished and coated with a carbon film. In other applications of SEM, individual mineral grains are required for the analysis of surface textures (Krinsley and Doornkamp 1973), at which point samples are disaggregated, washed in de-ionized water or sodium hexametaphosphate and mounted using double sided tape on stub surfaces.

6.3 DESCRIPTION OF THIN SECTION MICROMORPHOLOGY

Once produced, thin sections may be analysed through the use of standard petrological microscopes, using plane and cross-polarized light to identify textural and structural

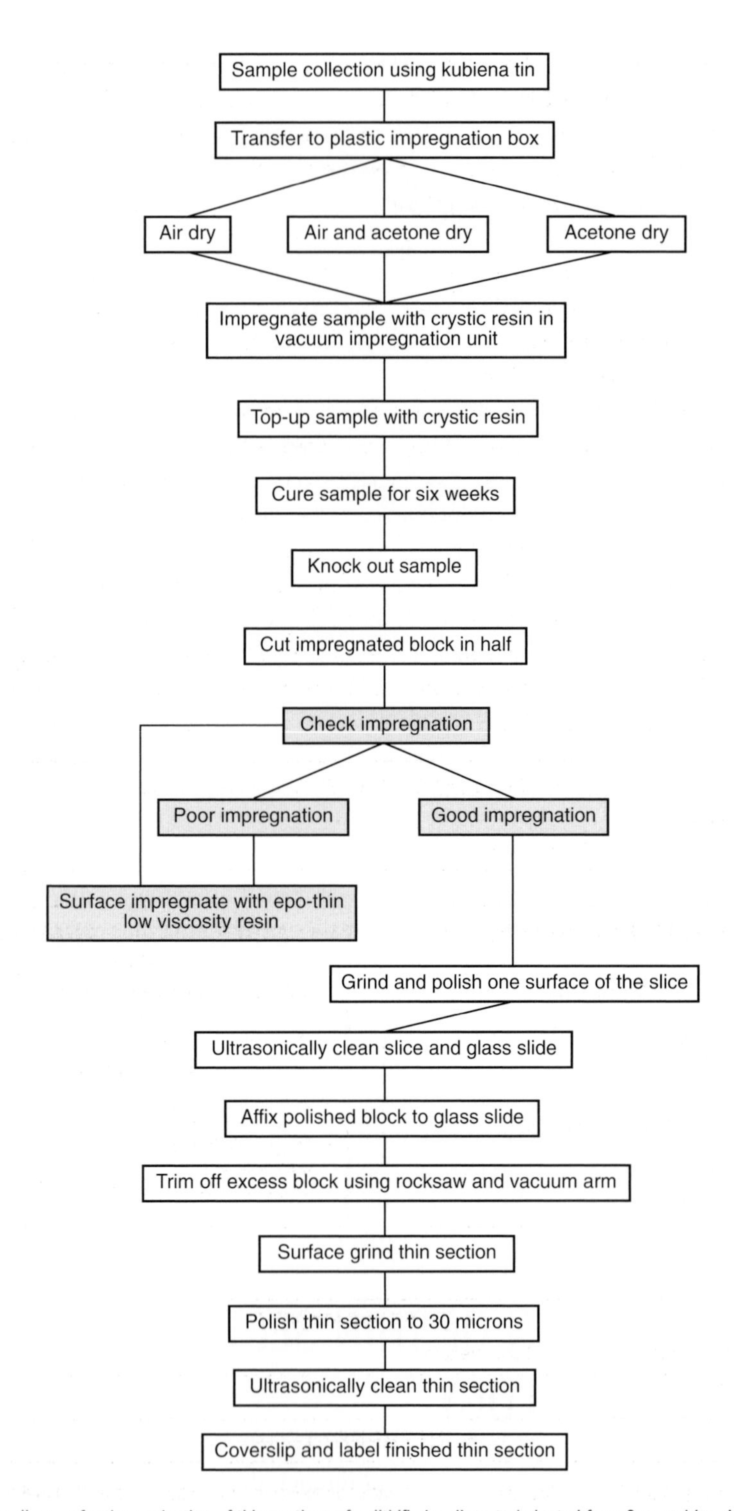

FIG 6.2 A flow diagram for the production of thin sections of unlithified sediments (adapted from Carr and Lee 1998).

characteristics of the sample. Thin sections are usually investigated at relatively low magnifications, of between 10× and 100×, because higher magnifications tend to focus the observer on individual grains, which may not be beneficial to interpretation. In some cases, the use of a projection macroscope provides a useful means of undertaking an initial evaluation of the large-scale structure of samples, due to a significantly larger field of view, also of value in microfabric analysis (Section 6.5). Increasingly sophisticated digital image capture and analysis equipment has been applied to the study of Quaternary sediments (Francus 1998; Tiljander *et al.* 2001; Zaniewski 2001) and whilst this approach is still in its infancy, it seems to offer significant potential in the drive towards quantifiable data collection within micromorphology.

The analysis of thin sections can only be applied meaningfully if description is systematic and researchers employ a 'standard' descriptive system. Without this it is almost impossible to communicate relevant data in a satisfactory manner. However, at present there is no universally accepted scheme for the description of thin sections from glacigenic sediments. This is possibly due to the restricted number of researchers using this technique, and the resulting limited exposure of the approach in the broader research literature. However, since the 1990s, a coherent scheme for the description of glacial sediments has been developed and refined (van der Meer 1987; 1993; 1996), partly based on the soil micromorphology nomenclature of Brewer (1976), providing a possible template for standardization (Fig. 6.3). However, variations on the scheme outlined below

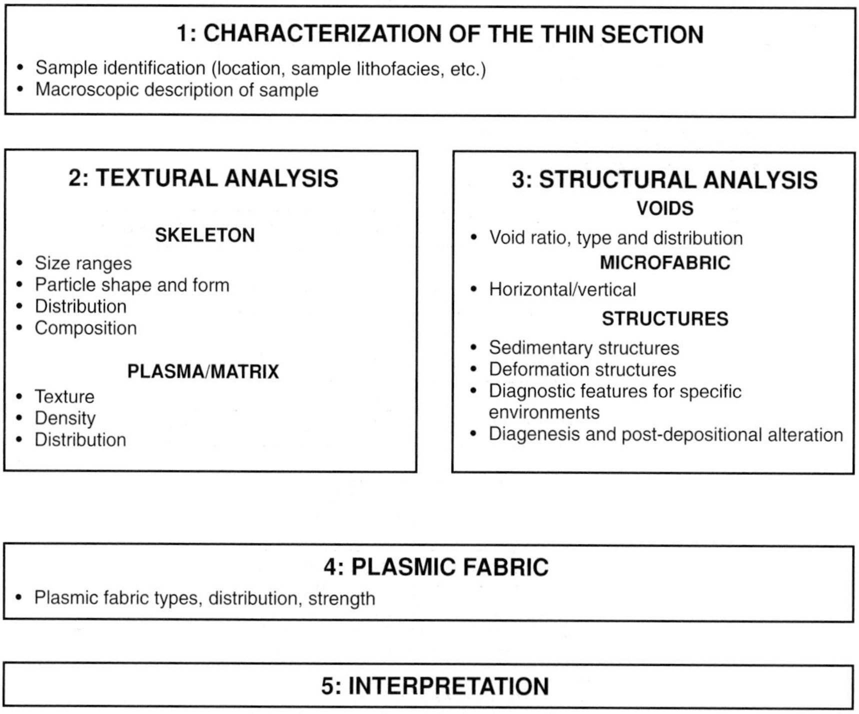

1: CHARACTERIZATION OF THE THIN SECTION

- Sample identification (location, sample lithofacies, etc.)
- Macroscopic description of sample

2: TEXTURAL ANALYSIS

SKELETON

- Size ranges
- Particle shape and form
- Distribution
- Composition

PLASMA/MATRIX

- Texture
- Density
- Distribution

3: STRUCTURAL ANALYSIS

VOIDS

- Void ratio, type and distribution

MICROFABRIC

- Horizontal/vertical

STRUCTURES

- Sedimentary structures
- Deformation structures
- Diagnostic features for specific environments
- Diagenesis and post-depositional alteration

4: PLASMIC FABRIC

- Plasmic fabric types, distribution, strength

5: INTERPRETATION

FIG 6.3 Suggested approach for describing thin sections of glacigenic sediments. This scheme is broadly based on the pedological scheme of Brewer (1976), as adapted by van der Meer (1993; 1996) and Carr (1999).

exist, with some researchers preferring the terminology of structural geology as a more appropriate means of description for subglacially-deformed sediments (Roberts 1995; van der Wateren *et al.* 2000). Despite the differences in nomenclature, the system outlined in Figure 6.3 provides a robust scheme by which a thin section may be comprehensively described. The contribution of structural geological approaches in micromorphology has yet to gain wide adoption, despite its clear potential contribution towards a better understanding of the kinematics of deformation processes.

Following the same approach as macro-scale analysis, micromorphological analysis examines both the structural and textural characteristics of individual lithofacies identified in a section. In many situations, techniques commonly used in field description may be applied at micro-scale, for example the analysis of grain shape (see Chapter 4). The strength of micromorphology, however, is in the identification of structures and textural characteristics in sediments where macro-scale analysis yields ambiguous data or identifies homogeneous sediment packages.

6.4 TEXTURAL AND STRUCTURAL ANALYSIS OF THIN SECTIONS

Textural analysis of thin sections consists primarily of breaking down the physical constituents of the sample into components identified on the basis of size, shape, distribution and lithology/composition. This approach may adopt the conventions and approaches taken by other forms of sedimentary analysis (Fig. 6.4), and is augmented by other techniques such as particle size analysis to quantify the particle fractions in a sampled lithofacies (Chapter 3). It is common in textural analysis of thin sections to discriminate between the *skeleton* and *plasma* (or *matrix*) of a sample. The former consists of all particles which may be observed individually, usually grains of coarse silt fractions and larger, whilst the latter is generally defined as the textural component below about 20 microns, within which individual grains cannot be observed. Textural analysis of skeleton grains can involve detailed investigation of the major components mentioned above, whilst description is much more limited for matrix material, often to an evaluation of coarseness (Fig. 6.4) and distribution of matrix material.

A *structural* description of thin sections forms the main aspect of the description of thin sections. While many glacigenic sediment facies appear to be homogeneous at a macro-scale, often providing ambiguous information regarding the process of deposition, at micro-scale these same sediments preserve a range of structures that may point to a particular origin. Structural elements of a thin section can be identified at a number of scales, from major structural elements often visible to the naked eye down to arrangements of individual particles only observed at higher magnifications (Menzies and Maltman, 1992). Whilst many of the larger structures visible in thin section are simply smaller versions of those observed in field sections, many of the smaller structures have few comparisons with macro-scale features.

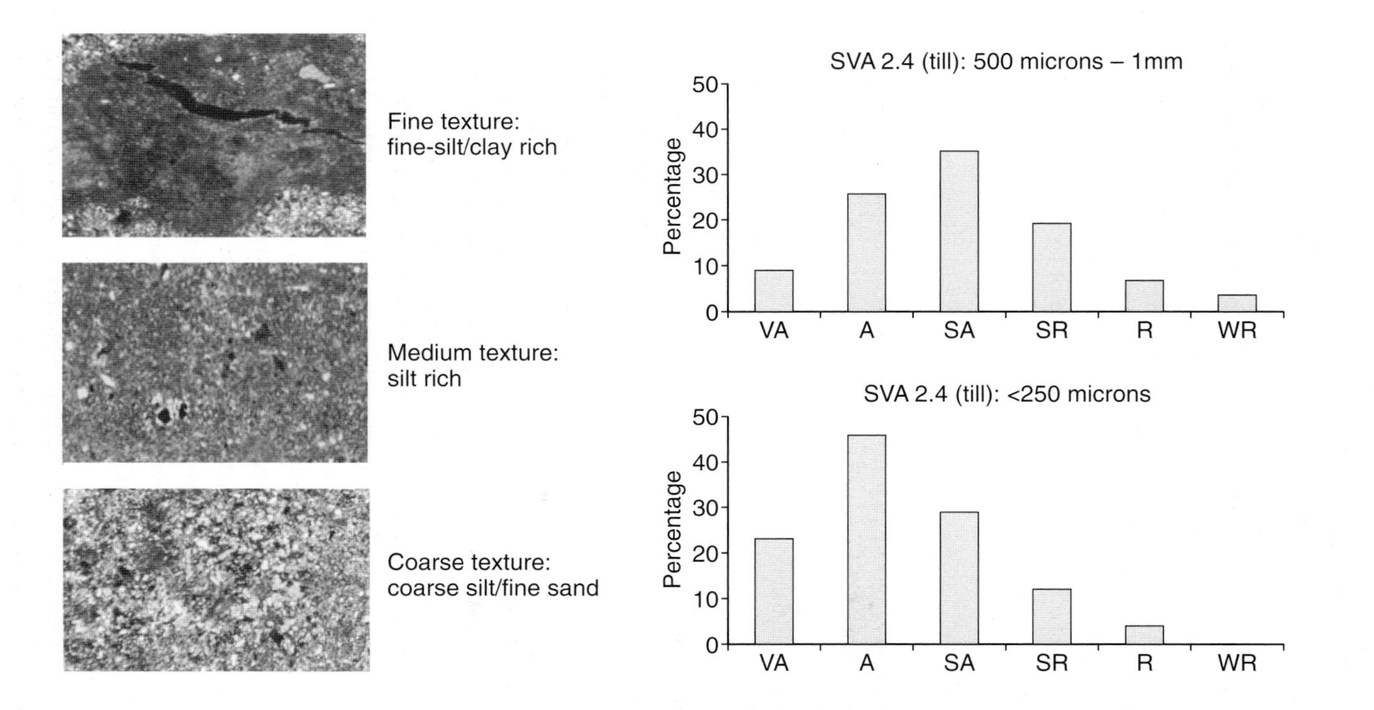

FIG 6.4 Presentation of textural data from a thin section. Left: identification of matrix coarseness in thin section. Right: in this sample, taken from a subglacial till sampled from a contemporary glacier in Svalbard grain roundness counts, based on the Powers roundness scale have been undertaken for the 500μm–1mm and <250μm size fractions. It is notable that the smaller grains are significantly more angular than the larger, and probably reflect edge-rounding of the larger grains in the subglacial bed.

6.5 STRUCTURES IDENTIFIED IN THIN SECTION

Depositional structures visible in thin section include many of the features often visible at a field scale, such as *bedding* and *lamination*. In the case of undeformed sediments preserving bedding structures at a field scale, the contribution of micromorphology is often limited by the size of samples obtained. However, at a microscopic scale it is usually possible to examine the internal structure of fine laminations (Fig. G41) providing further detail regarding the processes and environments of still or moving water bodies during deposition. The identification of *dropstone* structures within fine-grained bedded sediments provides an excellent indication of the supply of ice rafted material to a sediment body, not always visible at a macro-scale.

One of the key roles of micromorphology in recent years has been the discrimination between true sedimentary laminated sediments deposited in glaciaqueous environments and *pseudolaminations* produced by extreme attenuation of sediments through subglacial deformation (Roberts 1995; Carr 1998; 1999; Carr *et al.* 2000). As thin section micromorphology has developed over the past decade, studies of *deformation structures* have extended beyond simply those identified in subglacial sediments. Current research focuses on identifying associations of deformation structures that define specific strain regimes characterizing deformation occurring under sub-aerial and sub-aqueous mass-movement and density-driven mechanisms as well as glacitectonic mechanisms.

There have been two distinct approaches to the description of deformation features in thin section, depending upon the nature of the study being undertaken and the sample under investigation. Reconnaissance studies using thin sections and studies that involve the collation of significant numbers of thin sections have mainly followed an essentially descriptive approach, identifying and differentiating between structures indicative of rotational or planar deformation (van der Meer 1993; 1996; Menzies 1998, 2000; Fig. 6.5). Planar features (*augen, lineations, symmetrical pressure shadows*) tend to be associated with discrete shear (Fig. G42), whilst the more common rotational structures (*galaxy structures, soft intraclast 'pebbles', asymmetric pressure shadows*) reflect more pervasive deformation within a broad zone of deformation (Fig. G44). Other structures, such as *water escape structures* (Fig. G45) and *fractured quartz grains* (Hiemstra and van der Meer 1997; Fig. G46) provide an indication of the effective stress and pore water conditions of the sediment being deformed. Once identified, these associations of deformation structures may be related to specific styles of deformation (brittle/ductile) or to specific locations in a deforming sediment, and then to a process-based interpretation of the environment of deposition/deformation (Menzies 1998; 2000; Fig 6.6).

A second approach, where pre-existing sediments with a recognizable internal structure have been post-depositionally deformed, permits a more detailed approach to analysing deformation through the application of principles developed within structural geology (Phillips and Auton 1998; 2000; van der Wateren *et al.* 2000). The presence of deformed laminations, as illustrated in Figure 6.6, allows the detailed investigation of the geometry, kinematics and relative ages of different deformation structures (*tectonostratigraphy*), allowing the interpretation of complex 'polyphase' deformation sequences.

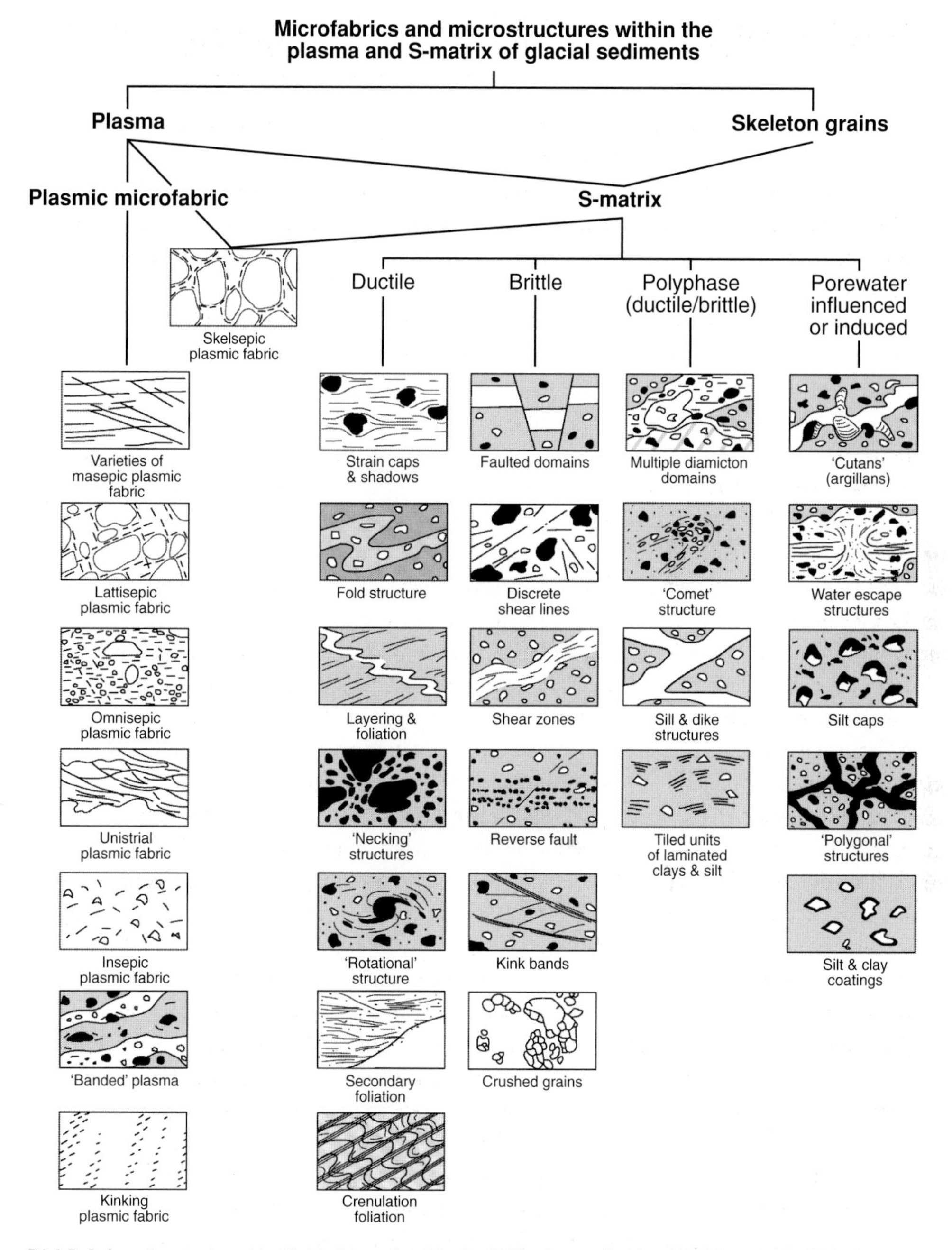

FIG 6.5 Deformation structures identified in thin section (Menzies 2000, after van der Meer 1993) (see also Fig 6.16). Note the grouping of structures into those that reflect brittle and ductile deformation.

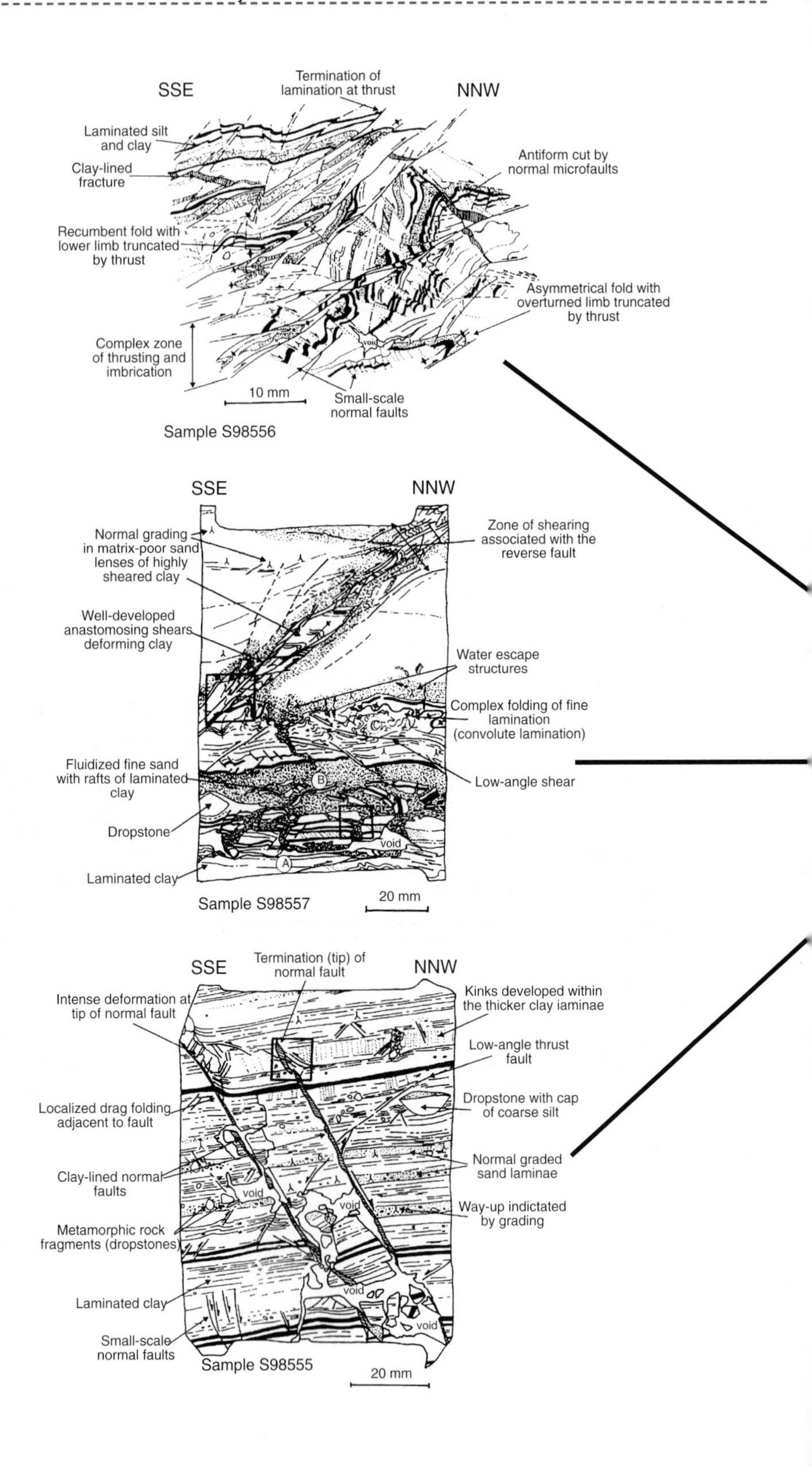
SSE
Termination of lamination at thrust
NNW
Laminated silt and clay
Clay-lined fracture
Antiform cut by normal microfaults
Recumbent fold with lower limb truncated by thrust
Asymmetrical fold with overturned limb truncated by thrust
Complex zone of thrusting and imbrication
void
10 mm
Small-scale normal faults
Sample S98556
SSE
NNW
Normal grading in matrix-poor sand lenses of highly sheared clay
Zone of shearing associated with the reverse fault
Well-developed anastomosing shears deforming clay
Water escape structures
Complex folding of fine lamination (convolute lamination)
C
Fluidized fine sand with rafts of laminated clay
B
Low-angle shear
Dropstone
void
A
Laminated clay
Sample S98557
20 mm
SSE
Termination (tip) of normal fault
NNW
Intense deformation at tip of normal fault
Kinks developed within the thicker clay iaminae
Low-angle thrust fault
Dropstone with cap of coarse silt
Localized drag folding adjacent to fault
Normal graded sand laminae
Clay-lined normal faults
void
void
Way-up indictated by grading
Metamorphic rock fragments (dropstones)
void
Laminated clay
void
Small-scale normal faults
Sample S98555
20 mm

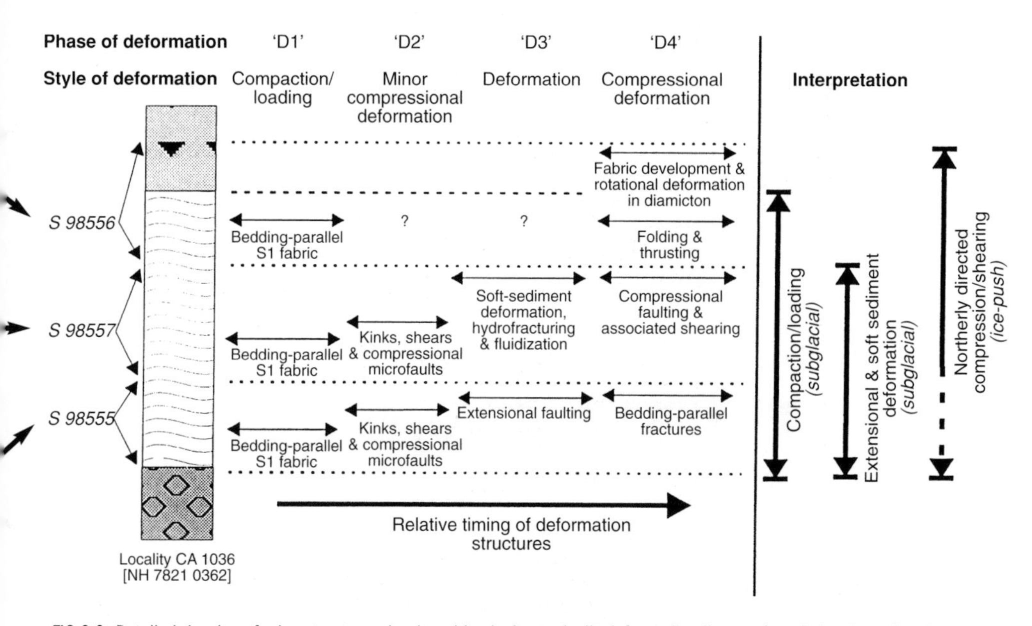

FIG 6.6 Detailed drawing of microstructures developed in glacitectonically deformed sediments from Raitts Burn, Strathspey, Scotland (Phillips and Auton 2000). Detailed drawings such as these allow the geometry and stratigraphy of deformation to be interpreted. Adopting a 'structural' approach, Phillips and Auton use these thin sections to reconstruct a four-stage record (right hand figure) of polyphase deformation at the site.

Voids found in thin sections may have a variety of origins, from natural voids inherent within the sediment that may provide an indication of sediment genesis, through to voids resulting from problems associated with the preparation of thin sections. Seven types of void are recognized within soil micromorphology (Kemp 1985: Fig. 6.7), although there has been little systematic investigation of void formation in glacigenic sediments. Studies of contemporary subglacial sediments identify parallel-walled *fissures* in tills that tend to reflect the plane of shear within sediments and may also be used to define structural elements in thin section, such as *marble bed* structures (van der Meer 1996; 1997; Hiemstra and van der Meer 1997) (Fig. 6.8), reflecting specific subglacial conditions. Other voids, such as rounded *vesicles*, may reflect beach cavitation (van der Meer *et al.* 1992a) or sediment dewatering processes (Bertran and Texier 1999). Irregular voids with indistinct boundaries (*vughs*) tend to reflect removal of sediment from a thin section during final polishing, as a consequence of incomplete impregnation of the sample, and are relatively common in glacigenic sediments, requiring careful identification.

Plasmic fabric may be defined as the arrangement of the plasma or matrix of a thin section, although in the context of thin section micromorphology the term is mainly applied to the examination of only the finest (<2mm) components of the plasma. Plasmic fabric is best investigated under cross-polarized light, under which interference colours produced by the interaction of the sample with the emitted light wave identify domains of similarly oriented clays and fine silts. Bundled domains of clays and silts may form a number of arrangements influenced partly by processes of sedimentation, but primarily plasmic fabrics are the result of

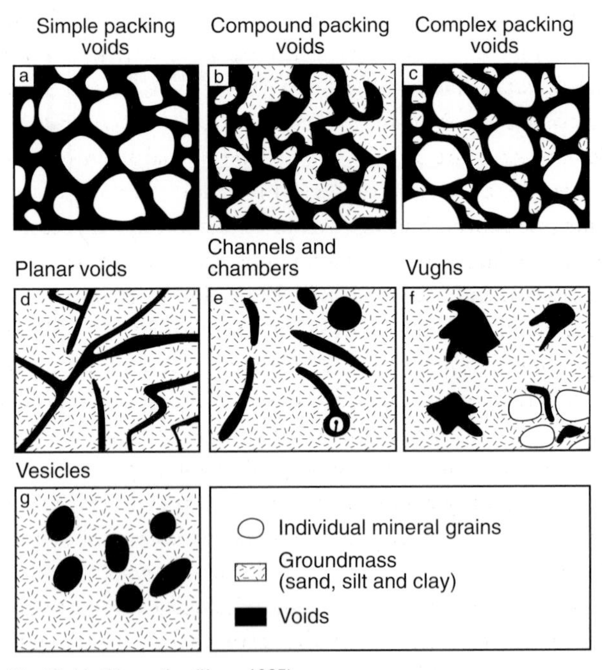

FIG 6.7 Common voids identified in thin section (Kemp 1985).

the re-orientation of domains through the application of stress during deformation, clearly demonstrated by experimental deformation of artificial clays (Hiemstra and Rijsdijk 2003).

The terminology of plasmic fabrics is somewhat complex, as a consequence of the adaptation of terminology from soil micromorphology. However, like the deformation structures outlined above, most plasmic fabrics may be grouped into two general classes. *Rotational plasmic fabrics* (*skelsepic, lattisepic, omnisepic*) indicate the rolling of particles in relation to each other, picking up and re-orienting domains of clays and fine silts against the surfaces of larger grains (Fig. G48). *Planar plasmic fabrics* (*masepic, unistrial*) reflect either the pervasive or discrete failure of sediment along shear planes (Fig. G49).

Additional plasmic fabrics provide an indication of other styles of deformation within a sediment sequence. For example, the presence of *kink-bands*, also known as a *kinking plasmic fabric* (Fig. G47) are an indication of lateral compression during deformation, and as such have been recognized in stacks of lake sediments that have undergone proglacial deformation under compressional stress regimes (Bordonau and van der Meer 1994).

Post-depositional structural changes to sediments may occur through the processes of weathering, or changes to the broader environment within which the sediment was originally deposited (Kemp 1985; 1998; 1999; Federoff *et al.* 1990). One of the most common post-depositional changes to recently deposited glacigenic sediments occurs through periglacial activity. Comprehensive investigation and reviews of the development of periglacial structures in soils and sediments are presented by Coutard and Mucher (1985) and van Vliet-Lanoe (1998). Often the microscopic imprint of periglacial activity is identified well below the lower limit of observed field cryoturbation features such as involutions or frost/ice wedges. Most commonly, the growth of interstitial ice results in lens-shaped voids in samples, often capped with a fine silt layer, producing a structure variously termed a *silt capping*, or *banded fabric*. Periglacial activity can disturb existing structures such as laminations, inducing what look like micro-scale ripples within laminations as a consequence of frost creep (Coutard and Mucher 1985).

The most problematic structure induced through periglacial activity is that of *skelsepic plasmic fabrics*, because these are often used as an indicator of rotational deformation in subglacial sediments (see above). Shrink-swell processes in clays are exacerbated by seasonal freezing of sediments, leading to the production of a skelsepic plasmic fabric within as little as 18 frost cycles (Coutard and Mucher 1985). The inducement of rotational plasmic fabrics such as those mentioned above may occur wherever shrink-swell processes occur, for example through groundwater fluctuation, and are not confined exclusively to periglacial settings. Where such structures are identified in sediments of potential subglacial origin, an assessment of clay mineralogy may indicate whether a skelsepic plasmic fabric reflects shrink-swell conditions, or grain rotation in a subglacial setting.

Other such processes are a key component of weathering, and may also include the translocation of clay, silt, iron and magnesium (hydr)oxides as well as carbonates. The re-growth of crystals from these solutes may also occur within glacigenic sediments, resulting in the production of *neoformations*. The micromorphological structures resulting from translocation and neoformation are beyond the scope of this chapter, but are covered in great detail by Kemp (1985; 1999).

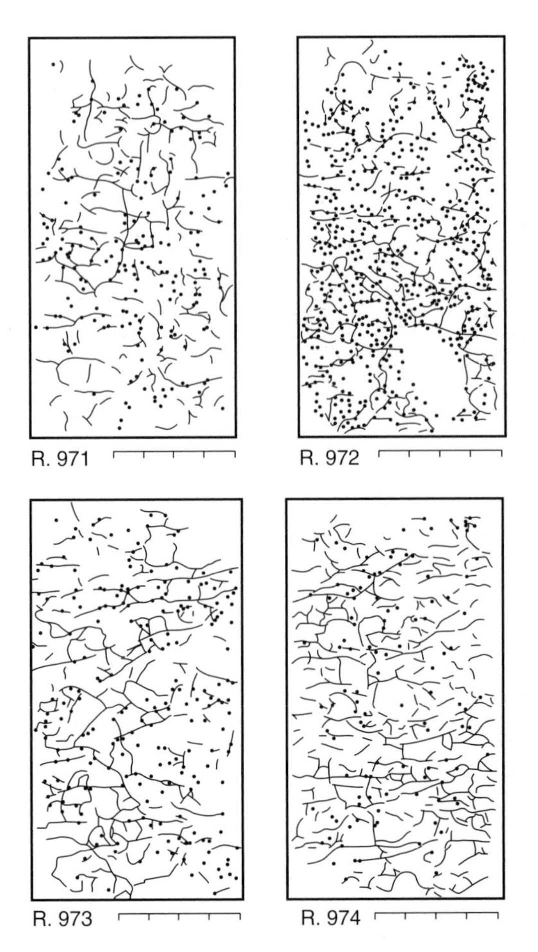

FIG 6.8 Marble-bed features (from Hiemstra and van der Meer 1997). In this group of samples from Saalian till, Wijnjewoude, the Netherlands, a series of parallel walled fissures identify a distinct structure in the sediment. Dots illustrate the position of fractured grains produced as a consequence of stress concentrations during 'marble-bed' deformation.

Recognition of *sampling/preparation-induced structures* is of key significance within micromorphology. As has been stressed throughout, the success of micromorphology is dependent on recovering and preparing a sample without any appreciable disturbance to the original sediment. In practice, some disturbance is unavoidable, and it is vital that the structures produced by such disturbance are recognized by the researcher. Field sampling of sediments may induce structures in two ways, either through compaction of samples due to excessive force being used during sampling, or through deformation induced by vertical compression during coring (Fig. 6.9). The transport of samples back to the laboratory may also induce the formation of structures, through dewatering of samples resulting in the formation of vesicles and density driven deformations, although fortunately these tend to be rare. Such features may easily be confused with *in situ* features, and further emphasize the need for careful sample packaging and transport, as well as adequate field description,

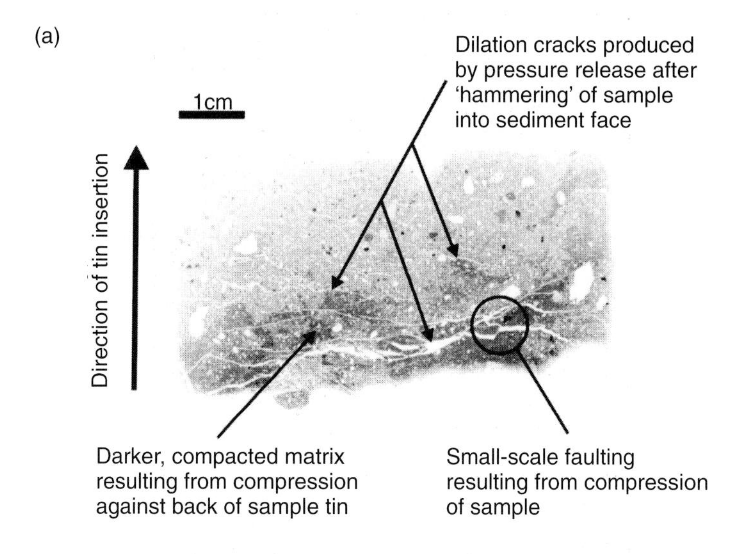

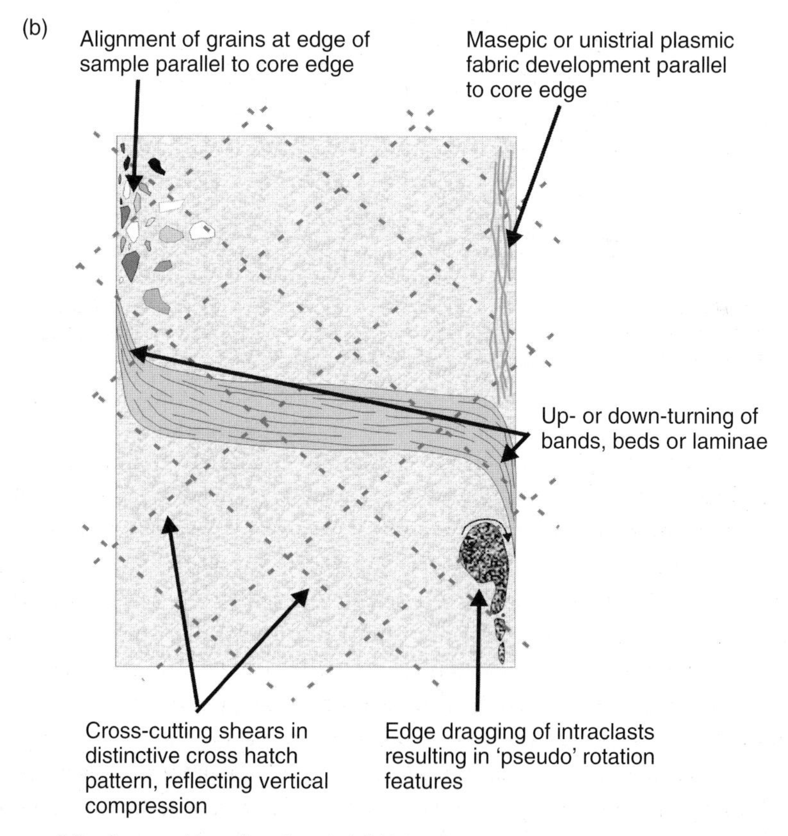

FIG 6.9 Damage and disturbance to thin sections through a) field sampling and b) coring-induced structures.

to identify whether a sample is at risk of dewatering. Finally, the processes of resin impregnation and thin section production can result in new structures produced through the partial removal of imperfectly impregnated samples or the presence of air bubbles trapped beneath slide coverslips, producing voids of distinctive structure (e.g. Fig G45).

6.6 APPLICATION OF SEM ANALYSIS TO SEDIMENTOLOGY

While there has been a significant volume of research on the application of thin section micromorphology only over the past decade, the application SEM to sedimentology extends back to the 1960s and 1970s. However, given the long history of research using SEM, there have been relatively few attempts in recent years to develop the application beyond areas that were identified at an early stage in the use of SEM techniques. A number of overviews of the role of SEM in the investigation of glacigenic sediments have been written (Whalley 1978; 1996; Bull 1981; Mahaney 1995) and this section summarizes the primary outcomes of such research. Two major approaches to sedimentary micromorphology using SEM have been adopted, looking either at *surface textures* of individual grains, or the *bulk properties* of sediment samples.

The *bulk analysis* of SEM samples is constrained by the small size of sediment samples that can be analysed. This, combined with the high degree of magnification involved in SEM analysis means it is often difficult to characterize the representative nature of an individual lithofacies unless a large number of samples is used (Whalley 1996). The nature of SEM studies is such that photo-montages need to be constructed to allow investigation of a sample larger than a few square millimetres (Fig. 6.10). As such, the bulk analysis of sediment micromorphology is probably better served by the analysis of thin sections. There are, however, a number of areas in which SEM analysis is of use within the analysis of bulk sediment micromorphology, as summarized by Whalley (1996). SEM images provide an enhanced depth of field and allow investigation at higher magnifications than thin sections, and may be of use in examining subtle sediment variation within fine-grained laminations for example. Additionally, the equipment of SEM allows a range of processing techniques that can enable much more detailed analysis of trace geochemistry through use of EDS probes. Owen (1994) has carried out a successful application of bulk analysis using SEM, in association with thin sections and macro-scale sedimentary analysis of glacial and non-glacial diamictons in the Karakoram and western Himalayas. Recent studies have illustrated the role of SEM in determining the nature of deformation at a microscopic level (Hirono 2000). Whilst describing pre-Quaternary sediments, Hirono (*ibid*) demonstrates how SEM analysis of micro-scale slip sense indicators provides information not evident at a macro-scale, which is of value to Quaternary and glacigenic sediment studies.

There has been significantly more analysis of samples of individual particles using the SEM within Quaternary and glacial sedimentology, primarily focusing on particle size, morphology and surface textures. As with thin section analysis, many of the standard

FIG 6.10 Photomontage of SEM photomicrographs of a lodgement till from the Karakoram Mountains (Owen 1994). The montage identifies a striated surface, with a sense of movement from the right to the left. Field of view is 4mm.

macro-scale approaches to describing particle form and morphology may be applied with SEM stubs. However, as Whalley and Orford (1982) and Whalley (1996) note, care must be exercised to ensure that viewing angle using the SEM does not skew assessment of particle shape. Although currently applied through the use of a human operator, it is likely that such assessments of particle shape and morphology will become automated through more effective digital image analysis.

One of the main areas of investigation using SEM on samples of individual grains is the analysis of surface textures, mainly of quartz grains, which are broadly ubiquitous in sedimentary bodies. Since the early work of Krinsley and Doornkamp (1973), many researchers have attempted to derive sediment history based upon the nature of the surface of individual particles (Bull 1981; Sharp and Gomez 1986; Whalley 1996). An individual particle surface texture will reflect the wear patterns derived from the processes of erosion, transport and deposition that it has undergone, and associations of different textural features should be indicative of specific depositional processes and environments. As Whalley (1996) notes, however, many of the surface features found may be produced in a range of different environments, resulting in some ambiguity in interpretation. Furthermore, quartz, while ubiquitous in most sediments, may be derived from different sources, resulting in different crystal structures, leading to development of distinct surface textures.

A range of surface textures and features of glacigenic sediments have been examined using SEM, assessing the nature of particle crushing and abrasion, formation of chattermarks and weathering processes (Fig. 6.11). These have been interpreted as evidence for erosional, transport and depositional processes as a key to interpreting sediment genesis and diagenetic processes. Key surface features, including conchoidal fractures, sharp angular edges and the presence of 'pre-weathered' and high-relief particle

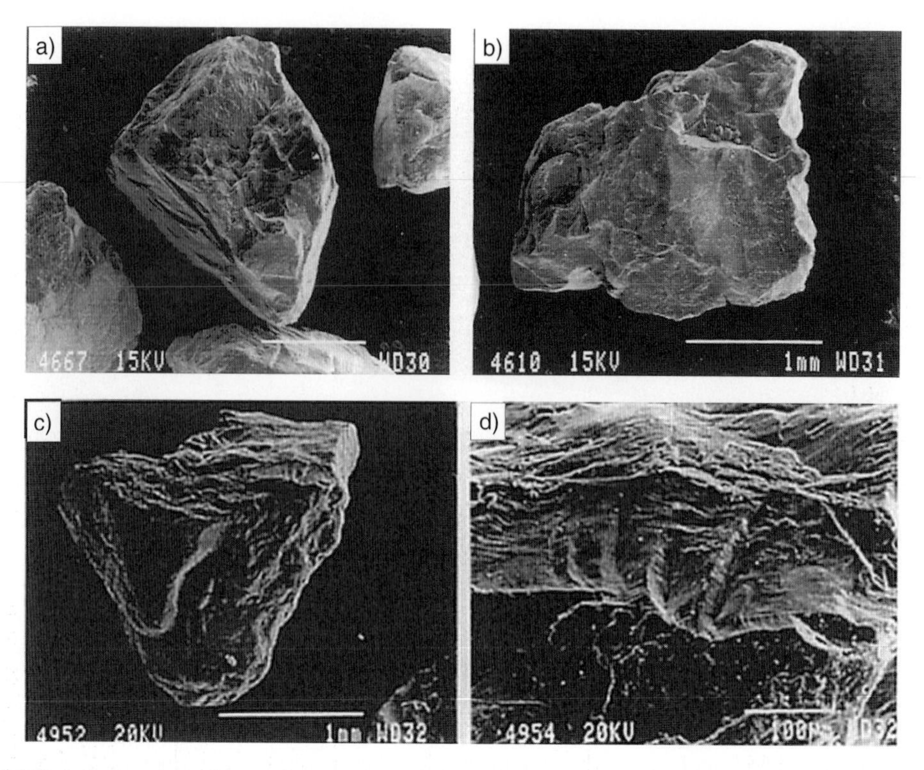

FIG 6.11 Surface textures and features of quartz grains recovered from subglacial tills in Estonia (Mahaney and Kalm, 2000; Mahaney *et al.* 2001). a): Angular quartz with sharp edges, high relief and striations (lower half of grain) and abraded fractures. b): Quartz grain with high relief, sharp edges, sub-parallel features and adhering particles (on step just above centre). c): Quartz grain with rounded edges (bottom) and sharp edges (top). d): Enlargement of centre of image C, showing fresh conchoidal fractures.

morphology are considered by Mahaney and Kalm (2000) and Mahaney *et al.* (2001) as diagnostic of subglacial environments (Fig 6.12).

While most applications of surface texture analysis of quartz grains have been used to aid genetic interpretation of sediments, a more controversial application of this analysis has been proposed by Mahaney *et al.* (1988; 1996). Mahaney (1995) suggests that within subglacial tills, characteristic surface features provide an indication of former glacier thickness. On the basis of analysis of quartz microtextures from modern and Quaternary tills, Mahaney (1995) and Mahaney *et al.* (1988; 1996) suggest that the presence of sub-parallel and *conchoidal fractures* and *grooves* imply ice thicknesses of in excess of 500m during transport and deposition. However, Clark (1989) notes that surface textural features do not necessarily reflect simply the thickness of ice, but rather the shear stress applied at the bed of a glacier, and the mode of sediment transport through the glacial system prior to deposition. It is also apparent that the relationship between ice thickness, basal shear stress and processes of crushing and abrasion is by no means clear, and that Mahaney's (1995) estimations should only be treated as speculative. However, as an application of SEM in attempting to examine the dynamics of former glaciers and ice sheets, the examination of surface textures of quartz clearly has significant potential.

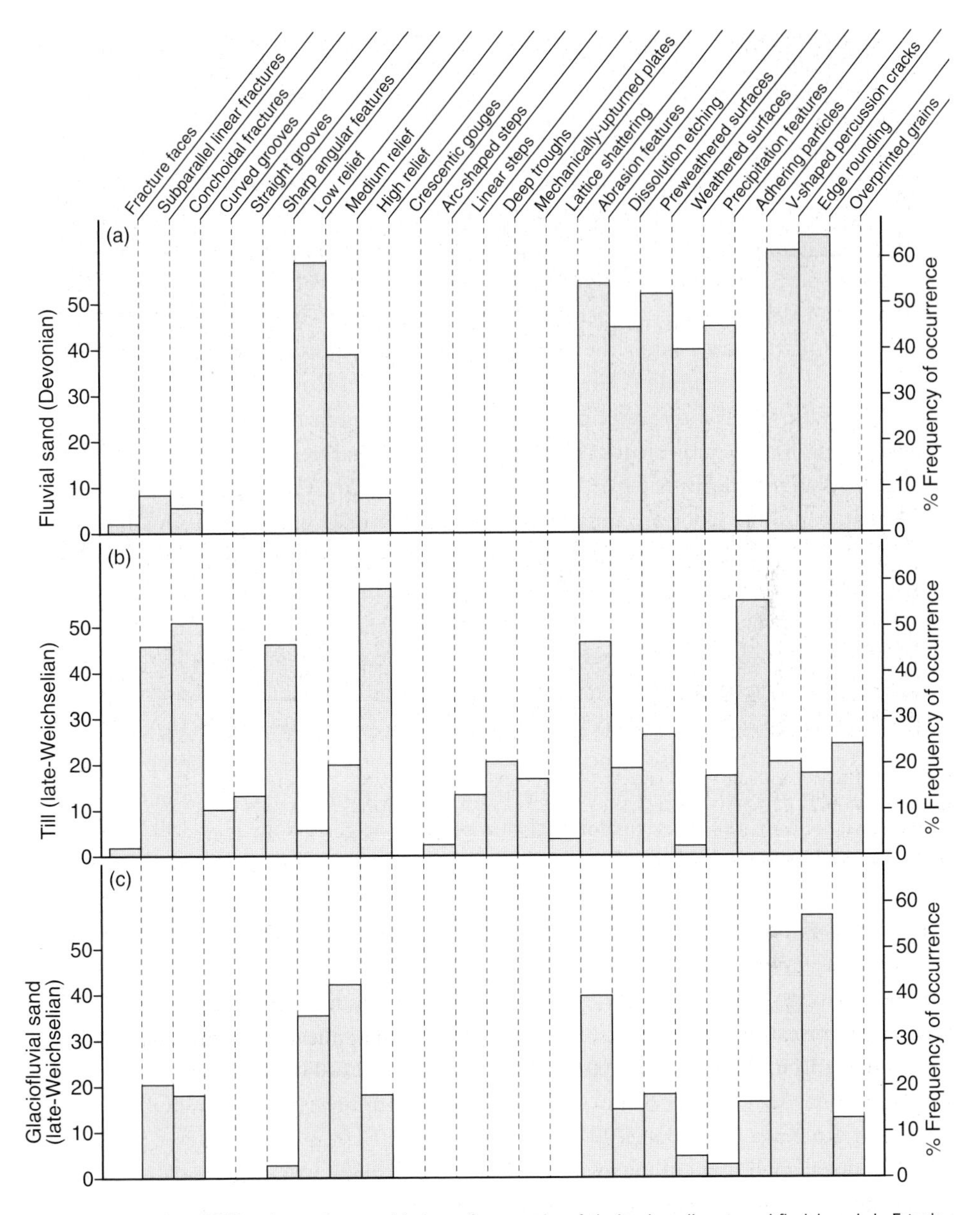

FIG 6.12 Presentation of SEM surface textures and features, from a series of glacigenic sediments and fluvial sands in Estonia (Mahaney and Kalm 2000). Note that this approach tries to quantify the presence of specific features from the samples, although it does not permit comparison between individual samples.

6.7 MICROFABRIC ANALYSIS

Whilst textural and structural description of thin sections and SEM samples often provides ample evidence to aid environmental reconstruction, much more information may be extracted from samples. One of the more useful techniques that may be applied to oriented samples is the examination of the arrangement of particles at a microscopic scale, or *microfabric*. In its strictest sense, microfabric includes all particle fractions, and thus includes the structure of plasmic fabric (Menzies 2000), but within glacial and Quaternary studies the term is mainly applied to the arrangement of sand-sized grains within a sample (Carr 1999; 2001; Carr and Rose 2003). Problems of data presentation and interpretation have limited the application of microfabric analysis using SEM, with most studies focusing on visual descriptive analysis of particle arrangements (Bull 1981; Derbyshire *et al.* 1985). More recently, with the increase in computing power, approaches adopting Fourier transformations have allowed some quantification of microfabric arrangement, reflecting the anisotropy of a sample (Derbyshire *et al.* 1992). However, very few studies of this type have been undertaken in recent years.

Microfabric analysis using thin sections has been a recognized approach in the investigation of glacigenic sediments since the middle of the twentieth century (Sitler and Chapman 1955; Harrison 1957; Ostry and Deane 1963; Evenson 1970; 1971; Johnson 1983), often in close association with macrofabric analysis, mostly applied to examining glacial sediments. The complexity and limitations of 3-D representation of microfabric using up to three orthogonal thin sections (e.g. Chaolu *et al.* 1993) resulted in the decline in use of microfabric in such investigations, concurrent with the significant advancement of macrofabric techniques of data presentation and analysis. In recent years there has been a resurgence in the use of 2-D microfabric analysis using thin sections, driven mainly by the desire to obtain more data from the limited volume of samples obtained through coring (Carr 1999; Carr *et al.* 2000). As Fig. 6.13 illustrates, microfabric investigation of vertically aligned thin sections from contemporary and Quaternary glacimarine and subglacial sediments identifies that the technique provides strong evidence allowing discrimination between fine-grained diamictons of either genetic origin (Carr 2001; Lee 2001).

An additional application of microfabric analysis relates to the investigation of overall sediment fabric, through the systematic measurement of fabric patterns at different particle size increments. As is apparent from Fig. 6.14, particles of different sizes do not respond in a uniform manner to a single applied stress (Kjaer and Krüger 1998). In many cases, there is a systematic phase-switch of particles that are aligned either parallel or transverse to the applied stress field. This is thought by Carr and Rose (2003) to reflect the nature of sediment strain in subglacial tills, by which different arrangements of size-specific microfabrics reflect distinct subglacial conditions during deposition. This application of microfabric analysis has not been

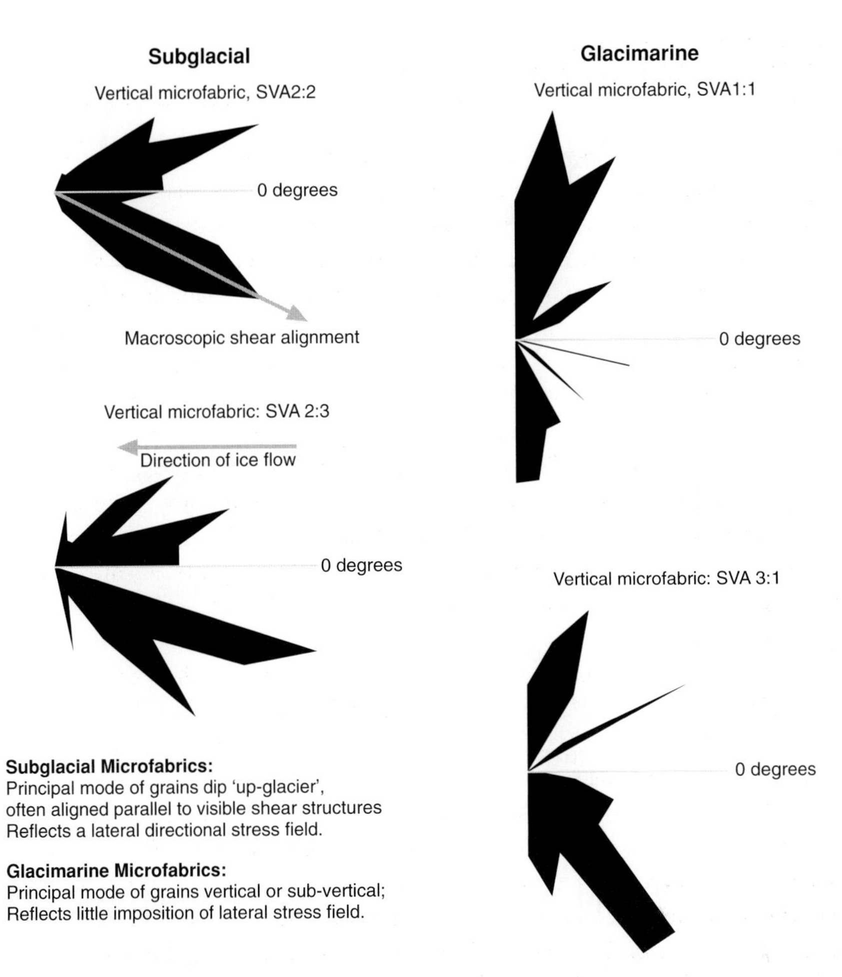

FIG 6.13 Vertical microfabrics of recently deposited glacimarine sediments compared with glacitectonized glacimarine sediments, St. Jonsfjorden, Svalbard (Carr 2001). The distinct vertical alignment of grains in the glacimarine sample reflects rain-out of sands from suspension or ice rafting with a low magnitude lateral stress field. As soon as the sediment has been deformed by subglacial glacitectonics, a very different microfabric signature is evident, reflecting the presence of a high magnitude, unidirectional lateral stress field.

repeated elsewhere as yet, and the results should be treated with extreme caution. However, such an approach may yield critical data regarding the nature of conditions beneath the Quaternary mid-latitude ice sheets essential to reconciling the modelled dynamics of former ice sheets with the preserved sedimentary evidence of glaciation.

It is clear that the potential of microfabric studies within micromorphology is significant, and the 're-discovery' of this approach in recent years is supplying significant new data, allowing greater understanding of key sedimentary and deformation processes.

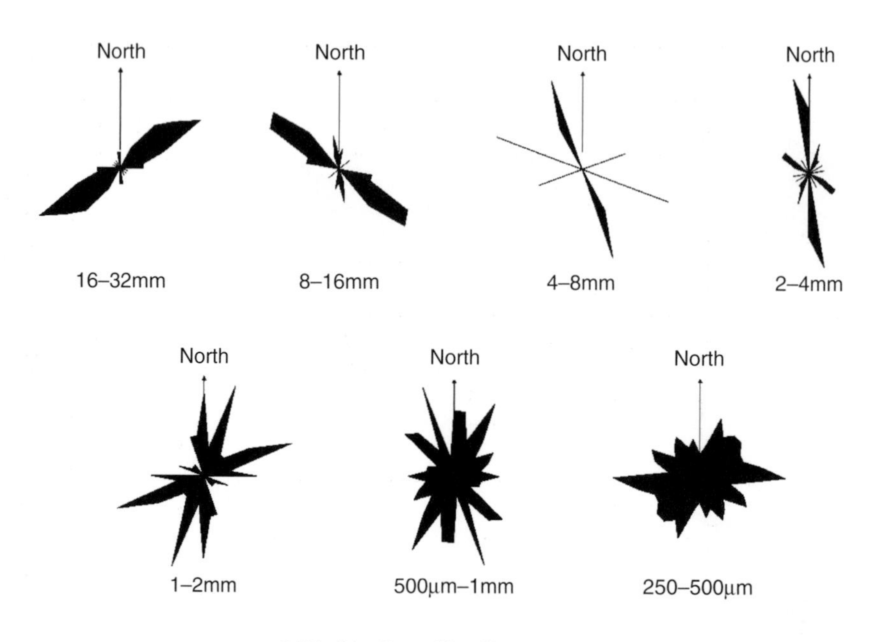

FIG 6.14 Macro- and microfabric orientation data from a recently deposited subglacial till at the margin of the Langjokull ice cap, Iceland. A former ice flow direction of 249° is independently provided by the orientation of subglacial bedforms. Whilst the strength of orientation tends to decrease with a decrease in particle size, it is apparent that there are systematic variations in grain orientation, switching between parallel and transverse phases.

6.8 PRESENTATION AND INTERPRETATION OF MICRO-SCALE FEATURES AND STRUCTURES

As noted earlier, there is currently no standard approach or taxonomy for describing the micromorphology of glacigenic sediments, although there are clearly close relationships between approaches taken by different researchers of both SEM and thin section techniques. This also means that there is no accepted format for the presentation and interpretation of micromorphological data. In some ways this is useful, because micromorphology has a broad application at many spatial scales, from the detailed investigation of sediment history at one location (e.g. Phillips and Auton 2000) through to regional interpretations of stratigraphic sequences (e.g. Carr 1999). As such, different styles of data handling and interpretation are controlled mainly by the spatial context of the micromorphological data.

At one end of the spectrum, micromorphology is used to discern the smallest scales of sedimentary or deformational processes. In this situation, detailed textural descriptions of thin sections, with accompanying sample drawings (Fig. 6.6) provide the essential level of detail required for interpretation of specific features or processes. This style of data presentation is

Micromorphological features →	Skeleton			Matrix	Voids		Deformation structures									Marine features		Plasmic fabric				
		Shape																				
Sample No. ↓	Grain size	>500µm	<500µm	Texture	Void ratio	Void type	Section elements	Rotation	Pressure shad	Crushed grains	Pepple I	Pepple II	Pepple III	Water escape	Lineations	Dropstones	Microfossils	Skelsepic	Lattisepic	Omnisepic	Masepic	Unistrial
SVA 1:1	250	SR	SR	C	M	P	–									•	••					
SVA 3:4	250	SR	SA	C	L	P	F									••	••					
SVA 3:3	250	SR	SA	C	L	P	F	•				•	•			•	••				•	
SVA 3:2	250	SR	SA	C	L	P	–						•				••					
SVA 3:1	250	SR	SR	C	L	P	–						•			•	••				•	
SVA 2:4	250	SR	A	C/M	L	F	–	•••	•	••		•	•		•					•	•	
SVA 2:3	250	SA	A	C/M	L	F	Ba/Bo	••	•	•					•			•			•	
SVA 2:2	250	SA	SA	C/M	L	F	F/S	•••	•••					•	••						•	•
SVA 4:6	250	R	SA	M	L	F	–	•	•	••							•			•	••	•
SVA 1:3	<250	R	A	C	L	F	–	•••	••								•				•	

FIG 6.15 Summary of thin section descriptions from Svalbard, from Carr *et al.* (2001). The shaded samples were interpreted as subglacially deformed tills, whilst the remainder were interpreted as different facies of glacimarine sediment. The style of presentation allows rapid comparison of samples, and facilitates the regional interpretation of a large body of sediment.

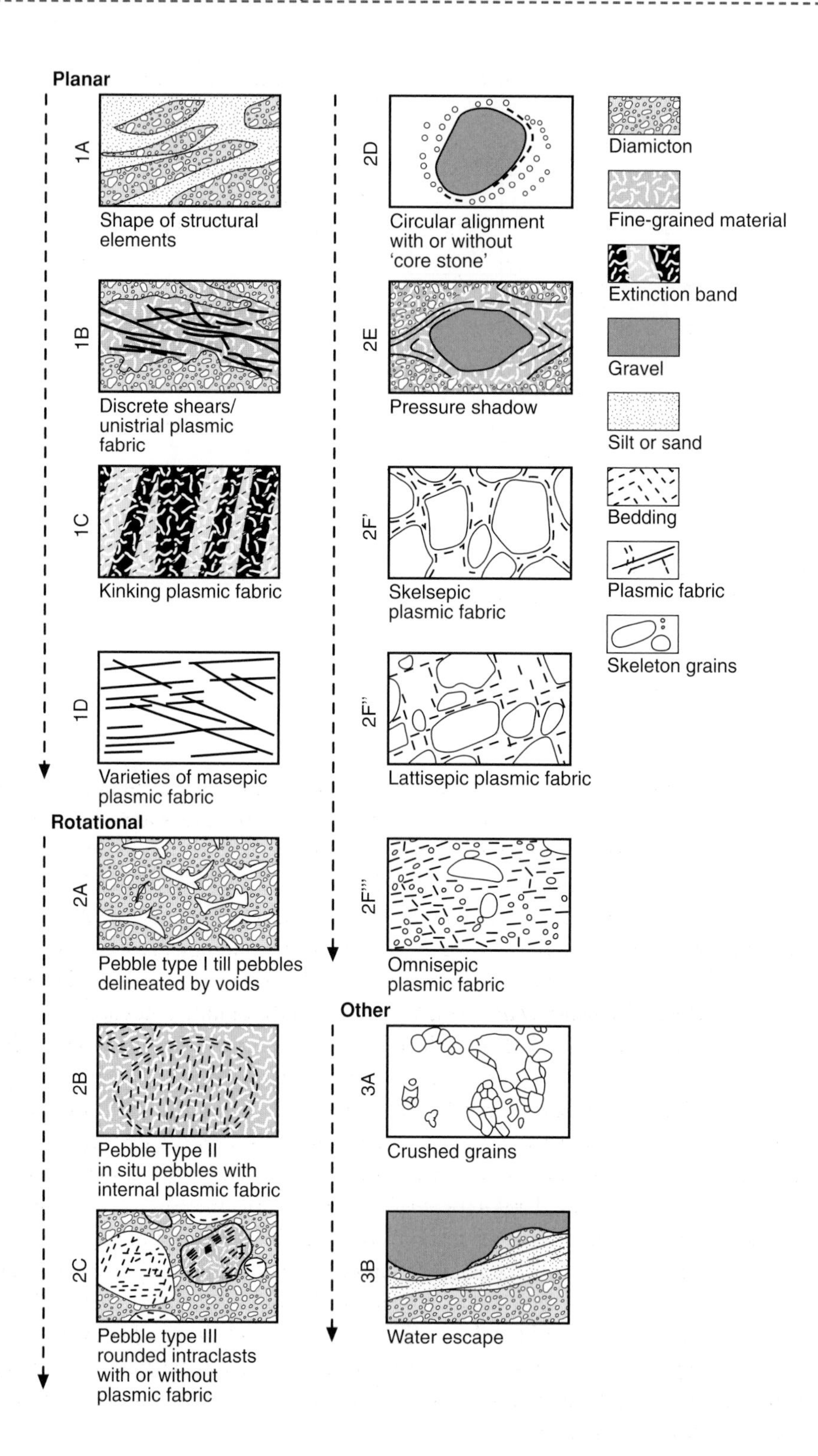

Planar
1A
Shape of structural elements
1B
Discrete shears/ unistrial plasmic fabric
1C
Kinking plasmic fabric
1D
Varieties of masepic plasmic fabric
Rotational
2A
Pebble type I till pebbles delineated by voids
2B
Pebble Type II in situ pebbles with internal plasmic fabric
2C
Pebble type III rounded intraclasts with or without plasmic fabric
2D
Circular alignment with or without 'core stone'
2E
Pressure shadow
2F'
Skelsepic plasmic fabric
2F''
Lattisepic plasmic fabric
2F'''
Omnisepic plasmic fabric
Other
3A
Crushed grains
3B
Water escape
Diamicton
Fine-grained material
Extinction band
Gravel
Silt or sand
Bedding
Plasmic fabric
Skeleton grains

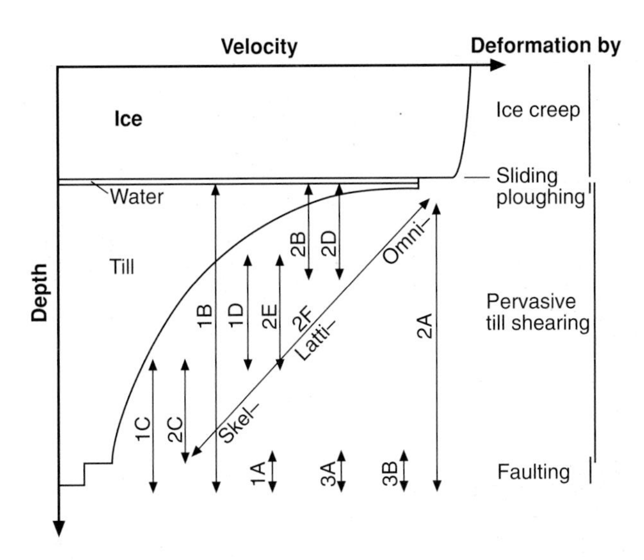

FIG 6.16 Left: Illustration of common microscopic deformation features identified in subglacial sediments. Right: Tentative locations for each type of structure in the subglacial deforming bed (van der Meer 1993).

common for the analysis of the micromorphology at a single site or section, and has been very important for the investigation of fundamental processes of, for example, subglacial sediment deformation (van der Meer 1993; Hiemstra and van der Meer 1997). Following this approach provides essentially qualitative data, which may provide exceptional depth of analysis, but does not readily support comparison between a large number of samples.

At the other end of the scale, there are an increasing number of regional studies of Quaternary sediment stratigraphies in which micromorphology provides critical data (van der Meer *et al.* 1992b; Mahaney *et al.* 1996; Carr *et al.* 2000). Studies that compile sedimentological and micromorphological data from many sites and samples tend to adopt a more structured approach to data presentation, using summary tables or diagrams to evaluate the relative abundance or strength of different micromorphological textures and structures (Fig. 6.15) with a minimum of text description. It is evident that these summary approaches also attempt to quantify aspects of the data, providing indications of the relative degree of structural development evident in specific samples. As such, these approaches provide a dataset that is more readily comparable between samples and studies. It should be noted, however, that the presentation of summary micromorphological data needs to be supported by some detailed investigation of samples to lend 'weight' to the summary data. Most studies tend to adopt a combination of both approaches in the presentation of micromorphological data, as well as other, issue specific methods, such as the quantification of subglacial microstructures to compare between till samples (Khatwa and Tulaczyk 2001), for example.

6.9 CASE STUDIES

There is an increasing body of research incorporating micromorphological analysis using thin sections or scanning electron microscopy, although until very recently this has been limited mainly to analysis of glacial sediments or the effects of soil formation. The case studies presented here attempt to illustrate the range of sediment types that have been investigated, to date, of significance in glacial and Quaternary sedimentology, and some of the principal characteristics of these sediments at micro-scales and the approaches taken to their interpretation.

6.9.1 Glacial sediments

Subglacial tills (van der Meer 1993)

On the left of Fig. 6.16 is the outline scheme of micromorphological structures identified in subglacial sediments by van der Meer (also see Fig. 6.6). On the right, van der Meer tentatively suggests specific parts of the subglacial deforming layer in which specific associations of features may be found. However, van der Meer does stress that the nature of the subglacial deforming bed is such that its character may vary in time and space, resulting in overprinting of planar and rotational forms of deformation.

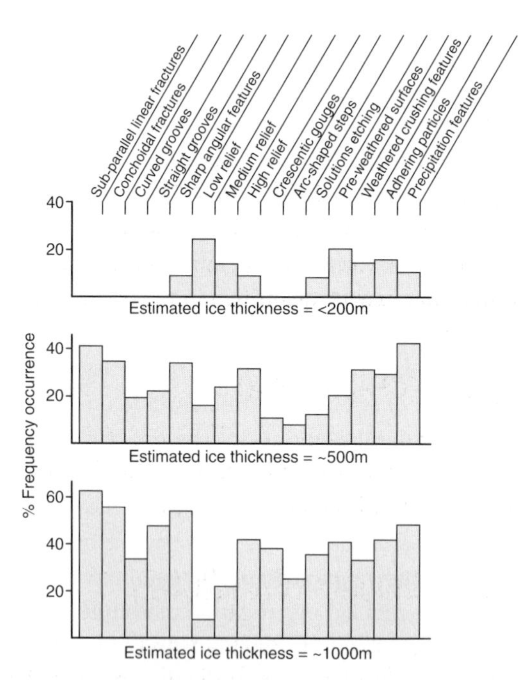

FIG 6.17 Variations of different microtextures on tills emplaced by glaciers of different thickness (Mahaney 1995). Samples used to construct this dataset were obtained from Quaternary and pre-Quaternary tills from Antarctica, Ellesmere Island, Svalbard, North America and the European Alps.

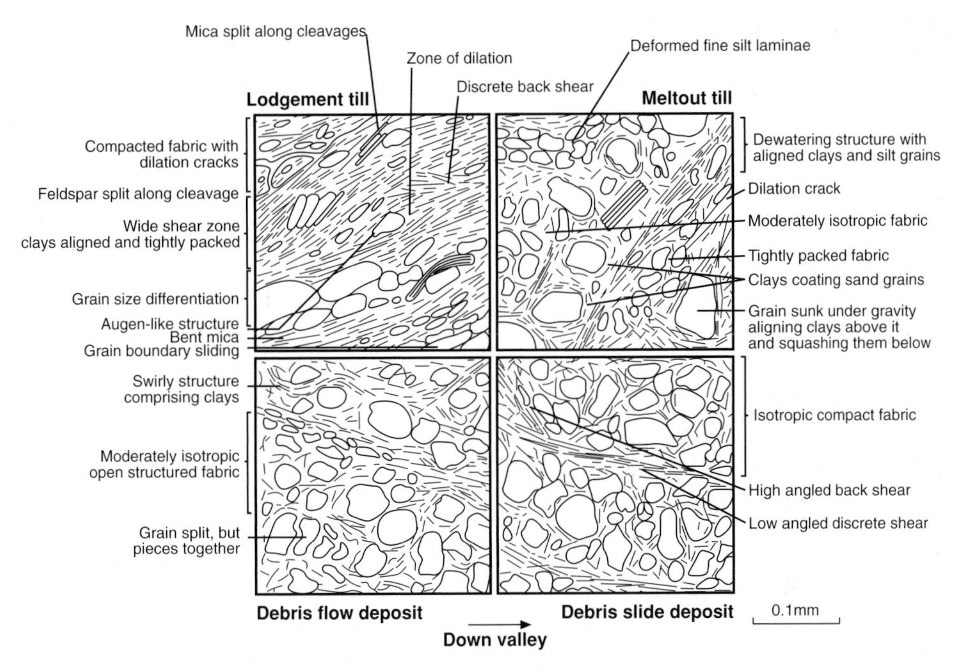

FIG 6.18 Summary diagrams of the characteristic micromorphological features of glacial and non-glacial diamictons in the Karakoram and western Himalayan mountains. This diagram is based on both thin section and SEM analysis (Owen 1994).

SEM grain surface textures (Mahaney 1995)

Figure 6.17 identifies the nature of surface microtextures of tills sampled from sites reflecting small valley glaciers less than 200m in thickness, expanded mountain glaciers up to 500m thick and continental ice sheets more than 1000m thick. It is clear that the range and development of microtextures increases with apparent ice thickness, although whether this is a simple relationship that may be quantified in the way Mahaney suggests is strongly disputed (Clark 1989).

SEM bulk description (Owen 1994)

Figure 6.18 illustrates the range of structures identified in glacial and non-glacial diamictons from the Karakoram Mountains and western Himalayas. This is one of the few studies to apply bulk structural analysis of SEM samples within glacial sedimentology.

6.9.2 Glacimarine sediments (Carr 2001)

Figure 6.19 presents a summary diagram of the characteristic micromorphology of recently deposited glacimarine sediments from western Svalbard. Carr uses this information derived from modern glacimarine sediments as an analogue for interpreting Late Quaternary sediments recovered from the North Sea Basin.

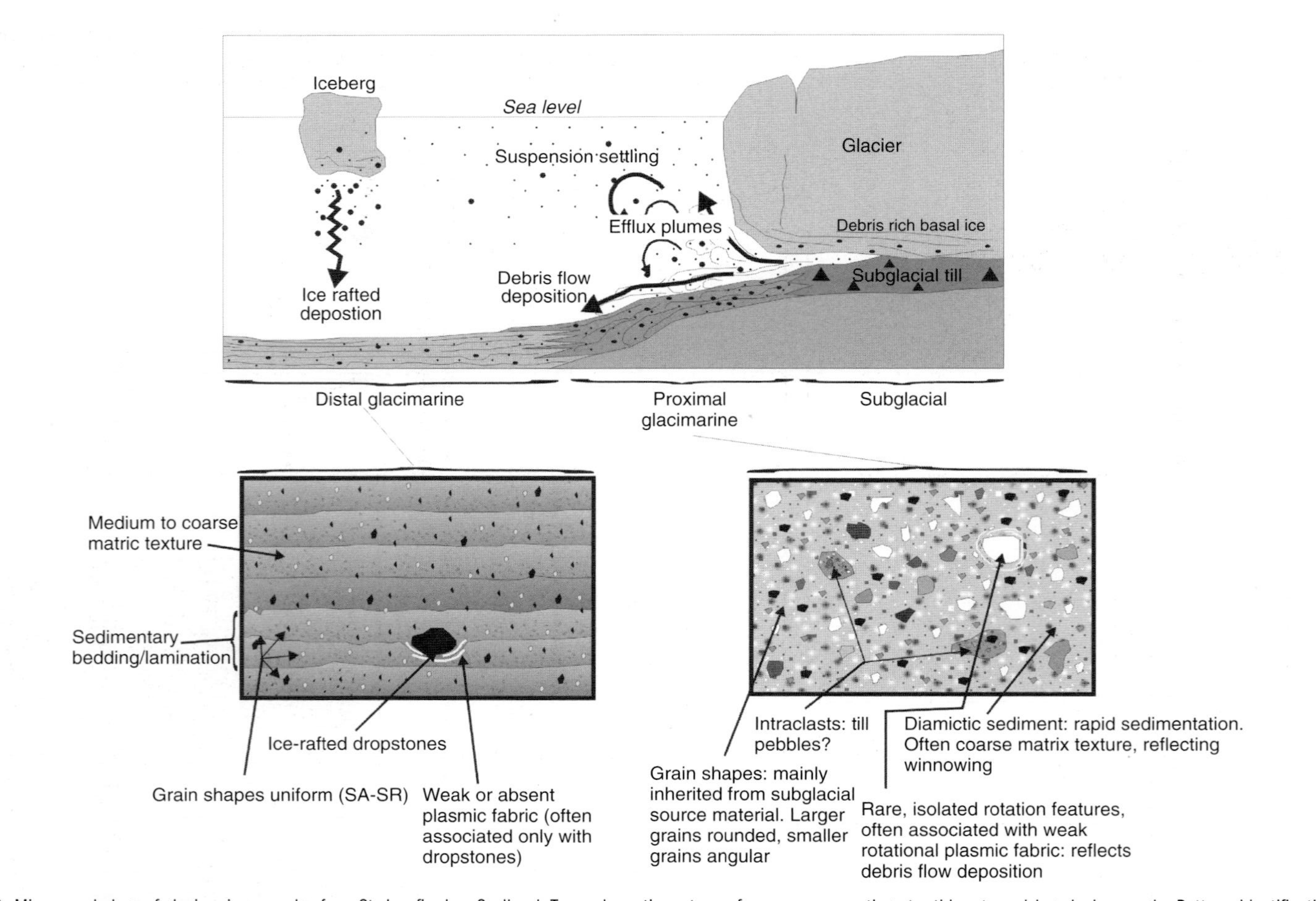

FIG 6.19 Micromorphology of glacimarine samples from St. Jonsfjorden, Svalbard. Top: schematic cartoon of processes operating at a tidewater calving glacier margin. Bottom: identification of key micromorphological features from proximal and distal glacimarine sediments.

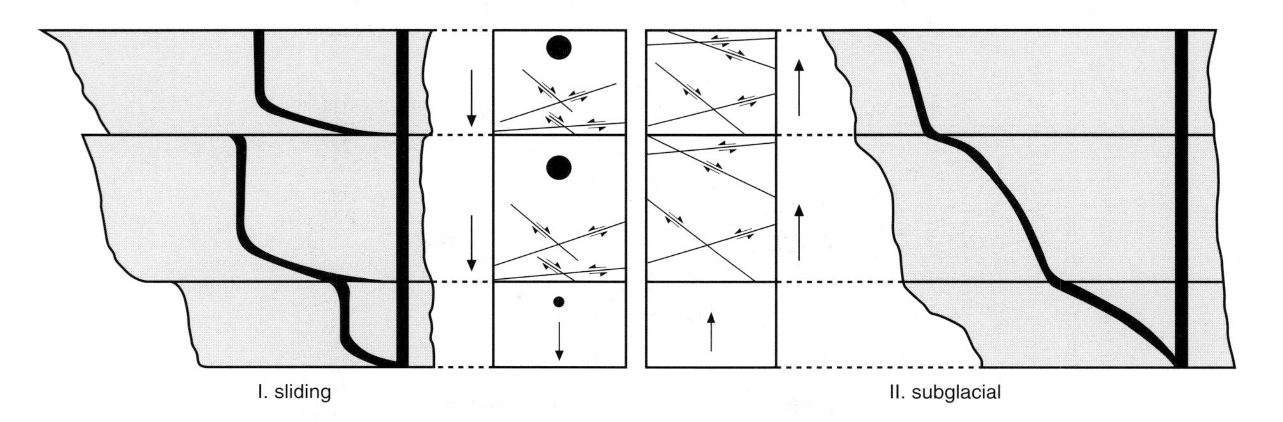

Micromorphological features →	Skeleton			Matrix	Voids		Deformation structures									Marine features		Plasmic fabric				
		Shape																				
Sample No. ↓	Dominant grain size (• m)	>500µm	<500µm	Texture	Void ratio	Void type	Section elements	Rotation	Pressure shadow	Crushed grains	Pebble I	Pebble II	Pebble III	Water escape	Lineations	Dropstones	Microfossils	Skelsepic	Lattisepic	Omnisepic	Masepic	Unistrial
HM 101–06																						
BER 6	250	–	SA	C	L	F	–	•				•				•	•					
BER 7	1mm	SR	A	C	L	F		•				••						•				

FIG 6.20 Top: photomicrograph of Late Weichselian debris flow deposit, North Sea Fan (Cross-polarized light, field of view 18mm). The soft sediment intraclast and limited rotation features are characteristic of proximal glacimarine debris flow deposits, as outlined by the summary table (Carr *et al.* 2000). Bottom: cartoon of deformation profiles, contrasting sub-aqueous gravity driven slides with subglacial deformation (Hiemstra 2001).

6.9.3 Submarine mass-movements (Carr *et al.* 2000; Hiemstra 2001)

The upper image of Fig. 6.20, from Carr *et al.* (2000) is a micrograph of Late Quaternary proximal glacimarine sediments deposited as a sub-aqueous debris flow on the North Sea Fan, with a description of key textural and structural features typical of proximal glacimarine (debris flow) sediments. The lower diagram, from Hiemstra (2001), is a schematic illustration comparing the likely deformation signatures of gravity driven sub-aqueous slides with subglacial deformation, and the predicted pattern of conjugate shear structures evident in thin section. The key contrast between the two environments reflects the location of the highest intensity of deformation, which in the submarine mass-movements is at the base of individually displaced units. Using such an approach, Hiemstra is able to discriminate between glacimarine deposits that have undergone gravity-driven or subglacial deformation in contemporary fjord deposits in Alaska.

6.9.4 Weathering and soil formation (Rose *et al.* 1999; Fig. G50)

Figure G50 illustrates the changes brought about by weathering and pedogenesis of Anglian till, deposited by a Scandinavian ice sheet. The top micrograph identifies silt coatings around many grains, as well as fragments of clay coatings (papules) indicating some disturbance of the sediment during soil formation, possibly through frost action. The middle photomicrograph identifies a layered void infill of clay, reflecting the passage of water and clay along a passage during soil formation. The lower image illustrates the formation of skelsepic plasmic fabric, that may indicate either rotational deformation of the sediment under subglacial conditions, or repeated freeze-thaw of clays in periglacial conditions: this example highlights the importance of basing micromorphological interpretations on a range of associated structures.

7 Particle lithology (or mineral and geochemical analysis)

John Walden

7.1 INTRODUCTION

As described elsewhere within this book, there are a wide range of approaches and various lines of evidence that can be used to reconstruct the environmental processes that have resulted in the deposition of a glacigenic sedimentary sequence. Many examples therefore exist in the research literature demonstrating the use of basic sediment description, particle-size analysis, clast fabrics or particle morphology. Detailed examination of the mineralogical or geochemical composition of glacigenic sediments is, however, less commonly done. As demonstrated by works such as Kujansuu and Saarnisto (1990), such compositional data can have applications in environmental reconstruction and may, in some circumstances, provide information of economic significance (e.g. tracing the bedrock source of commercially valuable minerals found within glacial sediment sequences).

In any research study of a Quaternary sediment unit or sequence, four fundamental questions usually need to be addressed:

- *correlation*; how are the sediments being studied related to sediments in other areas (correlation of sediments between sites)?
- *provenance*; where were the sediments derived from (their provenance) and therefore what was the direction of movement of the transporting medium?
- *process of deposition*; what mechanism (process; environment) was responsible for the deposition of a particular sedimentary unit?
- *chronology*; when were the sediments deposited?

By far the most obvious applications of compositional data are for sediment correlation and provenance indication. As in other sedimentary environments, the geochemical or mineralogical properties of the glacigenic sediments may be used as a basis to infer their likely source, or as a means of linking sediment bodies with similar compositions between locations. This approach is often referred to as chemical or mineralogical 'fingerprinting' of a sediment and all of the techniques introduced in Sections 7.3 to 7.8 can be used in this fashion. However, as illustrated in Section 7.9, compositional data can provide evidence that may help resolve each of the four questions listed above. Other applications therefore include studies on weathering history (which, in turn, may imply something about relative chronology) and processes of deposition.

A wide range of methods have been used for characterizing the geochemical or mineralogical properties of glacigenic sediments, and a comprehensive coverage would require a book in itself. This chapter aims to provide an introduction to a number of the more well-established and commonly used methods. These include:

- fine gravel analysis (clast lithological analysis) (Section 7.3);
- heavy mineral analysis (Section 7.4);
- bulk geochemistry (using X-ray fluorescence) (Section 7.5);
- mineralogy of the clay fraction (using X-ray diffraction) (Section 7.6);
- mineral magnetic analysis (Section 7.7);
- carbonate analysis (Section 7.8).

In each case, the basic principles of the methods will be introduced, followed by a consideration of the key practical issues facing potential new users. References to more detailed accounts of each method will also be provided. Prior to any such analysis being undertaken, samples need to be collected, and some recommendations for field sampling are made in Section 7.2. The chapter will conclude with a discussion of a number of case studies from the research literature that illustrate how the methods can be applied (Section 7.9).

7.2 FIELD SAMPLING AND SITE CONTEXT

Collection of appropriate samples in the field is clearly critical to the success (or otherwise) of any subsequent analysis of the mineralogical or geochemical composition of a sediment. The exact sampling procedure will obviously vary depending upon what analytical methods are to be used and the overall aim of the investigation, but standard 'good practice' in both field and la boratory sub-sampling, as described by Lowe and Walker (1997), Gale and Hoare (1991) and Gardiner and Dackombe (1983), should be followed.

In terms of site context, for a study that employs any of the compositional techniques described here, it is always preferable to have some background information on both the local and regional geology. Using lithology, mineralogy or elemental geochemistry to determine answers to the types of fundamental questions posed in Section 7.1 generally requires an understanding of the geological context within which the study site is situated. For unravelling questions about sediment provenance and correlation, such information is clearly critical.

Where possible, samples should be collected from open field sections. This allows a visual examination of the variability in the stratigraphy (both vertically and horizontally) and may, in turn, influence both the number of samples collected and the choice of specific sampling locations. Surface exposures should be cleaned of any loose or slumped material prior to sampling so that the material collected does not suffer from the obvious effects of surface weathering (such as discolouration). Careful notes should be taken on sample locations and, if multiple samples are taken from an exposure, sample depths should be recorded as well as a record of the site stratigraphy. The required detail recorded in the latter will depend upon the nature of the study (Jones *et al.* 1999; Chapter 2). In some cases a full site description, including vertical sediment logs, may be required but in others a simple field sketch of the stratigraphy may be sufficient.

Ideally, some sort of standardized procedure should be adopted for selecting specific sampling locations within a section (i.e. random or stratified). However, logistical constraints (such as accessibility or safety in vertical sections above a few metres in height) often require a more pragmatic approach to be adopted. Where accessibility allows, a series of vertically spaced samples, taken either at regular intervals or at intervals that ensure each unit is sampled in a representative fashion, is often employed.

Glacigenic sediments, and in particular glacial diamicts, can show considerable variation in their physical and chemical/mineralogical properties over relatively small spatial scales (<1m). In collecting samples for studies such as provenance indication or unit correlation, a single sample taken from a point location within even a modestly sized exposure (i.e. a few metres across) may not be fully representative of the composition of the unit as a whole. Where at all possible, multiple samples should be taken so that the relative magnitude of intra-unit variation can be quantified.

The nature of the specific sample collected will depend upon the subsequent analysis. For clast lithological analysis, suitable numbers of clasts need to be collected. Bridgland (1986) presents a detailed discussion of this topic but suggests that most workers would see *c.* 300 individual clasts as a suitable compromise between efficiency and statistical reliability. The volume of material required to obtain this number will depend upon the particle-size fraction chosen for the analysis (see Section 7.3).

For the majority of other forms of analysis described here, bulk matrix samples (i.e. material of sand grades and finer) need to be collected. Again, there is no absolute guidance that can be given other than to collect enough material to accommodate all the subsequent analyses, but *c.* 250g of matrix per sample might be considered a respectable minimum. Some consideration also needs to be given to the overall grain-size distribution (as assessed within the field) of the unit being sampled. This is particularly true for samples that are to be subjected to any of the analysis methods that are based upon a specific size fraction; enough bulk material needs to be collected in order to obtain a sufficient mass of the size fraction required. Field samples should be placed in air-tight plastic sample bags. These should be clearly labelled and, ideally, double bagged with additional labels between the inner and outer bag. Any trowels, spades or knives used in the sampling process should be free from rust, etc. and must be cleaned between the removal of each sample to prevent cross-contamination.

Once samples are returned from field to the laboratory, if it is not possible to prepare them immediately for analysis, they should be stored in a refrigerator or cold room (at *c.* 4°C). If samples were moist on collection or are thought to contain a significant proportion of organic matter, cold storage will reduce the possibility of any ongoing biological activity altering the chemical or mineralogical state of the samples.

If at all possible, all samples should be air-dried (preferably without too long a delay after collection). For magnetic analysis (Section 7.7) temperatures above 40°C can induce a change in the magnetic properties, and drying at high temperature is more likely to cause compositional alteration than air drying. Gale and Hoare (1991) provide a useful discussion of some of the problems encountered in drying samples prior to subsequent laboratory analysis and also provide detailed advice on the potential pitfalls involved in obtaining representative sub-samples from bulk field samples. Where at all possible, a sample splitter (also know as a riffle box) should be used in all sub-sampling. 'Grab' or 'dip' sampling, where a spoon or scoop is just dipped into an open bag of sediment, is unlikely to generate a representative sample as the bulk sediment may well have undergone particle-size sorting within the sample bag during transport and storage.

7.3 CLAST LITHOLOGICAL ANALYSIS

Bridgland (1986) provides an excellent review of the main aspects of clast lithological analysis, covering the ways in which the technique could be used, practical considerations, presentation and statistical analysis of the data and including some case studies using materials from a number of sedimentary environments.

As a tool for stratigraphic correlation and provenance studies, analysis of clast lithologies is a simple technique and requires little in the way of sample preparation. Identification of a suitable number of individual clasts allows the proportions of each lithology represented in the sediment to be calculated. The clast assemblages of each sample can be compared for correlation and related to particular source areas to determine provenance. The technique can also be used on a less quantitative basis, when particular 'indicator erratics' are present within a sediment that are associated with a distinct rock source, thus giving a simple estimate of sediment provenance.

Although in most cases the clast assemblage will be controlled by the provenance of the sediment, lithologies that are less resistant to mechanical or chemical weathering may be under represented due to abrasion during transport or post-depositional weathering. These factors must be considered when interpreting clast lithological data.

7.3.1 Sample preparation

In sampling for clast lithological analysis, two initial decisions have to be made. A suitable clast size range has to be selected for analysis and the sample size (number of clasts to be counted) needs to be specified. Bridgland (1986) suggests that the use of a consistent size range is necessary to enable comparisons between sample sets, but indicates that it is difficult to specify a single size range that is applicable to all work. Sizes commonly adopted are 8–16mm or 16–32mm, but some workers will work with both finer and coarser fractions than this. To a great extent, choosing a suitable size fraction may well be dictated by the types of sediment under study and the overall particle-size distribution. Sample sizes used by current workers vary from 50 to 1000 but counts of *c.* 300 give reasonable statistical validity. Apart from simple cleaning in distilled water, using an ultrasonic bath, no pre-treatment of samples prior to the analysis is required.

7.3.2 Analysis

Clasts should be examined using a hand lens and, where larger magnifications are required, using a binocular microscope and reflected light. Samples can be sub-divided into a number of classes based upon their lithological properties and, once the clasts have been classified, the data can be summarized as both raw count data and also as percentages. Such tables facilitate both a qualitative and quantitative (statistical) comparison between different sample sets.

Reference can be made to basic rock and mineral guides (i.e. Mondadori 1983) and standard geological texts for classification purposes. Fichter *et al.* (1991) and Fichter and Poché (1993) provide a collection of useful flow charts that can also be used to guide the novice user through the rock identification process, and examples are shown in Figure 7.1. Identification can be the main weakness of the technique as it is prone to subjectiveness, particularly when trying to classify clasts into groups when they may grade from one type to the other. Surface weathering of clasts can hinder identification. Such samples can be broken to reveal fresh surfaces. Production of flakes or thin sections can be undertaken but this removes what is one of the main advantages of clast lithological analysis – its speed.

Igneous rock texture and cooling history

Texture	Cooling history	Example
Glassy	Very fast cooling; non-crystalline.	Obsidian
Vesicular (cellular)	Very fast cooling with rapid gas escape forming bubbles in the non-crystalline rock.	Pumice, scoria
Aphanitic (fine grained)	Slow cooling; microscopic crystal growth.	Rhyolite, andesite, basalt
Phaneritic (coarse grained)	Very slow cooling; crystals grow to visible size.	Granite, diorite, gabbro
Porphyritic (two grain sizes)	Two stage cooling; one slow underground creating visible *phenocrysts*, the second fast at the Earth's surface producing a fine grained *groundmass*.	Any aphanitic rock with the adjective *porphyry*

FIG 7.1 Key for the identification of igneous rocks by texture and cooling history.

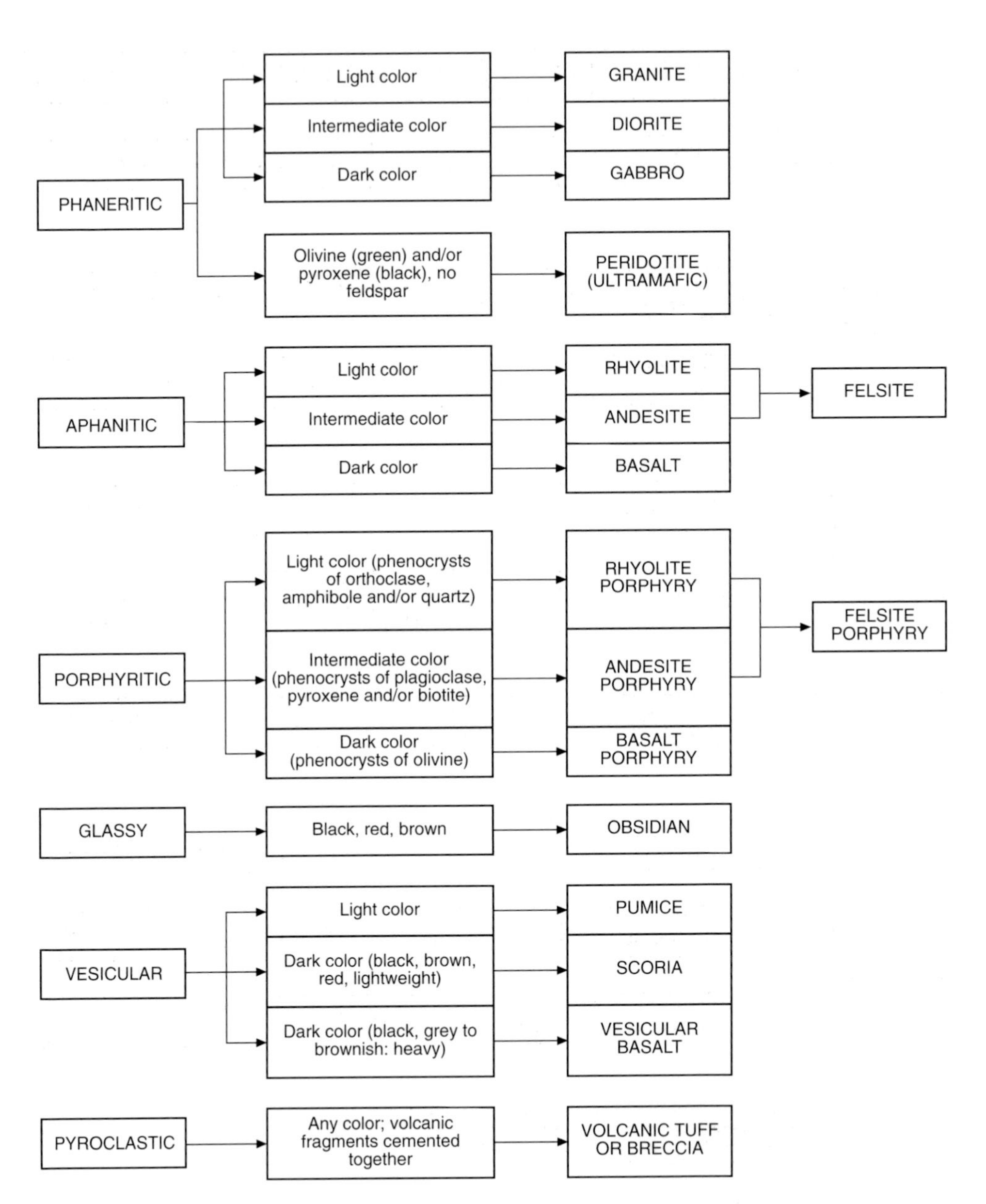

FIG 7.1 (b) Key for the identification of igneous rocks by texture and colour.

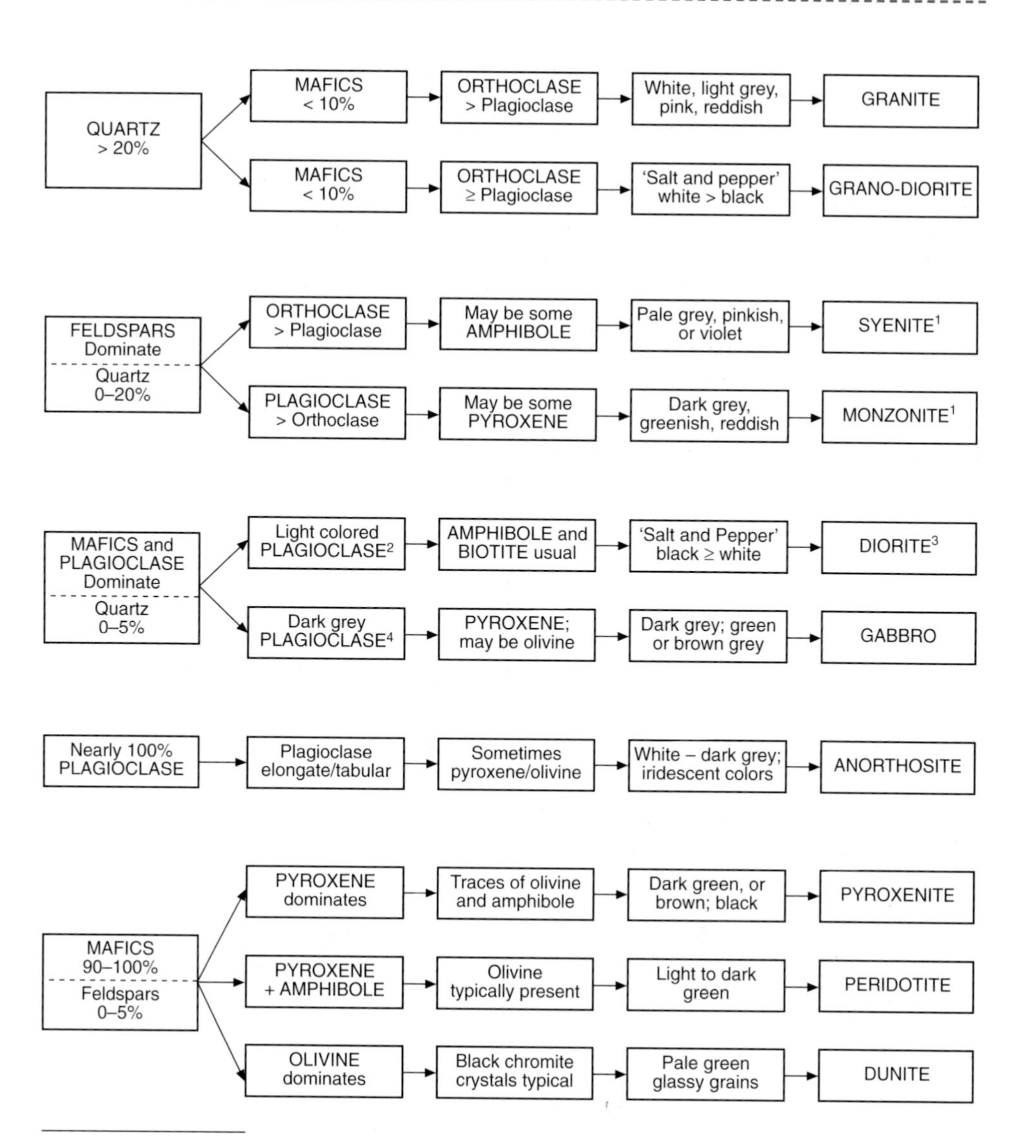

1 If quartz is present between 5–20% call the rock a **quartz syenite** or **quartz monzonite**.
2 Plagioclse is usually light-colored, but other colors are possible.
3 Quartz sometimes between 5–20% in diorites; call them **quartz diorites**.
4 This plagioclase is often grey; but other varieties are possible.

FIG 7.1 (c) Key for the identification of phaneritic rocks by composition.

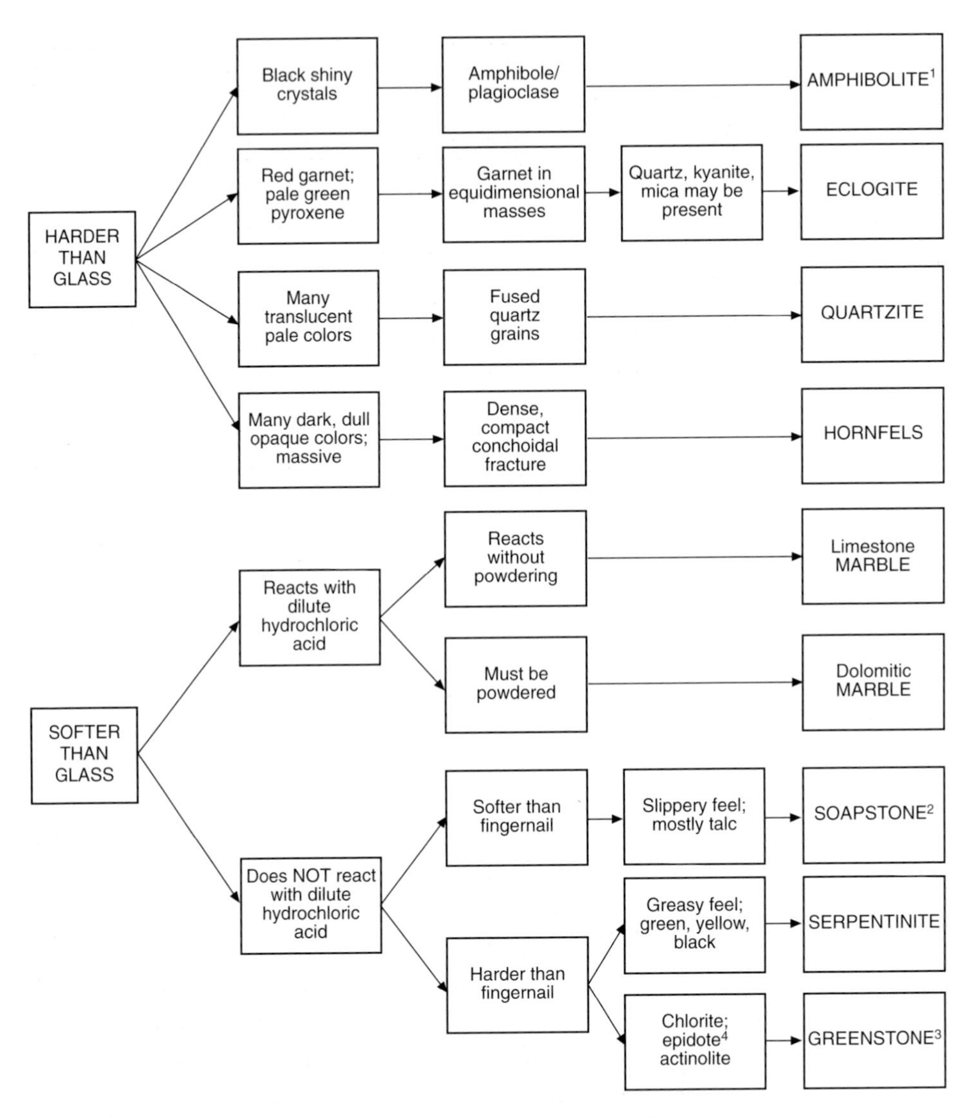

1 Amphibolite is usually foliated but some specimens may appear granular.
2 May be weakly foliated.
3 Greenstone is usually well foliated but massive varieties exist.
4 Epidote is a pale green. Often it is finely disseminated in the rock so individual crystals cannot be seen. Look for pale green patches within the darker green chlorite.

FIG 7.1 (d) Key for the identification of granular metamorphic rocks.

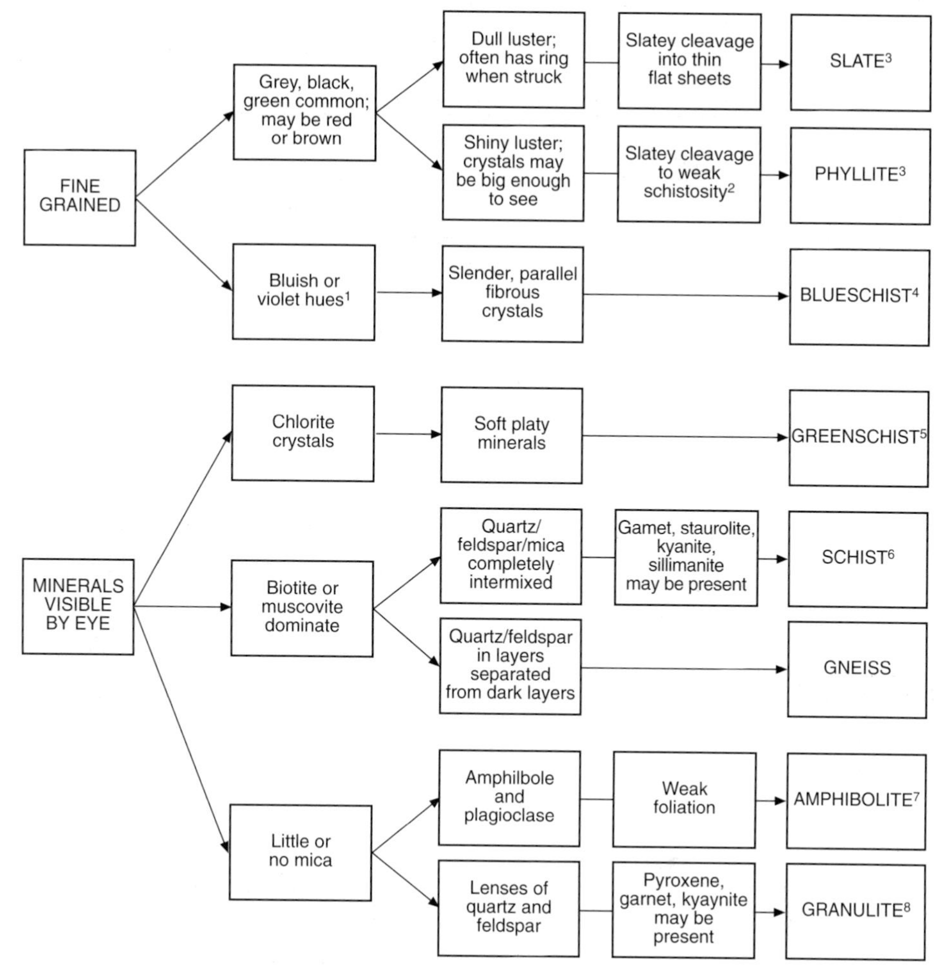

1 Under fluorescent light bluish hues may not be easy to detect. On the outcrop in full daylight, rock is usually distinctly blue in color.
2 Schistosity = coarse-grained foliation. Phyllites frequently have an undulatory surface and are not flat like slates and shales.
3 (Shale), slate, and phyllite completely intergrade with each other. Distinctions may be difficult. Ast your instructor.
4 Blueschist is also called glaucophane schist.
5 Greenchist may superficially look like slate/phyllite but has moderately developed schistosity (see note 2).
6 Rock may be called *garnet schist*, or *kyanite schist*, or *garnet-kyanite schist*, and so on depending on accessory minerals present.
7 Amphibolite may be granular in appearance.
8 Granulites may be crudely gneissic or granular in appearance.

FIG 7.1 (e) Key for the identification of foliated and banded metamorphic rocks.

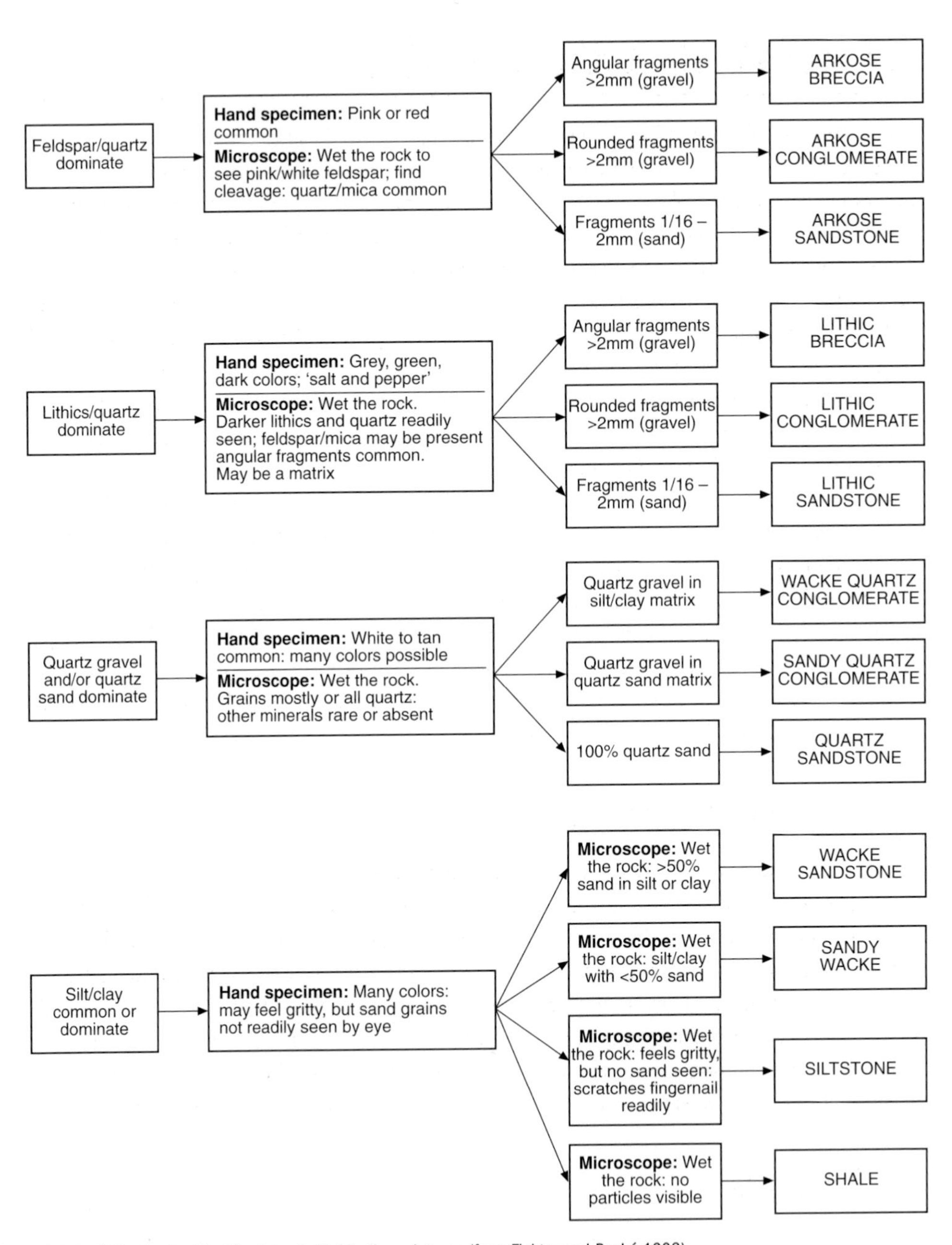

FIG 7.1 (f) Key for the identification of siliciclastic rock types (from Fichter and Poché 1993).

7.4 HEAVY MINERAL ANALYSIS

The technique of heavy mineral analysis has a long history in all branches of petrology and a number of authors have provided excellent discussions of its underlying principles, methodology and examples of its use for the study of sedimentary rocks (i.e. Krumbein and Pettijohn 1938;, Milner 1962; Carver 1971; Pettijohn 1975 and Lindholm 1987).

Several factors can influence the composition of the heavy mineral assemblage in a sedimentary rock, and all these need to be considered when interpreting the results of an analysis. Lindholm (1987) suggests that although the principal factor is source rock composition (and therefore it should be possible to correlate sediments derived from the same or similar source rocks), other influences include selective sorting (due to density variations) during transport or deposition and destruction of less stable species by post-depositional weathering or diagenetic processes.

As well as the care needed in interpreting the heavy mineral assemblages, a number of practical problems may present themselves. For instance, it can be difficult to extract the heavy minerals from lithified rock material (although this is not a major problem with most unlithified sediments of Quaternary age) and optical identification of the large number of grains needed to give a representative analysis of the heavy mineral assemblage within the bulk sample, can take considerable time. Despite these drawbacks, the value of the technique has been shown in many examples within the research literature (e.g. Derry 1933; Dreimanis *et al.* 1957 and Connally 1960).

7.4.1 Sample preparation

Standard procedures for heavy mineral sample preparation and subsequent analysis are documented in a number of sources (e.g. Carver 1971 and Lindholm 1987). Carver (*ibid*) states that identification of heavy minerals by petrological microscope can be a difficult task, but it is made easier if all grains are of approximately the same size, as diagnostic features (such as colour) are then more consistent. He also stated that these features tend to be more pronounced in larger grains and therefore suggested that the 2–3 phi fraction should be used, although other workers have suggested 3–4 phifractions can also be used (Lindholm 1987). The steps involved in sample preparation are shown in Table 7.1.

7.4.2 Analysis

Heavy mineral grains can be mounted in clove oil on standard microscope slides for identification using a petrological microscope capable of producing plane-polarized light. This form of temporary mount is less time consuming than mounting grains in Canada Balsam and also allows grains to be moved or rolled during examination, which can aid identification. The optical properties of the two mounting media are similar.

Identification is best achieved using a petrological microscope and a combination of reflected and transmitted light. For the 20 or 25 commonest opaque and non-opaque

Table 7.1 Step by step procedure for separation of heavy minerals (after Lindholm 1987; Mange and Maurer 1992).

Step 1	A sub-sample of the bulk matrix sample (enough to produce 3-4g of the 2-3 phi or 3-4 phi size fraction) is gently disaggregated using a pestle and mortar. No chemical treatments are usually needed for unlithified sediments.
Step 2	The disaggregated material is placed in distilled water in an ultrasonic bath to complete the disaggregation procedure. The 2-3 phi (or 3-4 phi) fraction is then separated by wet-sieving using suitable sieve sizes and dried in an oven at 40°C.
Step 3	The separation of the 2-3 phi (or 3-4 phi) fraction into its light and heavy components is achieved using a sodium polytungstate solution with a density of 2.85. This material is now preferred to traditional substances such as bromoform as it is non-toxic. A small amount of the chosen phi fraction (approximately 0.5-1.0g) is weighed and then placed into a 100ml centrifuge tube containing the sodium polytungstate solution and shaken vigorously. The tube is then placed in a centrifuge and spun for 15 minutes at 3000rpm. Overfilling of the tube can result in poor separation and repeat samples should be spun if insufficient heavy minerals are obtained initially.
Step 4	The heavy minerals (densities greater than 2.85) settle to the base of the tube leaving the lighter fraction floating at the top. A hypodermic syringe with a needle of a suitable aperture is inserted through the light minerals to the base of the tube and the heavy minerals are drawn by suction into the syringe. The contents of the syringe are then discharged into a Buchner Funnel fitted with filter paper. The heavy liquid can then be collected in a flask underneath the funnel ready for re-use.
Step 5	The light minerals remaining in the centrifuge tube should also be collected by using filter paper in a Buchner Funnel. Both the heavy and light mineral fractions are then washed in distilled water to remove any remaining traces of the heavy liquid.
Step 6	Finally, samples are dried in an oven at 40°C prior to analysis.

mineral types, only basic optical properties such as shape, colour and pleochroism (Table 7.2) are needed to identify most grains (Lindholm 1987). Krumbein and Pettijohn (1938) and Lindholm (1987) provide useful summaries of diagnostic criteria for the most common heavy minerals and Lindholm (*ibid*) also includes a simplified flow chart to aid identification (Fig. 7.2). However, the full colour images provided in Mange and Maurer (1992) provide one of the most useful comparisons if a set of 'standards' is not available for inspection.

For each sample analysed, counts are usually made until at least 200 non-opaque grains have been recorded (after Lindholm 1987). This figure is thought to give an adequate representation of the relative abundances of the various species present (Gwyn and Dreimanis 1979). As with clast lithological data, tables can then be compiled showing both raw count and percentage values for each mineral type. These data can then be subjected to both qualitative and quantitative analysis. For the latter, a wide range of statistical methods have been employed, from basic techniques such as χ^2 (Chi-squared) through to multivariate approaches including cluster and factor analysis (Davis 1986).

(a)

Isotropic

Colourless
Garnet
Spinel
Muscovite (biaxial "–," 2v = 29–42°)
Chloritoid (biaxial "+," 2v = 36–63°)

Coloured
Pale
Biotite (brown, biaxial "–," low 2v)
Garnet (pink, red, orange, yellow, green)
Chlorite (green to yellow green, low relief, biaxial "+" or "–")
Chloritoid (greenish grey, moderately high relief, biaxial "+")

Dark
Spinel (grass green or coffee brown)

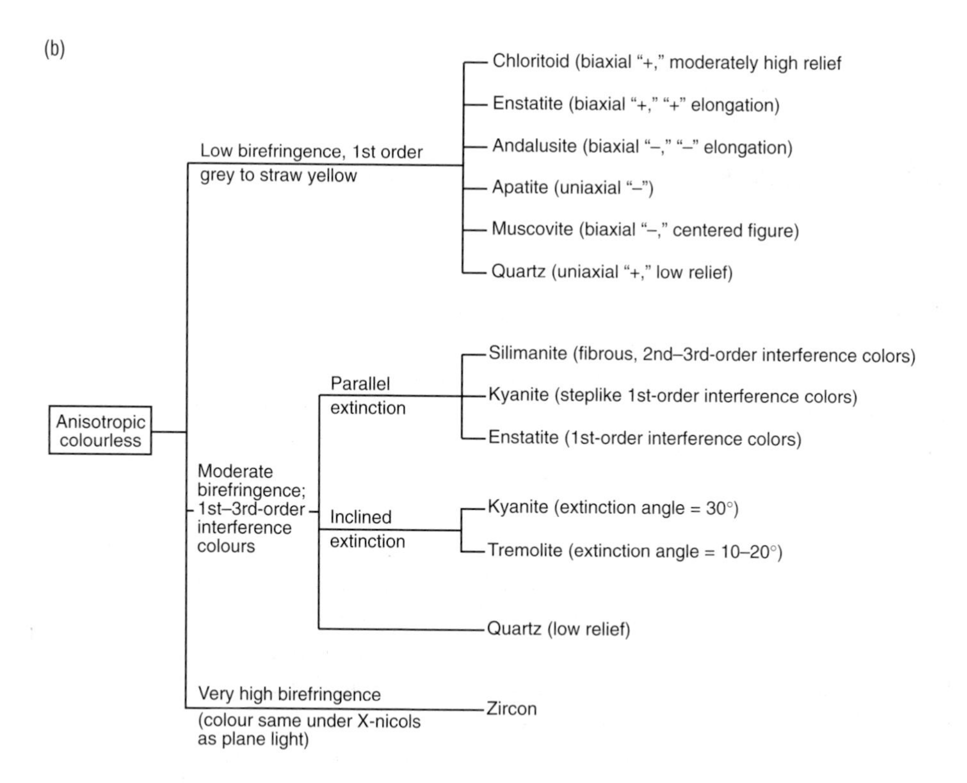

(c)

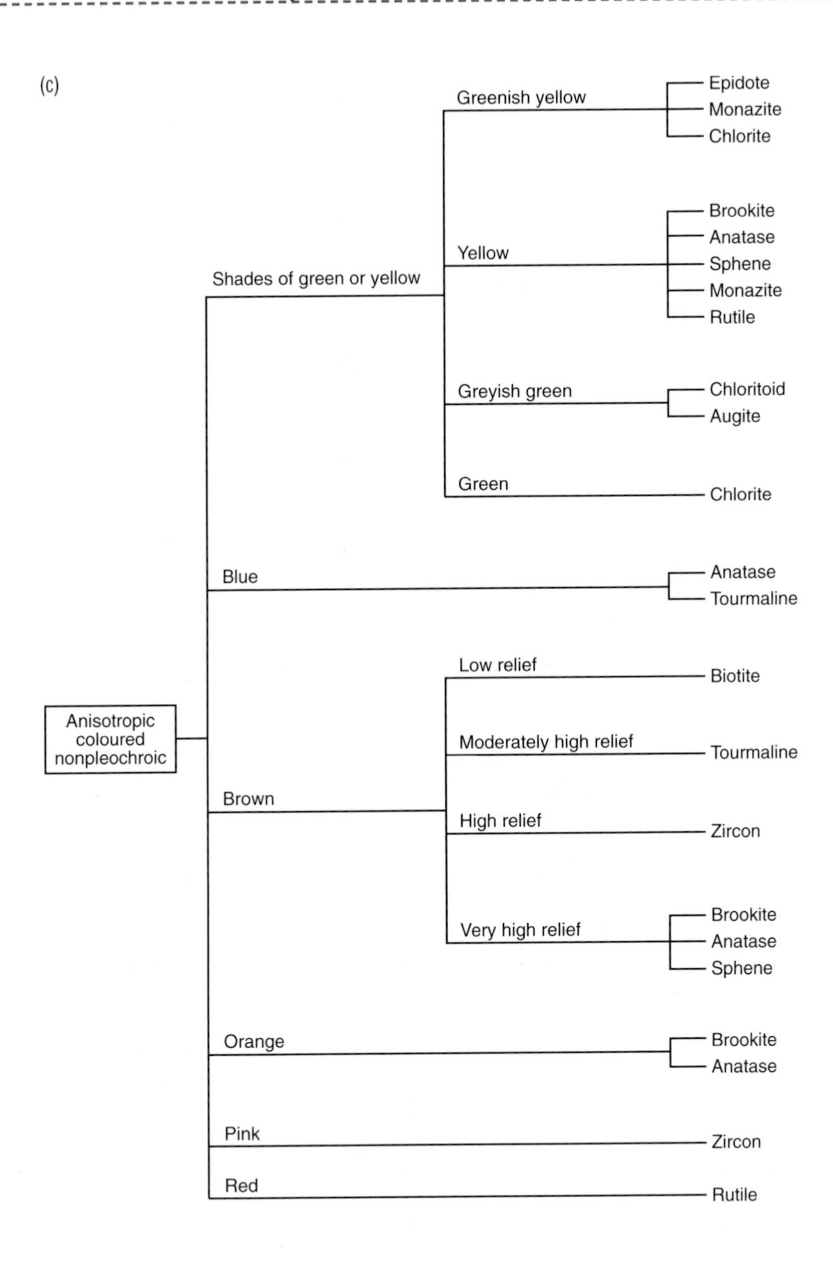
Anisotropic coloured nonpleochroic
Shades of green or yellow
Greenish yellow
Epidote
Monazite
Chlorite
Yellow
Brookite
Anatase
Sphene
Monazite
Rutile
Greyish green
Chloritoid
Augite
Green
Chlorite
Blue
Anatase
Tourmaline
Brown
Low relief
Biotite
Moderately high relief
Tourmaline
High relief
Zircon
Very high relief
Brookite
Anatase
Sphene
Orange
Brookite
Anatase
Pink
Zircon
Red
Rutile

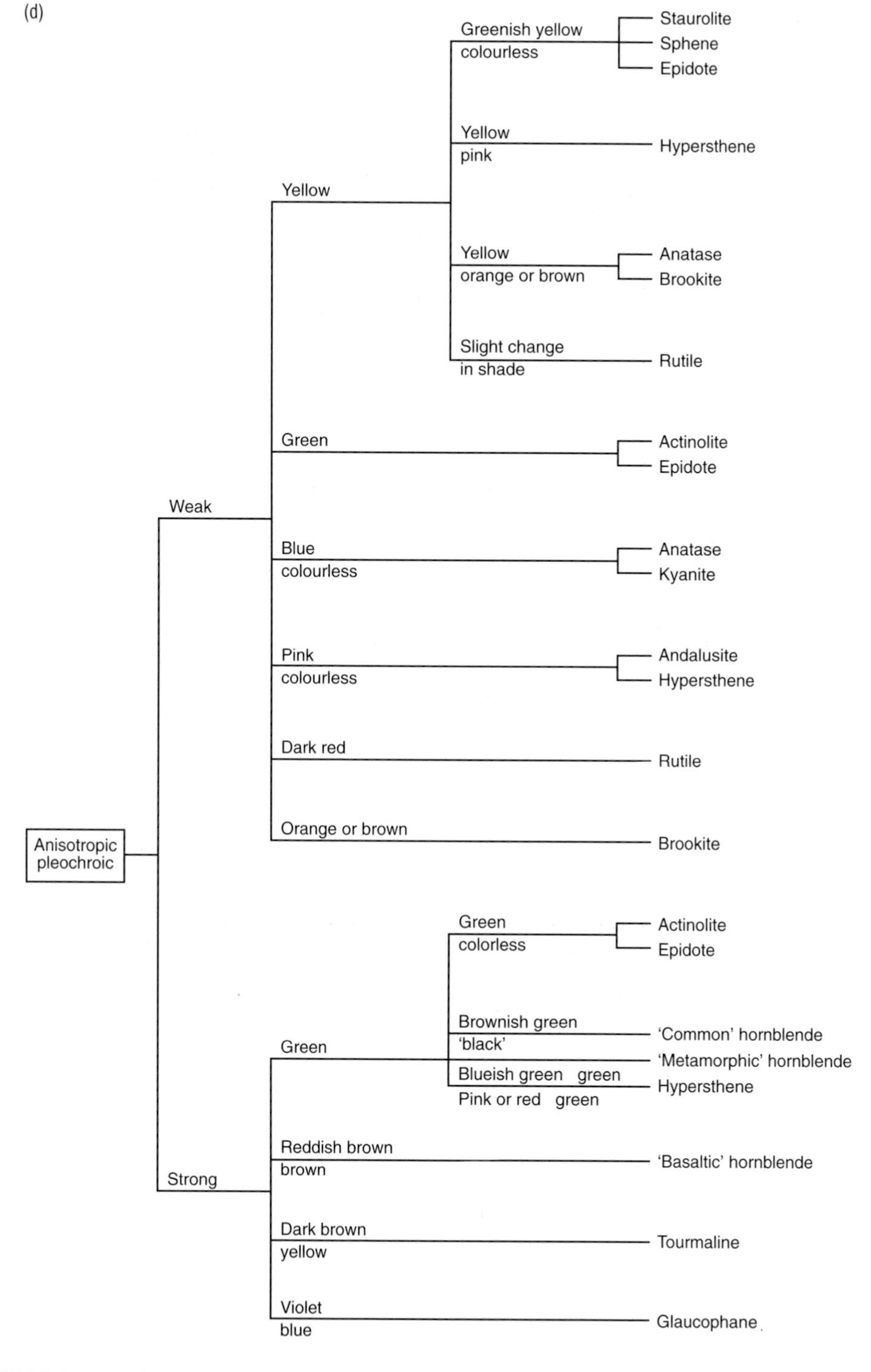

FIG 7.2 Flow charts for the identification of common heavy mineral species (from Lindholm 1987).

Table 7.2 Definition of the basic criteria used in the identification of heavy minerals (after Krumbein and Pettijohn 1938; Lindholm 1987; Mange and Maurer 1992).

Term	Definition
Opaque/ non-opaque	Opaque grains are those that do not allow light to pass through them and therefore appear dark when viewed via transmitted light (light from beneath the microscope stage). Opaque minerals can then be viewed via reflective light (a light source from above) where features such as shape or colour may be diagnostic of mineral type. In contrast, non-opaque minerals allow some transmission of light and may therefore appear as coloured or transparent mineral forms under transmitted light.
Isotropic/ anisotropic	Isotropic minerals remain dark when viewed under crossed nicols on a petrological microscope. Anisotropic minerals, because of a less symmetrical mineral structure, appear illuminated and, as the microscope stage is rotated, these grains may change from light to dark or from a colour to black. The point at which anisotropic minerals do not transmit light as the stage is rotated is known as *extinction* (see below) and this occurs four times as the stage is rotated through 360°.
Colour	Some non-opaque mineral types have distinctive colours and this can be used as a diagnostic criteria (see Fig. 7.2).
Pleochroism	Certain coloured mineral types exhibit variation in colour when rotated upon the microscope stage in plane-polarized light. This phenomenon is termed pleochroism. Isotropic minerals do not tend to show pleochroism.
Birefringence	The refracting properties of anisotropic substances produce two rays with different refractive indices when transmitting light (often labelled a 'fast' ray and a 'slow' ray). Due to velocity differences, these rays emerge from the mineral grain out of phase. When the components of these rays are resolved the microscope's prisms into the same plane interference results. This gives rise to bands of colour on the mineral surface and these are known as interference colours (normal spectrum colours; violet, indigo, blue, green, yellow, orange and red). A numerical measure of the difference in indices between the two rays is termed birefringence.
Extinction	Extinction describes the situation when an anisotropic mineral turns dark as it is rotated on a microscope stage under crossed nicols. This phenomenon occurs every 90° of rotation of the mineral. Straight extinction occurs when a mineral is in extinction with its cleavage parallel to the crosswires of the microscope eyepiece. In contrast, inclined extinction occurs when the cleavage is not parallel to the crosswires.
Shape	The shape or form of detrital minerals can be diagnostic of mineral type. Some minerals (e.g. quartz) do not show distinct or well-developed crystal form. However others, such as zircon or apatite have well-developed, distinct forms.
Cleavage	Many mineral grains, when broken, display a flat plane of breakage that is parallel to a possible crystal face. Cleavage planes are developed along planes of weakness in the atomic lattice and more than one cleavage may exist. Cleavage will influence the resulting shape of detrital mineral grains, but if the number and type of cleavage can be identified, it can be diagnostic of mineral type.
Relief	Some minerals show very strong edges (that is, they appear to have broad, dark boundaries). This is a result of the difference in the refractive index of the liquid the grain is mounted in and that of the grain itself. Such grains appear to 'stand up' on the microscope stage and these are said to display high relief. In contrast, when a grain displays less distinct edges it is said to have low relief.

7.5 GEOCHEMISTRY

X-ray fluorescence (XRF) analysis is one of several established techniques used to establish the elemental chemistry of a rock or sediment sample. Other modern methods that can also give elemental composition include atomic absorption spectrophotometry (AAS) and optical emission inductively-coupled plasma spectrometry (OE-ICP). Although these three techniques work on different principles, all can produce qualitative and quantitative elemental data.

XRF is considered here as an example of this type of approach as the equipment is perhaps more commonly available than other methods of 'whole-rock' elemental analysis and, on theoretical grounds, XRF is preferred to AAS for whole-rock analysis (as opposed to analysis of a few individual elements) as sample preparations are simpler and all elements can be analysed at the same time. Also, larger quantities of sample tend to be used, generally ensuring that the data are more representative of the bulk sample.

The theoretical basis of XRF, as applicable to the analysis of geological materials, has been outlined by various authors (e.g. Johnson and Maxwell 1981; Goudie 1981; Potts 1987; Thompson and Oldfield 1986; Fairchild *et al.* 1988). Discussion here will therefore concentrate upon some practicalities and representative examples of its use in the study of glacial diamicts.

7.5.1 Sample preparation and analysis

Preparation of samples for XRF analysis can take various forms. Table 7.3 describes a common procedure, following methods outlined by Smith (1985), Potts (1987), Fairchild *et al.* (1988) and with some modifications suggested by Peachey *et al.* (1985).

Table 7.3 Step by step procedure for preparation and analysis of diamict matrix by XRF (after Smith 1985 and Peachey *et al.* 1985).

Step 1	A sub-sample of approximately 30g is extracted from the bulk sediment matrix using a suitable sample splitter and then air-dried prior to gentle disaggregation in a pestle and mortar.
Step 2	Peachey *et al.* (1985), suggest that this material should then be dry-sieved through a 3.25 phi (0.105mm) sieve to reduce the dilution effect of quartz sand on other, hopefully diagnostic, elements.
Step 3	The portion of the sample reaching the sieve pan is then weighed (in grams to three decimal places) and placed in a furnace at 550°C for eight hours to remove any organic material (calcined). On removal from the furnace, the sample is cooled in a desiccator and then re-weighed so that percentage weight loss on ignition could be calculated.
Step 4	Most modern XRF instruments can accept samples in a number of forms. A common form for sediments and soils is to make a hard 'diskette' or 'pellet'. However, the calibration of the results is dependent upon the samples having a uniformly fine particle size, and thus all calcined samples are usually ground in a mill (i.e. Tungsten barrel mill) prior to forming the diskette.
Step 5	A 10g mixture of sample and binding agent (i.e. Hoerch wax or Movial) in appropriate proportions (advice will be needed here from the specific laboratory undertaking the analysis) is thoroughly mixed and finally pressed into a diskette at a pressure of 12 tonnes/m^2.

Depending upon the equipment available, analysis of major elements (i.e. Fe, Si, Ti, etc.) and minor (trace) elements (i.e. Pb, Zn, Cu, Ni, etc.) can be made. Providing sample preparation is consistent, modern XRF systems allow all the major elements to be assessed to within 200–300ppm (and some with considerably greater accuracy). Trace elements can often be estimated to an accuracy of less than 5ppm.

If required, the multivariate data generated can then be analysed using a wide range of statistical methods (Davis 1986). As with the other forms of analysis introduced in this chapter, the elemental composition of a sediment can be used for both provenance and correlation studies. Elemental data can also be used in a commercial context, where trace element composition within glacigenic sediments can be used to identify sources of economically valuable rock materials (Kujansuu and Saarnisto 1990). However, changes in the elemental composition with depth (e.g. based upon a series of samples taken from a vertical profile through a unit or set of units) can also be used to infer the process of deposition or to identify the impacts of post-depositional weathering (see Section 7.9.3).

7.6 CLAY MINERALOGY

Various authors have outlined the physical principles of the XRD technique (Klug and Alexander 1974; Jenkins and de Vries 1970) and this theory has been summarized by a number of other workers dealing specifically with its application to sediments (Carroll 1970; Brindley and Brown 1980; Starkey *et al.* 1984; Lindholm 1987; Wilson 1987; Hardy and Tucker 1988). Unlike XRF, which detects the elements present within a sample, XRD reveals how those elements are combined, giving the main crystalline phases within the sample. The principles of the technique mean that greater accuracy is achieved if the samples consist of finer particle size, explaining why it is the most widely used technique for the identification of clay minerals (Wilson 1987).

7.6.1 Sample preparation

A variety of sample preparation techniques are used for analysis of clays by XRD. However, most workers agree that for a full analysis of the clay fraction, samples should be presented to the diffractometer in the form of 'orientated' mounts on ceramic tiles (Gibbs 1968 and 1971; Carroll 1970; Starkey *et al.* 1984; Wilson 1987; Lindholm 1987 and Hardy and Tucker 1988), although other mounts, such as frosted glass slides, are also commonly used. The clay fraction can be mounted onto an unglazed ceramic tile by suction, ensuring that the majority of the individual clay plates lie parallel to the surface of the tile. Alternatively, a sediment slurry can be placed onto a glass slide and left to dry. A typical sample preparation procedure is shown in Table 7.4.

7.6.2 Analysis

As shown in Table 7.5, the diagnostic peak angles obtained from standard air-dried samples for some clay minerals can interfere with one another. As such, even for

Table 7.4 Step by step procedure for sample preparation of the diamict matrix clay fraction (finer than 9 phi) for analysis by XRD (after Carroll 1970; Lindholm 1987 and Hardy and Tucker 1988).

Step 1	A sub-sample is taken from the bulk field sample using a sample splitter, air-dried and then gently disaggregated in a pestle and mortar. The amount of bulk sample used is dependent upon the overall particle-size distribution, but only a very small quantity of the clay-sized material (finer than 2µm or 9 phi) is actually needed for each mount.
Step 2	The bulk sample is then placed in a beaker with distilled water and full disaggregation achieved either by vigorous shaking or placing samples in an ultrasonic bath. Flocculation can be a problem in the separation of clays and therefore, if required, a very small amount of a deflocculant (i.e. ammonium hydroxide NH_4OH) can be added to the suspension prior to treatment in the ultrasonic bath.
Step 3	The resulting suspension is then transferred to a settling column and allowed to stand until all material coarser than 9 phi has settled out. The time taken for this will depend upon the depth of the water column and is calculated using Stoke's Law (Folk 1974). The clay-sized material remaining in suspension (finer than 9 phi) is then decanted and forms the suspension used for the subsequent stages of the sample preparation.
Step 4	The suspension for each sample is thoroughly shaken and then dropped by pipette onto a disk cut from unglazed ceramic tile held in a modified Buchner Funnel. The water is drawn through the tile by suction from a vacuum pump, leaving a coating of clay upon the tile surface. This process is repeated until the tile surface is covered with a reasonable thickness of clay. In fact, only a very thin layer is needed as the X-ray beam does not penetrate the sample surface beyond a few microns (4–12µ, Gibbs, 1968). The tiles are allowed to air dry and then placed in a desiccator for storage prior to analysis.

Table 7.5 Diagnostic angles for the major clay minerals and non-clay mineral species commonly found in the finer than 9 phi fraction by XRD (after Carroll 1970; Lindholm 1987 and Hardy and Tucker 1988).

Group	Mineral	Degrees 2θ	D spacing (Å)
Clays			
Kaolinite	Kaolinite	12.34, 24.88, 20.34	7.17, 3.58, 4.37
Mica	Muscovite	8.87, 26.76, 17.77, 35.00, 19.77	9.97, 3.33, 4.99, 2.56, 4.49
	Illite	8.84, 19.82, 26.77, 34.36	10.00, 4.48, 3.33, 2.61
	Biotite	8.75, 26.45, 33.69	10.10, 3.37, 2.66
	Vermiculite	6.22, 19.42	14.2, 4.57
Smectite	Montmorillonite	5.89, 19.73	14.2, 1.53
Chlorite	Chlorite	12.40, 6.20, 24.9, 18.78	7.13, 14.20, 3.58, 4.74
Non-clays			
Quartz		26.67, 20.85, 36.56	3.34, 4.26, 2.46
Feldspar	Orthoclase	26.94, 23.60	3.31, 3.77
	Albite	27.92, 23.54	3.20, 3.78
Calcite		29.43	3.04
Dolomite		30.99	2.87

qualitative work, three analyses are traditionally carried out on each sample. Most workers agree that the series of steps shown in Table 7.6 will allow a qualitative identification of the main clay minerals present (Starkey *et al.* 1984). The requirement for heating to 450° or 550°C, is one reason why ceramic tile mounts are superior to other methods (Gibbs 1971) such as mounts on glass slides. Modern XRD systems, operating under full computer control and calibrated with reference to suitable standards, can provide database searching to match the diffraction patterns with those of specific mineral types.

The information from the four treatments shown in Table 7.7 allows a more certain identification of the clay mineral types within the sample than by simply using an

Table 7.6 Steps used to enable identification of clay minerals by XRD (after Starkey *et al.* 1984 and Lindholm 1987).

Step 1	An initial trace is obtained from the clay samples (each sample on a ceramic tile mount and having been stored in a desiccator overnight).
Step 2	The tiles are then placed in a desiccator in which ethylene glycol has replaced the normal desiccating agent and left at a temperature of 60°C for approximately 4 hours.
Step 3	A second trace is obtained from the samples.
Step 4	The tiles are then heated to 450°C for at least 1 hour. On removal from the furnace, samples are immediately air quenched in a standard desiccator and allowed to cool.
Step 5	A third trace is obtained from the samples.
Step 6	The tiles are then heated to 550°C for at least 1 hour. On removal from the furnace, samples are immediately air quenched in a standard desiccator and allowed to cool.
Step 7	A fourth trace is obtained from the samples.

Table 7.7 Identification of clay minerals from XRD traces using four treatments (after Johns *et al.* 1954; Carroll 1970; Starkey *et al.* 1984 and Lindholm 1987). All angles of diffraction are given as degrees 2θ.

Mineral	Untreated	Ethylene glycol	Heated to 450°C	Heated to 550°C
Illite	Reflections at 8.8, 17.7 and 26.7 degrees 2 theta	No change	Becoming a little more intense	Becoming a little more intense
Montmorillonite	Reflection between 6.8 and 5.9 degrees 2 theta	Moves to 5.2 degrees 2 theta and can become more intense	Moves to between 9.9 and 8.8 degrees 2 theta	No change
Chlorite	Reflections at 6.3, 12.6, 18.9 and 25.2 degrees 2 theta	No change	Peak at 6.3 degrees 2 theta increases. Other peaks disappear	No change
Kaolinite	Reflections at 12.4 and 24.9 degrees 2 theta	No change	No change	All peaks disappear

air-dried sample. Clay minerals that produce peaks at similar angles in an air-dried state, respond differently to glycolation and heating, causing both shifts and changes in intensity of peaks. Interpretation of the traces from these various treatments is summarized in Figure 7.3 and Table 7.7 (after Carroll 1970; Starkey *et al.* 1984 and Lindholm 1987).

By comparison with sample standards, modern XRD analysis systems can generate quantitative data, however, Johns *et al.* (1954) suggested a procedure that has given satisfactory semi-quantitative results based upon the qualitative data obtained purely from treatments 1 to 3 (Table 7.8). This method uses relative peak heights or peak areas and can give semi-quantitative estimates of the respective proportions of montmorillonite, chlorite, illite and kaolinite.

As illustrated in Section 7.9.4, the types and relative abundance of particular clay mineral types can be used for both correlation and provenance indication studies. However, another application includes the post-depositional weathering history of a glacigenic sediment, where clay mineral alteration may take place as a result of weathering/pedogenic processes, and examples of this type of study are also introduced in section 7.9.4.

7.7 ENVIRONMENTAL MAGNETISM

Environmental magnetic measurements are now gaining a wide acceptance as a means of characterizing a rock, sediment or soil for such purposes as provenance indication or stratigraphic correlation. All substances, including rock forming minerals, have magnetic properties and, with appropriate instrumentation, these properties can be measured.

Table 7.8 Computational procedures used for the calculation of semi-quantitative estimates of montmorillonite, illite, chlorite and kaolinite composition of the clay fraction (after Johns *et al.* 1954). All angles of diffraction are given as degrees 2θ.

Step 1	Using illite (8.8 untreated peak) as a base, this is compared directly with montmorillonite (5.2 glycolated peak) to get the ratio of illite:montmorillonite.
Step 2	A quartz peak at 26.6 may interfere with the illite peak at 26.7 (untreated). The component of the peak at this angle due to illite is needed for calculation of the illite:chlorite+kaolinite ratio. The quartz component at 26.6 can be found by comparing the quartz peak at 20.8 (untreated) with the quartz peak at 26.6 (untreated). The 20.8 peak should be 'pure' quartz and the ratio of quartz (20.85) to quartz (26.6) is approximately 1:2.85.
Step 3	The component of the peak at 26.7 (untreated) due to illite is found by applying the above correction for quartz at 26.6 (untreated).
Step 4	The ratio of illite (26.7 untreated) to the combined kaolinite and chlorite peak at 25.0 to 25.5 (untreated) is found.
Step 5	The ratio of chlorite and kaolinite can be found by comparing the size of the untreated and the 450°C 25.0 to 25.5 peaks. This gives the chlorite+kaolinite to kaolinite ratio.
Step 6	From the above, the relative proportions of illite, kaolinite, chlorite and montmorillonite can be compared and these ratios normalized to 100% to give a semi-quantitative estimate of their relative abundances.

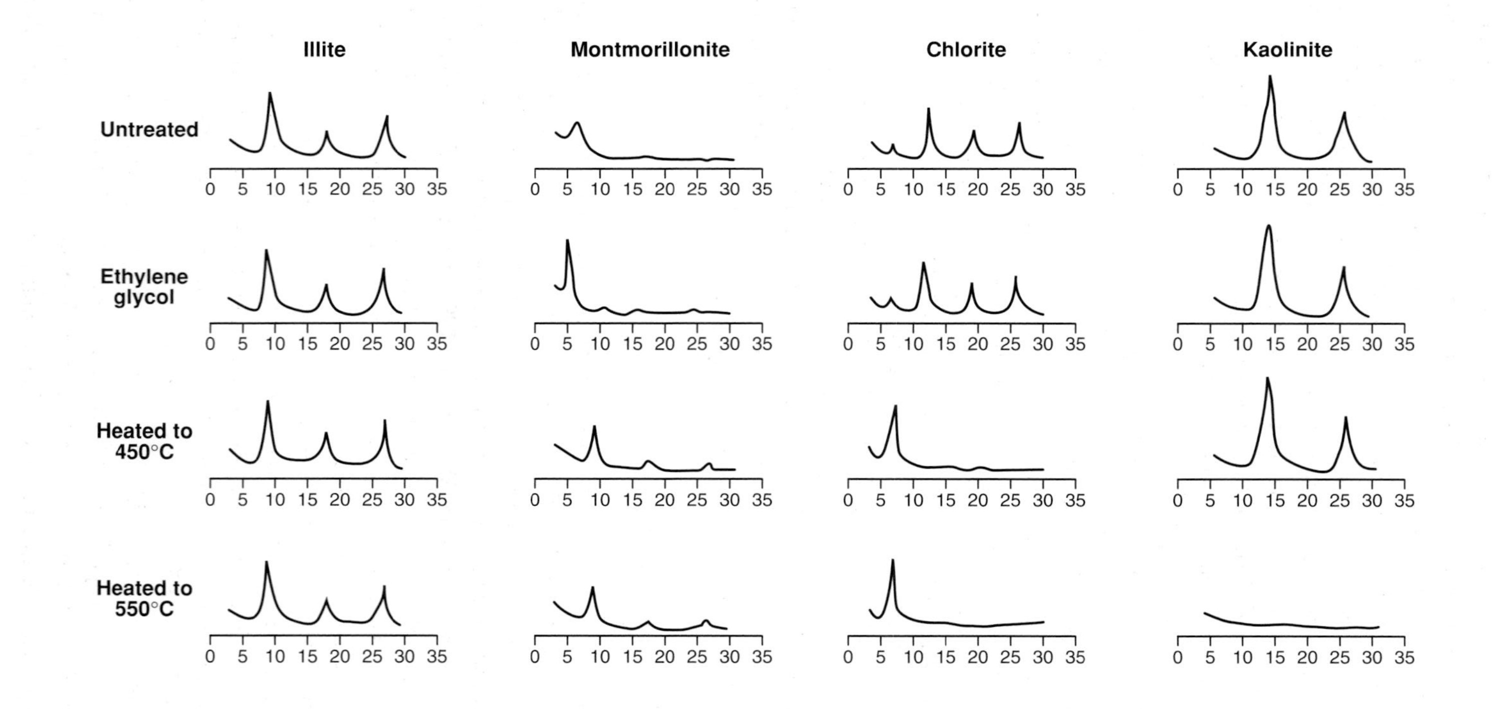

FIG 7.3 X-ray diffraction patterns for common clay minerals in response to various treatments (after Johns *et al.* 1954 and Lindholm 1987).

Different mineral types have distinct magnetic properties allowing identification and differentiation of rock and sediment types based purely on their magnetic characteristics (Thompson and Oldfield 1986; Walden *et al.* 1999; Maher and Thompson 1999).

In the absence of other influences, the assemblage of minerals (and their magnetic behaviour) within a sediment represents a mix of all the source rocks from which it is derived. As such it should be possible to use the mineral magnetic character of a sediment to relate it to its source or sources. In a similar manner, it should be possible to correlate sediments that are separated spatially, but derived from the same source. In this sense, mineral magnetic methods of sediment correlation or provenance determination use the same underlying principle as other geochemical or petrological approaches; that is, some measurable compositional property of the sediment is compared to a stratigraphically equivalent sediment for correlation purposes, or compared to potential source rock to establish provenance.

Also, as with other geochemical techniques, in reality a number of other factors can affect the primary magnetic character of a sediment and alter this purely provenance derived signal (e.g. weathering and selective sorting). However, in most cases, the influence of these other factors can be assessed and interpretation of mineral magnetic data adjusted accordingly. Indeed, in some studies (e.g. Maher 1986; Maher and Thompson 1999) it is these 'secondary' effects that are of most interest.

In comparison with methods such as XRF or XRD, the equipment is relatively inexpensive and preparation and analysis are time modest. The measurements are, however, extremely sensitive and, while the method only provides information on the iron mineralogy within the material, it is possible to detect differences between samples that would be outwith the resolution of many other analytical methods.

7.7.1 Sample preparation

Walden (1999) provides a summary of sample preparation methods for magnetic analysis. In studies of glaciagenic materials both clast and matrix fractions can be analysed, although the latter is far more common. In both cases, however, sample preparation for basic analysis is relatively speedy and requires little in the way of sophisticated laboratory equipment.

Standard magnetic instrumentation (Walden *et al.* 1999) can deal with samples up to *c.* 25mm in size. Clasts of this size or smaller can therefore be cleaned in distilled water and, where necessary, placed in an ultrasonic bath to remove any matrix from the clast surface. Once clean, clasts should be dried at 40°C.

Sample preparation for the matrix is somewhat more involved but requires no specialized equipment and therefore several sample sets can be prepared simultaneously. As with other compositional methods, particle size can influence the bulk mineralogical or geochemical composition of a sediment. Many workers recommend that magnetic measurements should be made on a particle size specific basis (Oldfield *et al.* 1985; Smith 1985; Thompson and Oldfield 1986; Walden and Slattery 1993). This can involve either working with a single particle size fraction (i.e. finer than 2mm or finer than 63μm) or a series of size fractions. For the former, bulk samples can be either dry- or wet-sieved through a suitable aperture sieve. Wet-sieving should be performed using distilled water and all sample drying should take place at low temperature (< 40°C) or by air-drying.

In many sediment types, particularly those that demonstrate limited variability in their particle-size distributions, working with a single size fraction is sufficient. However, for a more thorough analysis or where considerable particle-size variability is present (as in many diamict sediments), splitting the bulk matrix sample into a small number of discrete size fractions can be helpful. A combination of sieving and settling provides an adequate method (Walden and Slattery 1993) and the main steps involved in this process are described in Table 7.9.

Samples should finally be weighted (in grams to two decimal places) and then packed into 10cc plastic pots suitable for use with the mineral magnetic instrumentation. It is important that the sample is immobilized within the sample pot. To this end, matrix samples can be wrapped in standard kitchen cling film, and this also means the pots themselves are kept clear for re-use.

7.7.2 Analysis

A full range of environmental magnetic measurements are described by Thompson and Oldfield (1986) and the practical issues involved in the analysis are outlined in Walden *et*

Table 7.9 Preparation of the diamicts matrix samples for mineral magnetic analysis (after Smith 1985; Walden and Slattery 1993).

Step 1	A 200g sub-sample from each bulk matrix sample is obtained using a sample splitter. The sub-sample is placed in distilled water and left in an ultrasonic bath for 1 hour in order to disaggregate the sediment. A few drops of a weak ammonia solution are added to all samples as a dispersant. Smith (1985) suggested that this has a negligible effect on the resultant magnetic properties of the sediment.
Step 2	The number of size fractions required will depend upon the overall particle-size distribution of the sample in question. Walden and Slattery (1993) suggest that four size fractions should be adequate for most work (0–4 phi (sand), 4–6 phi (coarse silt), 4–9 phi (fine silt) and finer than 9 phi (clay). A bulk sample should also be retained for subsequent magnetic analysis to provide a comparison between the bulk and particle-size specific measurements.
Step 3	The bulk sample is wet-sieved through 0 and 4 phi sieves to a) remove material coarser than 0 phi, b) obtain the sand fraction and c) obtain the silt+clay fraction for further separation.
Step 4	The pan fraction from the wet-sieving procedure (finer than 4 phi) is then subjected to a series of decantations, using settling times based upon Stoke's Law (from Folk 1974). This size separation procedure is simple but has proved to be both rapid and easily repeatable (Walden and Slattery 1993).
Step 5	Each size fraction is then dried at a temperature no greater than 40°C in order to avoid thermal effects on the magnetic properties (Oldfield *et al.* 1981 and Collinson 1983).
Step 6	Known weights of each size fraction (and a bulk matrix sample) are wrapped in kitchen cling film and packed (immobilizing grains within each sample) into 10cc plastic pots to give particle-size and weight-specific samples suitable for use with the mineral magnetic instrumentation.

al. (1999). As described by these authors, a basic, modern environmental magnetic 'kit' might consist of four instruments; a susceptibility meter, an a.f. demagnetizer with anhysteretic magnetization capability, a magnetometer and a pulse magnetizer. This combination of equipment allows a wide range of common, room temperature, mineral magnetic parameters to be calculated (i.e. magnetic susceptibility (χ), Susceptibility of Anhysteretic Remanent Magnetization (χ_{arm}) and Saturation Isothermal Remanent Magnetization (SIRM)). Common parameters, and their basic interpretation, are listed in Table 7.10. Thompson and Oldfield (1986) outline the range of values for these parameters (and others) exhibited by various magnetic mineral types and natural rock

Table 7.10 Common room-temperature mineral magnetic parameters and their basic interpretation (after Thompson and Oldfield 1986; Maher and Thompson 1999; Walden *et al.* 1999).

Parameter	Interpretation
χ (10^{-6} m^3 kg^{-1})	Initial low field mass specific magnetic susceptibility. This is measured within a small magnetic field and is reversible (no remanence is induced). Its value is roughly proportional to the concentration of ferrimagnetic minerals within the sample.
χ_{fd} (10^{-6} m^3 kg^{-1})	Frequency dependent susceptibility. This parameter measures the variation of magnetic susceptibility with the frequency of the applied alternating magnetic field. Its value is proportional to the amount of magnetic grains whose size means they lie at the stable single domain/superparamagnetic (< 0.1μm) boundary.
χ_{ARM} (10^{-6} m^3 kg^{-1})	Anhysteretic Remanent Magnetization (ARM) is proportional to the concentration of ferrimagnetic grains in the 0.02 to 0.4μm (stable single domain) size range. The final result can be expressed as mass specific ARM per unit of the steady field applied (χ_{ARM}).
SIRM (10^{-6} Am^2 kg^{-1})	Saturation Isothermal Remanent Magnetization is the highest amount of magnetic remanence that can be produced in a sample by applying a large magnetic field. The value of SIRM is related to concentrations of all remanence-carrying minerals in the sample but is also dependent upon the assemblage of mineral types and their magnetic grain size.
Soft IRM (10^{-6} Am^2 kg^{-1})	The amount of remanence acquired by a sample after experiencing an applied field of 40mT. At such low fields, the high coercivity, canted-antiferromagnetic minerals such as hematite or goethite are unlikely to contribute to the IRM, even at fine grain sizes. The value is therefore approximately proportional to the concentration of the low coercivity, ferrimagnetic minerals (e.g. magnetite) within the sample, although also grain-size dependent.
Hard IRM (10^{-6} Am^2 kg^{-1})	The amount of remanence acquired in a sample beyond an applied field of 300mT. At fields of 300mT, the majority of ferrimagnetic minerals will already have saturated and the value is therefore approximately proportional to the concentration of canted antiferromagnetic minerals within the sample.
IRM Backfield Ratios	Various magnetization parameters can be obtained by applying one or more magnetic 'reverse' or 'backfields' to an already saturated sample. The magnetization at each backfield can be expressed as a ratio of IRM_{field}/SIRM and can discriminate between ferrimagnetic and canted antiferromagnetic mineral types.

samples, while Walden *et al.* (1999) discuss the practical details of the analysis in more detail. Maher and Thompson (1999) demonstrate a number of applications of the method within the Quaternary sciences.

7.8 CARBONATE ANALYSIS

Dreimanis (1962) provided an excellent summary of both the applications and precedures for analysis of carbonates within glacial sediments. While advances in analytical technology mean other methods are now available, as a rapid, low-cost approach, the gasometric methods using the Chittick apparatus described by Dreimanis (1962) are still perfectly capable of producing results to an acceptable level of accuracy. Gale and Hoare (1991) describe a similar gasometric method based upon the Bascomb Calcimeter and which could also be performed in the majority of basic sediment laboratories without the need to resort to specialised analytical equipment.

As with many of the other methods introduced above, quantitative analysis of the carbonate content of a diamict sediment may provide evidence to support stratigraphic correlation or provenance studies. However, given the susceptibility of carbonate minerals to chemical weathering processes, vertical profiles of carbonate concentrations can also give useful insights into the depth and degree of post-depositional weathering of a sediment unit.

7.8.1 Sample preparation

Dreimanis (1962) recommends analysis using the Chittick apparatus should be performed on the very fine sand and finer fraction. Approximately 20g of dry sediment should be gently disaggregated using a pestle and mortar. This is then dry-sieved through a 3.75 phi (74μm) sieve. Of the material passing through the sieve, 1.70g is weighed out to be used in the analysis. If it is known that the carbonate content of the sediment is high (>40%), then 0.85g of sample is used.

7.8.2 Analysis

The Chittick apparatus can be constructed from basic laboratory glassware. A full description is given in Dreimanis (1962) and the references therein. The apparatus is designed to determine the volume of CO_2 gas evolved from carbonates within the sample when it reacts with acid. If it is assumed that calcite is the carbonate mineral present, then a simple conversion formula can be applied to calculate the weight percentage of carbonate present in the sample. However, sediment samples may also contain other carbonate mineral forms. While many minor forms can be ignored due to low abundance, some attempt must be made to distinguish between calcite and dolomite. Failure to do so can lead to a systematic over-estimation of the total carbonate content.

Dreimanis (1962) suggests a two stage measurement process that allows the proportions of both calcite and dolomite to be estimated. Experimentation has shown that all the calcite present within a sample reacts within the first 30 seconds of the analysis while less than 5% of the dolomite reacts during this initial stage. The dolomite continues to react for between 15 to 45 minutes, after which the evolution of CO_2 stops. Two readings of the CO_2 developed are therefore taken; the first after approximately 30 seconds (to allow calculation of the weight percentage calcite) and a second after approximately 15 minutes (to allow the calculation of the weight percentage dolomite).

Dreimanis (1962) details the full procedure, which also takes into account the temperature within the apparatus and the barometric pressure to improve the accuracy of the results. Errors of less than 0.5% are possible with the method and results have been shown to be highly repeatable.

7.9 CASE STUDIES

7.9.1 Clast lithological analysis in the study of glacial diamicts

The majority of applications of clast lithological analysis have examined gravel deposits, but the technique has also been used in the study of glacial diamicts (i.e. Shetsen 1984). Bell *et al.* (1989) used clast lithology data to distinguish between glacial diamicts derived from local and regional sources, allowing them to infer ice flow patterns and limits in the Nachvak Fjord in northern Labrador, North America (Fig. 7.4). The fjord is thought to have acted as a major flow path for ice originating from the interior of Labrador and moving east into the Labrador Sea. Other workers (Clark and Josenhans 1986 and Josenhans *et al.* 1986) have proposed that the last regional ice sheet terminated at least 30km to the east of the current coastline, and that within the area of the Nachvak Fjord,

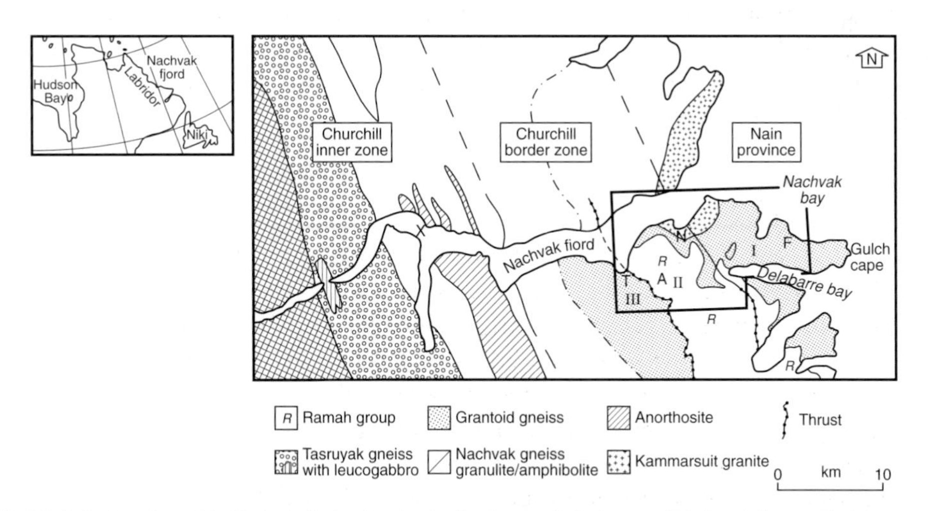

FIG 7.4 Bedrock geology of the Nachvak Fiord region showing the three geological zones within the study area. Clast lithological data were used as evidence to suggest less extensive regional ice cover during the last major glacial phase than had been proposed on the basis of other evidence (adapted from Bell *et al.* 1989). F = Flies valley; A = Adams Lake; T = Tinutyarvik valley; N = Naksaluk valley.

ice thickness of 200 to 300m must have existed. This would mean that regional ice must also have filled the subsidiary Tinutyarvik, Naksaluk, Adams Lake, Kammarsuit and Flies valleys to the south of the fjord, all of which have altitudes well below 200m. In contrast, Bell *et al.* (1989), on the basis of clast lithological analysis, considered the Nachvak Fjord to be an ice marginal area at the maximum extent of the last regional ice advance.

Bell *et al.* collected samples suitable for clast lithological analysis from diamict exposures throughout the area with the purpose of establishing the extent of samples with a westerly (as opposed to a predominantly local) provenance. Three potential source areas were recognized; local sources were split into two 'zones' (I and II) and a further zone (III), lying to the west, represented a more distal source. A westerly provenance was assumed to represent the influence of a major regional ice advance and local sources to indicate smaller scale valley glaciation. Clasts were placed into one of several lithological classes and the proportions in each class were then calculated. Cluster analysis was used to group the samples based upon their clast composition. Four dominant clusters were identified (labelled M, N, O and P), and for each of these clusters, the mean percentages for four of the lithological types were calculated from all sample sets (Table 7.11). Cluster P showed by far the largest input of clasts from zone III and Bell *et al.* (1989) therefore inferred that samples in cluster P represent the deposits of the last regional ice advance. Clusters M, N and O represent various combinations of locally derived clasts from zones I and II.

When the spatial distribution of the samples in each cluster were examined, it was found that, although samples in cluster P were present in the Tinutyarvik and Naksaluk valleys, they were absent from the Adams Lake and Flies valleys where locally derived material dominates. From this, Bell *et al.* (1989) concluded that the last regional glaciation was not as extensive as proposed by Clark and Josenhans (1986) and Josenhans *et al.* (1986) and that the relatively low lying Adams Lake and Flies valleys were free of regional ice, and proposed a revised regional interpretation of events during the Middle and Late Wisconsinan.

Table 7.11 Mean values of lithological composition for sample groups. Samples were placed into clusters M, N, O and P on the basis of cluster analysis (from Bell *et al.* 1989).

	Mean cluster value (%)				
Cluster	Class A	Class B	Class C	Class D	Number of samples in cluster
M	0.0	94.1	1.6	3.4	11
N	19.4	53.8	11.5	8.6	22
0	59.5	13.8	2.4	23.0	6
P	0.0	7.2	64.8	16.4	20

Key: – Class A clasts derived predominantly from Zone I
Class B clasts derived predominantly from Zone II
Class C clasts derived predominantly from Zone III
Class D clasts derived from all zones

7.9.2 Heavy mineral analysis in the study of glacial diamicts

Gwyn and Dreimanis (1979) used heavy mineral assemblages in glacial diamicts to distinguish between sediments derived from various glacier lobes in the Great Lakes region of North America (Fig. 7.5). Several major ice streams were thought to have existed in the region, and the sources for the sediments deposited by these ice streams fall into three major structural provinces; Superior, Southern and Glenville, each with their own discrete range of rock types.

Gwyn and Dreimanis obtained samples of diamicts thought to belong to each of the major ice lobes and performed heavy mineral analyses on over 100 samples. 17 minerals and mineral groups were counted. Initial inspection of the data, in graphical form, suggested that sediments derived from the Superior and Southern Provinces exhibited low total heavy mineral percentages and large proportions of both hornblende and the clinopyroxenes (diopside and various augites). Samples derived from the Grenville Province showed higher total heavy mineral percentages, less clinopyroxenes (but relatively higher orthopyroxenes – enstatite and hypersthene) and higher garnet proportions. On a visual basis, the authors also recognised sub-sources within some provinces.

The existence of both the major and minor distinctions between the provinces was examined by multivariate statistical methods. Cluster analysis divided the samples into two main groups as expected (Superior-Southern and Grenville types). Further sub-division was most successful with five subgroups, three within the Superior-Southern samples and two within the Grenville samples. These groupings confirmed the qualitative interpretation of the data and produced a sensible spatial distribution within the study area. R-mode factor analysis was used to see which mineral types were most diagnostic for each of the source areas. Effective separations of the various source groups was achieved by considering the abundance of epidote, purple garnet and red garnet (Fig. 7.6). Discriminant function analysis was used to provide a method by which samples of unknown origin could be identified based upon their heavy mineral assemblages.

Gwyn and Dreimanis (1979) concluded that the heavy mineral assemblages of glacial diamicts within the Great Lakes region provided an effective provenance indicator, and that multivariate statistical analysis of heavy mineral data provided a useful tool by which samples of unknown origin could be related to a particular ice lobe.

7.9.3 XRF analysis in the study of glacial diamicts

Peachey *et al.* (1985) showed how XRF data could be used for both provenance indication and geochemically-based correlation of Quaternary and pre-Quaternary deposits (including diamicts) in Suffolk, England. The relationships between the various units within the sequence had been debated by a number of workers (i.e. Lake *et al.* 1977 and Rose and Allen 1977) and studied using sedimentological and palaeontological techniques. In particular, the Kesgrave Sands and Gravels have been linked to both the stratigraphically lower crag deposits and the stratigraphically higher glacial deposits. Peachey *et al.* (1985) hoped that a geochemical approach might help to resolve this problem.

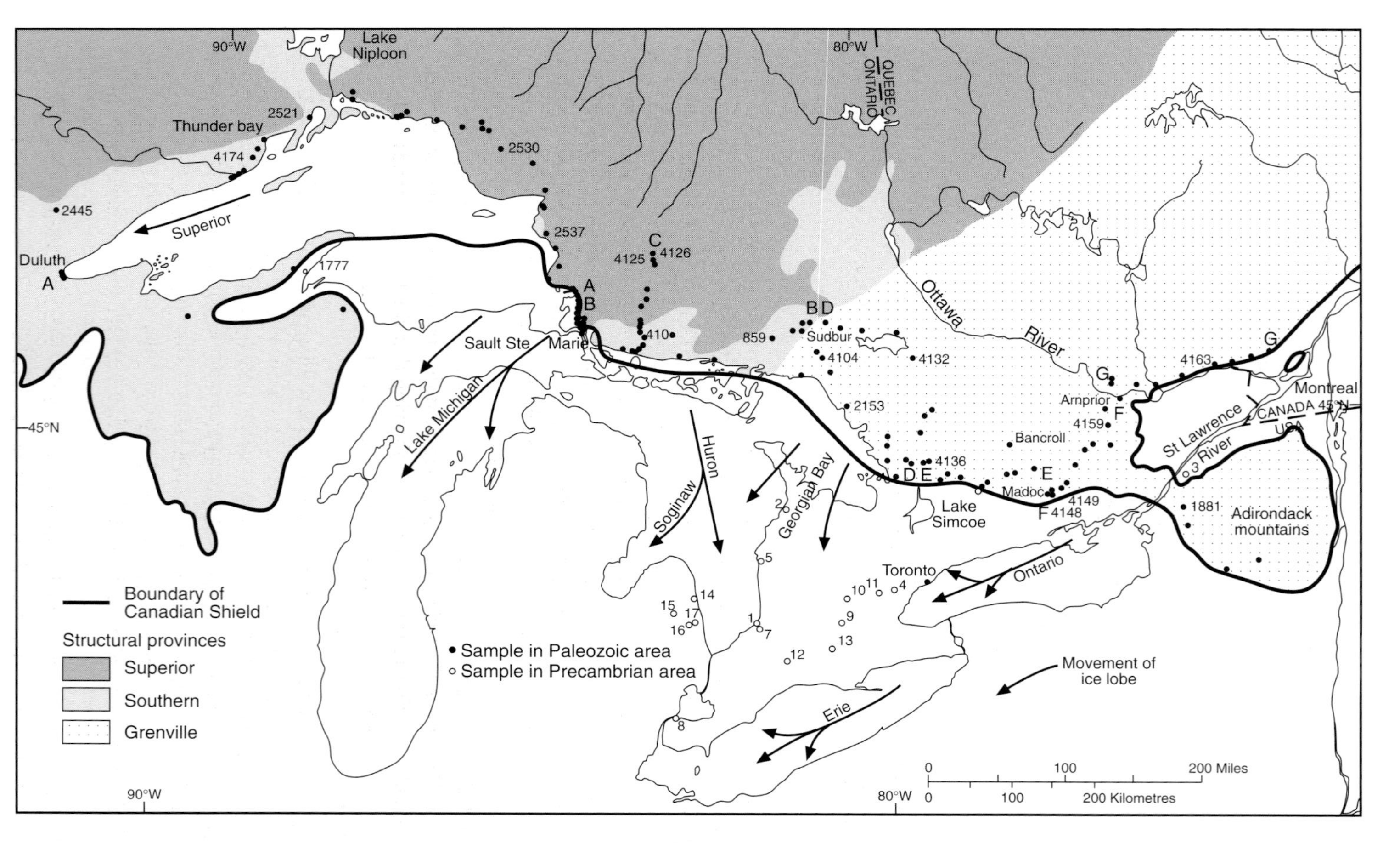

FIG 7.5 Till sample locations, flow directions of main glacial lobes and outline of structural provinces in the Canadian Shield in the Great Lakes Region (from Gwyn and Dreimanis 1979). On the basis of heavy mineral data, two main ice-source areas were defined; a Superior-Southern Province source and a Grenville Province source.

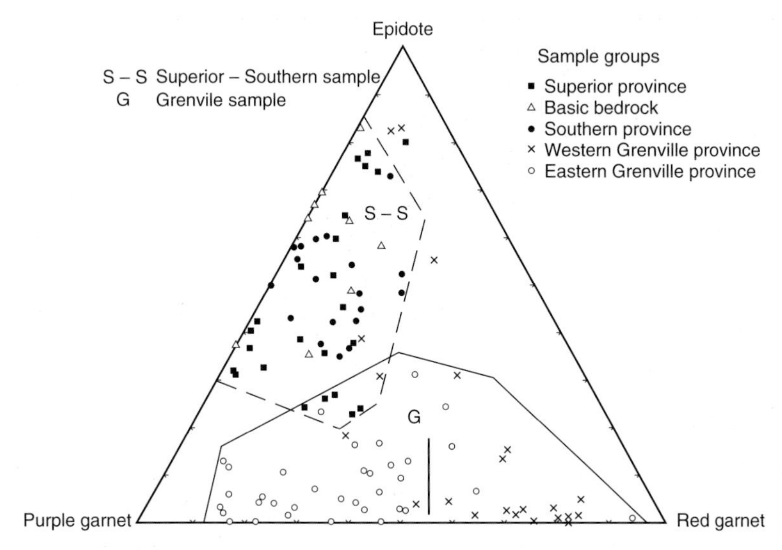

FIG 7.6 Ternary plot of selected heavy minerals, showing separation of samples into different source compositions (from Gwyn and Dreimanis 1979).

Samples from the major units within the stratigraphy were taken at a number of localities and subjected to examination by both XRF and DC-arc emission spectrometry. A comparison of the data collected by the two techniques showed that both methods produced similar results (Roberts and Peachey 1983). Bivariate scatter diagrams were used to show patterns in the concentrations of particular elements. For example, plots of Ti and Zr (thought to be reflecting levels of rutile and zircon) showed distinct patterns. The Kesgrave Sand and Gravel samples showed a similar distribution to the Red and Norwich crag deposits. However, the diamict samples formed a distinct group. These distinctions between the various deposits were supported by data for other elements such as Cr, V, Rb, Sr, Ca and Cu, although no statistical treatment of the geochemical data was presented.

From this, Peachey *et al.* (1985 and 1986) suggested that the geochemical data clarified the stratigraphic relationships in the sequence, showing that the Kesgrave Sands and Gravels were derived from the crag and the provenance of the diamict sediments was distinctly different.

Broster (1986) showed how data on elemental composition obtained by XRF and AA could be used to differentiate between two Late Wisconsinan diamict units in Port Albert, Ontario, Canada (Fig. 7.7). The units were exposed in a single 300m long section, the underlying unit, the Lower St. Joseph Till, separated from the higher Upper St. Joseph Till by a distinct stratigraphic break from which two separate glacier advances were inferred (although not necessarily of vastly different ages). Fabric measurements suggested that both diamict units had been derived from ice flowing from the north west (out of the Lake Huron basin).

Samples were collected from vertical profiles passing through both diamict units and data for ten major and seven trace elements were obtained. Plots of each element with

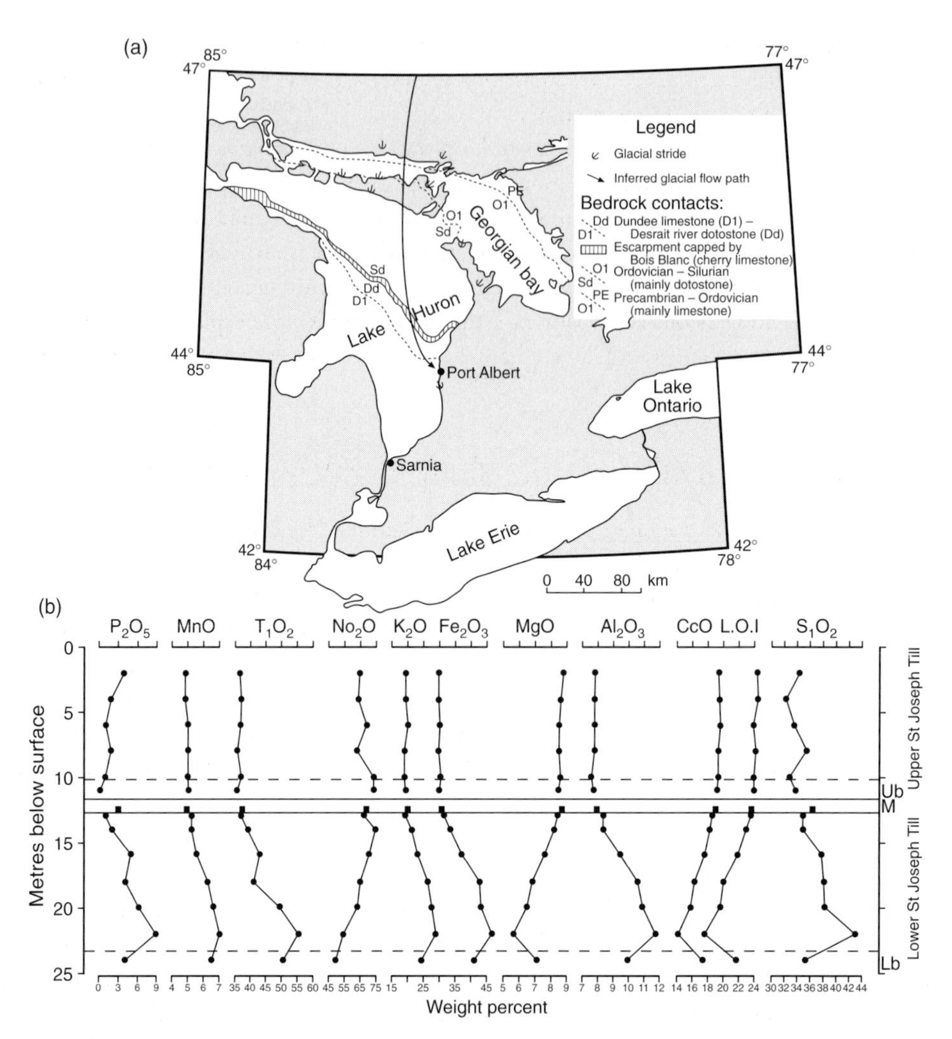

FIG 7.7 XRF data from tills in Port Albert. a) study area and inferred ice-flow path. b) Vertical variation in elemental composition of the Upper and Lower St. Joseph Till. As both till units are derived from the same ice-flow directions, Broster (1986) inferred that the compositional variation in the lower unit reflects variation in up-ice source rocks that have been essentially preserved through the entrainment-transport-deposition system (redrawn from Broster 1986).

depth in the profile showed little lateral variation between profiles but consistent vertical variations were present (Fig. 7.7b). The upper diamict showed very little vertical compositional variability, as did its clast content. In contrast, the lower unit showed strong vertical trends in a number of both major and trace elements. Analysis of the clast content of this unit reveals that there is an increase upwards in the abundance of clast types from successively up-glacier rock formations.

The data were also subjected to a rigorous statistical analysis and the composition of the two units was found to be significantly different for almost every element analysed. A multivariate analysis of the same data (using principal-components analysis, cluster analysis and discriminant functions) confirmed the distinctions and allowed the dominant trends in the data to be seen more clearly.

Given that the fabric data suggested a similar provenance for the two diamict units, Broster argued that depositional processes were responsible for the compositional stratification. The vertical variation of the lower unit reflects the same vertical stratification of debris layers in the glacier ice, as debris from more distant sources is gradually raised further from the ice-bedrock interface by continued entrainment of fresh material. The homogeneous composition of the upper diamict was explained by increased mixing of the basal debris layers due to thrusting and shearing during the second glacier advance.

7.9.4 Clay XRD analysis in the study of glacial diamicts

Clay XRD has been used in studies on a variety of sedimentary rock types including glacial diamicts (i.e. Perrin 1957; Glentworth *et al.* 1964; Beaumont 1972). In studies on glacial diamicts this technique has been used for two main purposes; first, for correlation and provenance indication and second, to establish the weathering history of the sediment being studied. This second application is also of relevance here. A knowledge of the likely products of weathering within the clay fraction of the diamicts to be studied will enable qualitative statements to be made about the effect of post-depositional chemical change upon the underlying geochemical character of each diamict, when used for correlation and provenance studies.

Correlation and provenance studies

Glentworth *et al.* (1964), in studying the parent materials for various soil associations in North East Scotland, used clay XRD data as one of a range of techniques to differentiate between glacial diamicts. Five soil associations were described, developed upon a red glacial drift. Previous work had suggested that all the red drift material (both glacial diamicts and lacustrine deposits thought to be associated with the diamicts) found in the coastal plain of Aberdeenshire had been derived from Old Red Sandstone sediments in the Vale of Strathmore, giving an ice-flow direction from the south west.

Samples from the basal horizons of all five soil types were analysed and Glentworth *et al.* made various deductions from these data. First, the Tipperty lacustrine deposit appeared similar in composition to the glacial diamict. This suggests that the lacustrine deposit is derived from the diamict or from the same source as the diamict. Second, the drift deposits that form the parent materials for the Laurencekirk and Stonehaven soil associations appear distinctly different from those of the Tipperty, Peterhead and Hatton soil associations and the lacustrine and diamict sediments from the Tipperty quarry. The Laurencekirk and Stonehaven associations are closest to the Strathmore source area and would therefore be expected to most closely reflect its geochemistry. Glentworth *et al.* therefore suggested that the Tipperty, Peterhead and Hatton area had been influenced by ice from an additional source. Using supplementary data from examination of the light and heavy mineral suites of the same samples, it was suggested that ice from the Scandinavian sheet, moving across the present coastline from the North Sea, had deposited sediments derived from red argillaceous rocks on the bed of the North Sea.

Weathering studies

Droste (1956) showed how clay minerals in glacial diamicts can be altered by weathering. Using samples taken from profiles in diamict units in Wisconsin, USA, he showed that where weathering of the upper surface of the profile had taken place (i.e. during pedogenesis), a change in the relative proportions of particular mineral types could be seen with decreasing depth in the profile.

The data showed how samples at the base of the profiles contained predominantly chlorite and illite. The upper samples, taken from below the zone of soil development, showed illite, but the chlorite had diminished and been replaced by vermiculite. The weathering of chlorite to vermiculite is part of one of two main weathering sequences in clay minerals (Melkerud 1984) related to the conditions within the sediment or soil. These sequences are:

1 mica > illite > vermiculite > smectite

2 ferromagnesian minerals > chlorite > vermiculite > smectite

Snäll (1985) showed a similar example of the other sequence using diamict sediments from east-central Sweden. Illite was gradually replaced by vermiculite towards the surface of the profile (Fig. 7.8). Snäll stated that all samples were taken below levels where superficial weathering was visible (below the obvious area of soil development), with the assumption that illite, chlorite and kaolinite were primary clay minerals and vermiculite a secondary product of weathering. Snäll concluded that the XRD data could be used to indicate that weathering had occurred even when visual evidence is absent.

7.9.5 Mineral magnetic analysis in the study of glacial diamicts

Until relatively recently, work that has applied mineral magnetic measurements has used either a very restricted set of parameters or only worked on bulk samples of the sediment matrix (Vonder Haar and Johnson 1973; Gravenor and Stupavsky 1974; Barendregt *et al.* 1976 and Chernicof 1983).

However, Walden *et al.* (1987; 1992a) have demonstrated that a more comprehensive mineral magnetic analysis can provide extremely useful data to aid provenance and correlation studies in glacigenic sediment sequences. This work quantified the intra- and inter-unit variability of diamict units within the Quaternary sediment sequence of the Isle of Man (Fig. 7.9). The data produced were consistent with the broad field-stratigraphic relationships between the various units sampled, and supported sediment provenance inferences based upon earlier clast lithological and fabric analysis (i.e. Dackombe and Thomas 1985). The data also showed that the magnetic analysis had considerable potential for identifying intra-unit variability (i.e. through vertical exposures within a particular stratigraphic unit). Walden *et al.* (1996) compared the magnetic data obtained with that

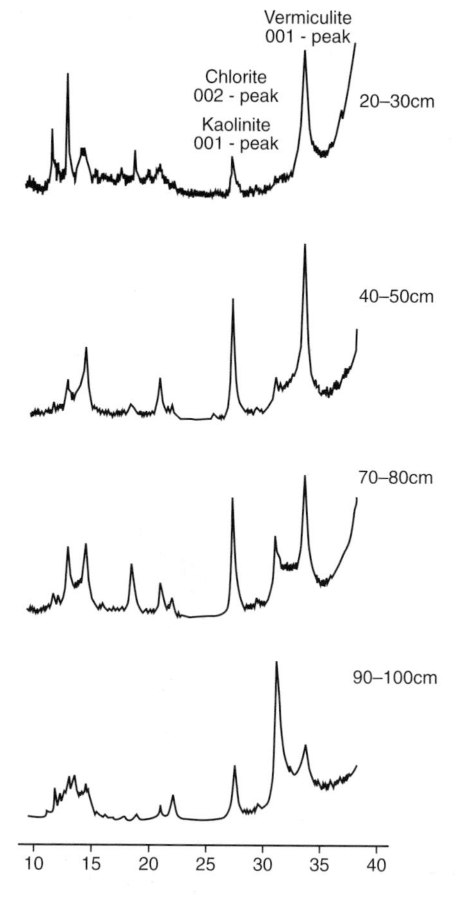

FIG 7.8 XRD traces for the clay fraction separated from samples taken from a vertical profile through a soil developed on glacial sediments in Uppland, Sweden. The changes in clay mineral distribution with depth are due to weathering, with illite being replaced by vermiculite. Similar changes are found in weathered till within the region (after Snäll 1985).

from other compositional analyses applied to the same sediments. They concluded that the information provided by the magnetic analysis was comparable in quality to that obtained from XRF, XRD, clast lithological analysis and heavy mineral analysis.

Other studies have also suggested that mineral magnetic measurements can offer some insights in studies of post-depositional weathering of glacial sediment sequences (Walden and Addison 1995) and depositional environments (Walden *et al.* 1995).

7.9.6 Carbonate analysis in the study of glacial diamicts

A considerable body of diamict compositional data has been accumulated by the Geological Survey of Canada (Shilts 1993) and this includes carbonate analysis (e.g. Dreimanis and Terasmae 1958; Kettles and Shilts 1989). While much of these data have

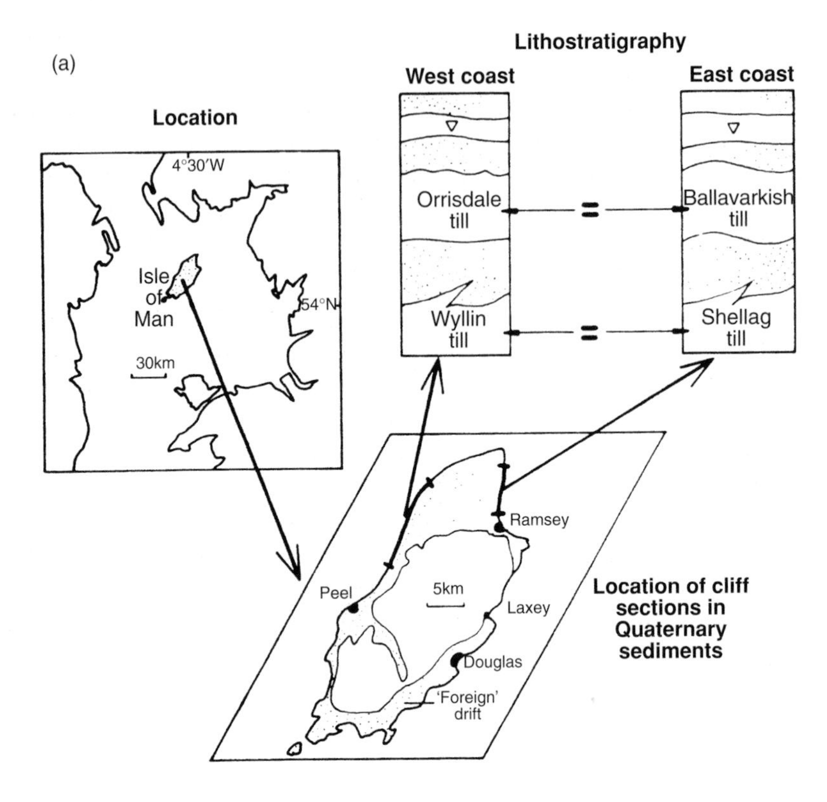

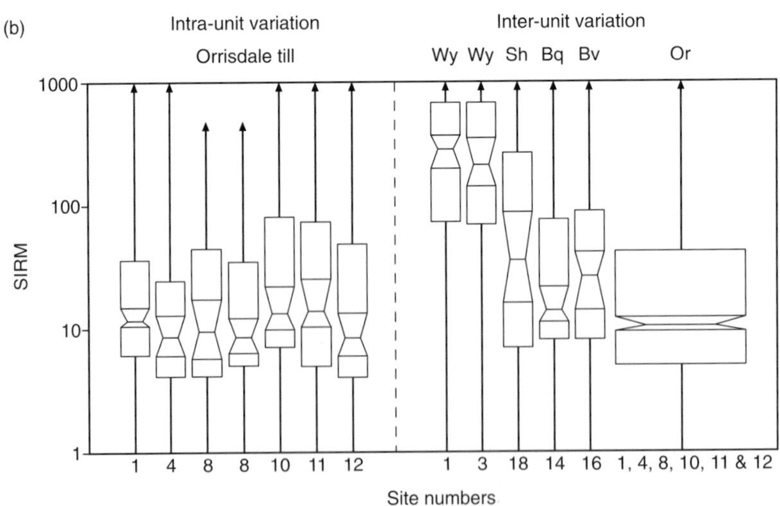

FIG 7.9 a) Field location and summary stratigraphy. Diamict matrix samples for the same sediment units show similar compositional relationships on the basis of magnetic, heavy mineral and XRF analyses; b) Notched box plots of SIRM (10^{-5} Am^2 kg^{-1}) values from the clast sample sets from four diamict units from the Isle of Man, showing intra-unit variation (lateral and vertical) of the Orrisdale Till (left side) and inter-unit variation between all diamicts sampled (right side). Notches that overlap suggest no significant difference between the median values of the two sample sets at approximately the 5% significance level (after Walden *et al.* 1992a; 1996).

been applied to studies of sediment correlation and provenance, in other cases post-depositional weathering has been studied.

As an example of the latter approach, Shilts and Kettles (1990) used carbonate analysis as part of a comprehensive geochemical and mineralogical study through a vertical glacigenic sediment exposure of the Lennoxville Till, Thetford Mines, Quebec, Canada. This exposure contained both fresh and weathered materials. The results demonstrated that compositional alteration due to post-depositional weathering can be significant, particularly in the finer silt and clay fractions. While visual evidence such as colour change may provide an early indication of such alteration, any studies attempting to determine stratigraphic relationships between exposures or sediment provenance must clearly establish whether the compositional signal is detritial. To this end, evaluation of the degree of post-depositional weathering is an essential pre-requisite.

Shilts and Kettles (1990) also compared carbonate determinations made using the Chittick method of Dreimanis (1962) with two other forms of analysis; the Leco Carbon Analyser (Foscolos and Barefoot 1970) and total carbonate dissolution. They found that the Chittick and Leco methods produced comparable results. However, the total carbonate dissolution method tended to over-estimate carbonate concentration as other silt and clay sized mineral forms such as Fe-oxides can also be dissolved and removed from the sample.

8 Engineering properties

Brice R. Rea

8.1 INTRODUCTION

The engineering (geotechnical) properties of glacial sediments refers to the fundamental physical properties that can be used to describe the response (strain) of sediments (referred to as *soils* in engineering literature) to applied stresses. The term 'engineering properties' reflects the extensive range of procedures, equipment, relationships and terminology developed in soil mechanics. Understanding the response of glacial sediments to applied stress is important for a number of reasons, from the pure scientific ends of interpreting the dynamics of glaciers (e.g. subglacial sediment deformation) to implementing the correct piling procedures for construction (geotechnical engineering). As in many fields of earth sciences, collaboration between the earth scientist and the engineer is one of mutual benefit. As the rigour and repeatability of engineering tests, required for safe construction, excavation, etc, provide information useful to the earth scientist, so an understanding of the formation and deformation history of the sediments is useful to the engineer. In this chapter, details on, and the use of, various engineering properties and the methods for their measurement, will be provided. While many *in situ* engineering tests exist for measuring the physical properties of sediments, a number of these are only feasible in commercially funded operations and, as such, lie outside the scope of this chapter. It should be noted that, in the following sections, *in situ* refers to undisturbed deposited sediments as opposed to *in situ* 'subglacial' sediments, except in Section 8.8. Only tests that are likely to be undertaken by the glacial geologist with access to (easily portable field and lab-based) soil testing equipment will be discussed.

As has been demonstrated in earlier chapters of this book, glacial materials come in a range of guises, from fine-grained matrix-dominated lowland tills to bouldery mountain tills, through glacifluvial and glacilacustrine deposits to complex glacitectonized sequences of mixed origin sediments. These sediments all have intrinsic physical properties that provide information on their strain behaviour and strain history. However, it must be remembered that the goals of the glacial geologist and the engineer are rather different. The engineer is interested in developing standardized tests and criteria that are applicable and repeatable in order to assess the requirements of the specific engineering problem at hand. Even more fundamental is the (general) desire of the engineer to know the peak shear strengths of sediments in order to correctly design engineering solutions for foundations, etc. For the glacial geologist this is important, but so also is the residual or ultimate strength (the quasi steady-state value the shear strength attains after the peak shear strength is overcome), given the significant strains many glacially deformed sediments undergo (Hooyer and Iverson 2000a).

8.2 PARTICLE SIZE

Particle size is probably the most fundamental property of all sediments, and is a routine analysis undertaken by engineers and geologists alike. The determination of particle size may provide information regarding the characteristics of the source material, transport distance, mixing and incorporation of different sediments and post-depositional processes. Particle sizes and their physical ordering exert a major control on the mechanical properties of sediments. For example, in response to an applied stress, a coarse-grained diamict has a significantly different stress-strain response than a silt/clay-rich diamict (Müller and Schlüchter 2001). One significant point, which must be made at this stage, is the importance of clays. They are fundamental in understanding the mechanical properties of sediments, because clays have very specific properties, especially in the presence of water. Clay minerals have a platy-layered structure, and these platy particles can align as a result of shear strain, leading to a reduction in the strength of sediments. The platy surfaces carry a negative charge and have a great affinity for water, so clay minerals can hold large amounts of 'adsorbed' water bound to them (e.g. montmorillonite may contain up to 40% adsorbed water), which leads to significant swelling and shrinking upon wetting and drying. The combination of the platy minerals and affinity for water combine to produce plasticity – the ability to take up and retain a new shape when they are deformed. All of these properties are 'diluted' by the presence of 'inactive' clay-sized particles, e.g. quartz and feldspars, which are often found in glacial materials. While particle-size analysis can provide information on the size range of the sediments it does not identify clays *sensu stricto*, only clay-sized material. For the identification of clay minerals geochemical analysis using, for example, XRD (x-ray diffraction) equipment is required. A full description of the use of particle-size analysis has been given in Chapter 3.

8.3 GENERAL PHYSICAL PROPERTIES

Some general physical properties are required to characterize the physical state of sediment and are pre-requisite for computation of the engineering properties. These relate the proportions of different material phases present in the sediment, i.e. solids (inorganic and organic), which form a skeleton having pore spaces filled with liquid (water) and/or gas (air). The most widely used phase model for soils is the unit solid volume, where the volume under consideration is composed of 1 volume unit of solid material, which is assumed to remain constant (i.e. the solid particles are considered incompressible given the range of loads under consideration), and a volume of voids (water and/or air filled) which will vary depending upon loading and drainage. Figure 8.1 displays conceptually the unit solid volume model, from which a number of important quantities are derived.

The void ratio (e) is given as:

$$\text{void ratio} = \frac{\text{volume of voids } (V_v)}{\text{volume of solids } (V_s)} = \frac{e}{1} = e \qquad (1)$$

and the porosity (n) is:

$$\text{porosity } (n) = \frac{\text{volume of voids } (V_v)}{\text{total volume } (V)} = \frac{e}{1 + e} \qquad (2)$$

and also:

$$e = \frac{n}{1 - n} \qquad (3)$$

It should be noted that these are both ratios and as such dimensionless. Another important quantity is the moisture content (m), which is given by:

$$\text{moisture content} = \frac{\text{mass of water}}{\text{mass of solids}} = \frac{M_w}{M_s} \qquad (4)$$

Details of the determination of these quantities can be obtained from many general soil mechanics and soil-testing books and manuals (e.g. Bowles 1981; Whitlow 1983).

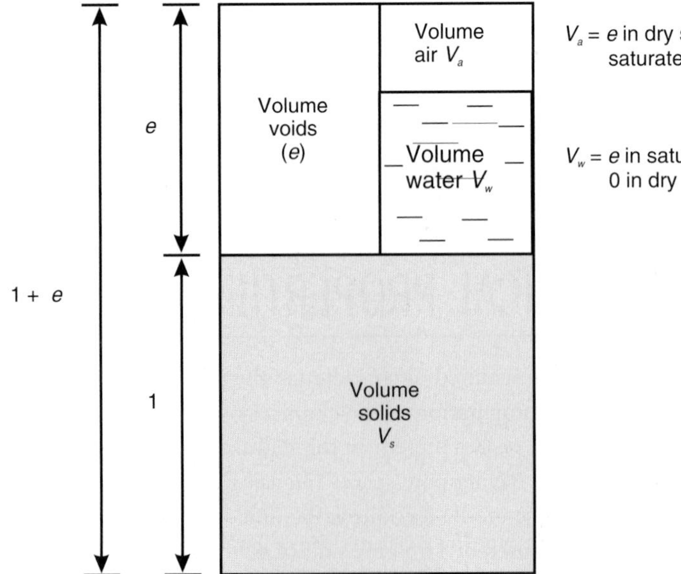

FIG 8.1 The unit solid volume model and phase relationships found in sediments.

8.4 ATTERBERG LIMITS

The Atterberg limits, sometimes referred to as index properties, are used to classify fine-grained soils in engineering, and define the limits at which the sediment acts as a solid, plastic or liquid (Fig. 8.2) and are given below in order of decreasing moisture content:

1| Liquid Limit (LL), is the moisture content at which the sediment will flow under its own weight and represents the transition from viscous fluid to plastic behaviour.

2| Plastic Limit (PL), is the moisture content at which the sediment can be rolled into a thread of 3mm diameter without breaking and represents the point at which the material passes from plastic to pseudo-plastic.

3| Shrinkage Limit (SL), is the moisture content at which no further decrease in volume is experienced even with continued moisture loss, and represents the transition from pseudo-plastic to solid.

For determination of the Atterberg limits a sample of the sediment is air dried, as oven drying may reduce both the LL and the PL by 2–6% (Bowles 1981), and broken up with a pestle and mortar, taking care not to crush individual particles. The sample is then passed through a 425μm sieve and the coarse material discarded. The sieved sample is mixed with water to form a thick paste and allowed to stand for 24 hours. This paste is then used in the different tests required to obtain the Atterberg limits. In the following sections, a description of the relevant test procedures is provided, but further, more

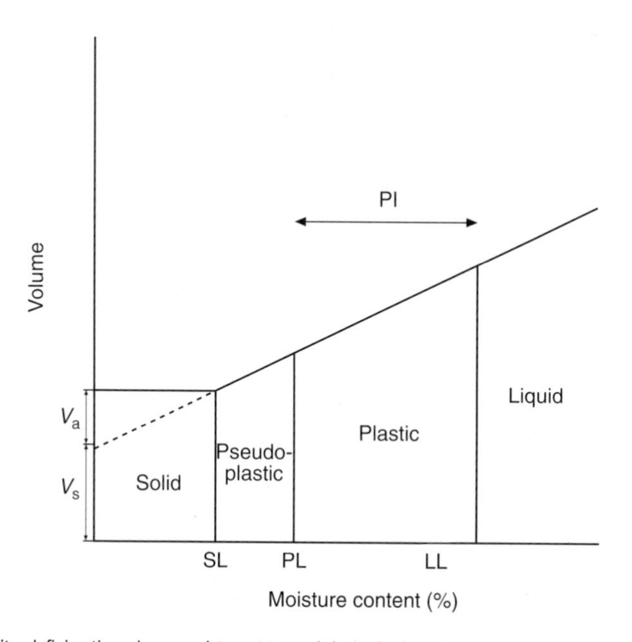

FIG 8.2 Atterberg limits, defining the volume, moisture state and thus, physical state of sediment.

detailed information may be found in relevant engineering texts (e.g. Bowles 1981; Vickers 1983).

8.4.1 Liquid Limit (LL)

The LL can be measured by two methods a) the cone penetrometer and b) the Casagrande apparatus.

a) The cone penetrometer measures the penetration of a standard cone (apex angle 30°) under a load of 80g for a set, 5 second, time period (Fig. 8.3a). A sample (from the preparation described above) is remixed for 10 minutes, filled into a metal cup (taking care to avoid trapping air) levelled and placed below the cone. The cone is lowered, so that it just touches and marks the surface of the sample, and is then locked in place with the dial gauge reading noted. The cone is then released and penetrates the surface of the sample for a duration of 5 seconds. The test is repeated a few times on samples from the same paste mix and the moisture content determined for each sample, with an average taken. The test is repeated for 5 or 6 other paste mixes with different water contents, and penetrations should all lie between 15–25mm. Cone penetration vs. moisture content is plotted on a graph, a best fit straight line plotted and the moisture content at 20mm penetration is taken as the LL (Fig. 8.3b).

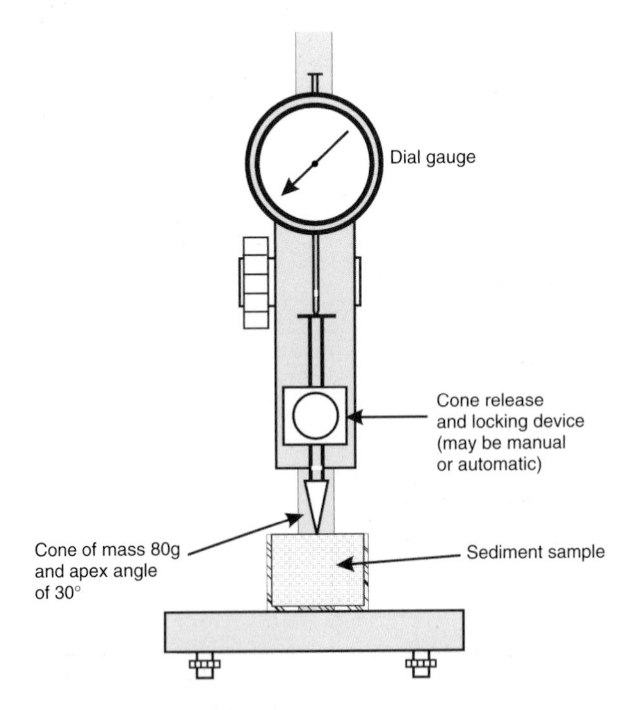

FIG 8.3 a) Standard design for the cone penetrometer.

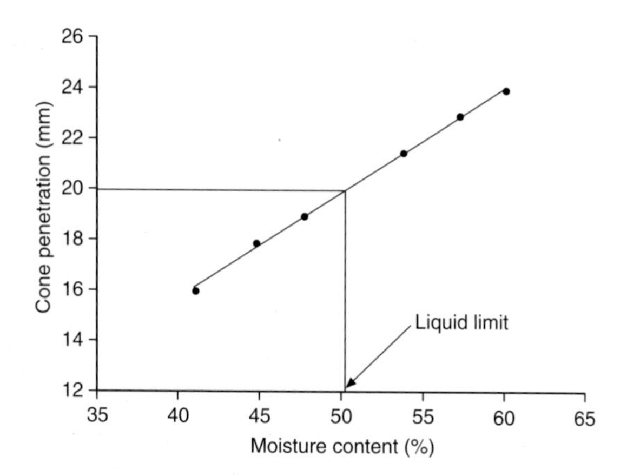

FIG 8.3 b) Graphical method for defining a sediment's LL from cone penetrometer data.

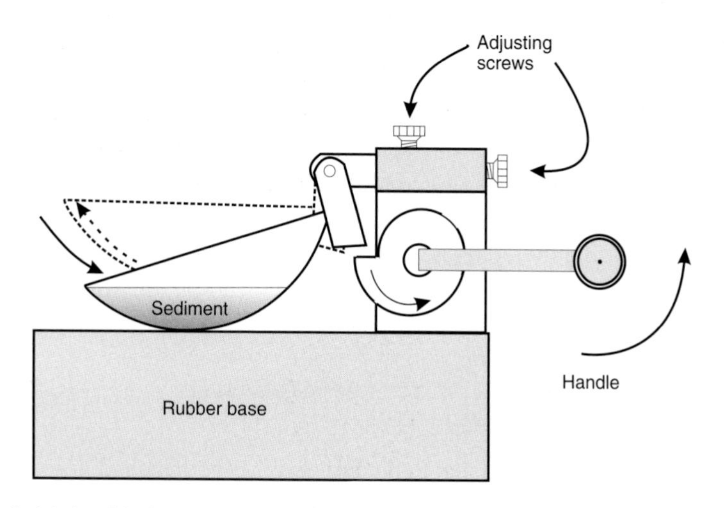

FIG 8.3 c) Standard design of the Casagrande apparatus.

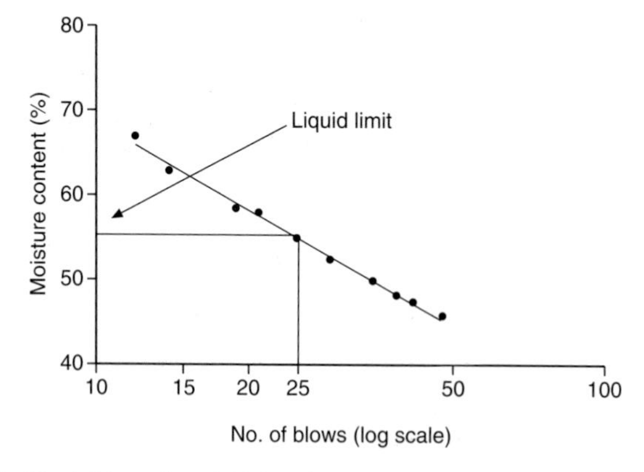

FIG 8.3 d) Graphical method for defining sediment's LL from Casagrande apparatus data.

b) The Casagrande apparatus consists of a rotating cam that raises and drops, by 10mm, a metal cup containing a sediment sample (Fig. 8.3c). The sediment sample (from the preparation described above) is remixed for 10 minutes and filled into the cup and levelled to a horizontal surface. A groove is cut through the soil sample using the standard grooving tool, supplied with the apparatus. The handle of the apparatus is then turned (at ~2 revs/s), raising and dropping the cup, gradually closing the groove. Once the groove has closed along a 13mm section the number of turns is noted. The test is repeated on more of the same sample and an average recorded. Further tests on samples with different moisture contents complete the procedure. The results are plotted as moisture content vs. log of number of blows and the moisture content at 25 blows is taken as the LL (Fig. 8.3d).

The cone penetrometer method is favoured over the Casagrande method due mainly to the test repeatability, less potential for operator error and its simplicity in terms of test procedure, result plotting and equipment maintenance.

8.4.2 Plastic Limit (PL)

A golf ball sized sample of the paste (from the preparation described above) is taken and divided into eight equal parts. The first part is rolled on a glass plate at a rate of 80–90 strokes per minute (1 stroke = forward and back) until a 3mm thick thread is formed. This thread is broken up, rolled again into a 3mm thread, with the procedure repeated (the handling process dries the sample) until the sample breaks up and can no longer be rolled into a thread, i.e. it has passed from plastic to psuedo-plastic. At this point the moisture content is determined.

8.4.3 Shrinkage Limit (SL)

The SL is infrequently calculated due to the fact that it is not required in soil classification. It is most useful for identifying sediments that undergo large volume changes with changing water content, in other words, where swelling clay minerals are present in quantity. A known volume of sediment sample (from the preparation described above) is mixed with water until close to its LL. This sample is slowly dried with measurements of mass and volume (often done using a volume displacement vessel) taken periodically during this process. The point at which volume decrease ceases with further decreases in moisture content is the sediment SL.

8.4.4 Plasticity Index (PI)

The Plasticity Index PI is the range of the plastic phase of the sediment and is given by LL–PL. Note that for cohesionless sediments (i.e. sediments with no clays), the PI is zero as the PL and LL are equal. For soils with small clay contents it is often difficult to determine the LL and PL. In such instances the *Linear Shrinkage Test* provides a method of estimating the PI for these types of soils. A sample of the paste (from the preparation described above) is filled into a semi-circular mould (vibrated to remove air) and the surface levelled. The sample is then air dried at 60°C until it has shrunk clear of the mould, and subsequently may be oven-dried at 105°C, to speed up the drying. The sample length is then measured (taking a mean if required) and the LS calculated from:

$$\mathrm{LS} = \left(\frac{L_0 - L_f}{L_0}\right) \times 100 \qquad (5)$$

where L_0 is the initial length and L_f is the length after drying. The PI is then obtained from:

$$\mathrm{PI} = 2.13 \times \mathrm{LS} \qquad (6)$$

Once the PI and LL are known the data can be plotted onto a plasticity chart (PI vs LL), which provides a method of sediment classification (it is used to establish the sub-groups of fine-grained soils in the British Soil Classification System). Figure 8.4a shows the standard format of the plasticity chart with the Casagrande A-line providing an arbitrary division between clays (above) and silts (below).

Boulton and Paul (1976) plotted data from a number of tills onto a plasticity chart and found that modern glacial sediments with a LL >22% plotted close to a straight 'T-line' that lies above the Casagrande A-line (Fig. 8.4b). The modern glacial sediments lie above the A-line due to the presence of coarser material than that found in fine-grained soils. As suggested by Boulton and Paul (1976) the T-line is useful for indicating the nature (clay-sized as opposed to true clays) and quantity of the clay-size fraction in glacial sediments. Greater use of the plasticity chart would perhaps prove useful in studies of glacial sediments, for example, in characterizing which sediments would be most prone to accommodate strain from glacier induced thrusting (e.g. Benn and Evans 1996; Christiansen and Suaer 1997; Harris *et al.* 1997). Sladen and Wrigley (1983) plotted data from 'supraglacial, flowed, melt-out and lodgement' tills on a plasticity chart and showed a progression up the T-line from the 'coarse-grained supraglacial', through the 'melt-out' into the 'lodgement' (Fig. 8.4b). The location of different tills on the T-line represents the change in content of clay-sized material as a function of transport distance and transport mode; supraglacial and melt-out tills are likely to be coarse and plot in the bottom left of the T-line while 'lodgement' tills will have undergone more deformation and thus, comminution, increasing their clay-sized fraction and position up the T-line. Interestingly the flowed tills plot away from the T-line and start approaching the A-line, indicating a preferential increase in the fine-grained material as a result, presumably, of sorting during multiple phases of reworking during 'flow'. Theoretically, in a subglacially-transported deformation till unit where mineralogy is 'constant' and there is no mixing with other sediments, increasing down-glacier transport distance should relate to increasing comminution and thus an increasing position up the T-line.

8.4.5 Liquidity Index (LI)

The Liquidity Index LI is calculated as follows:

$$\mathrm{LI} = \frac{m - \mathrm{PL}}{\mathrm{PI}} \qquad (6)$$

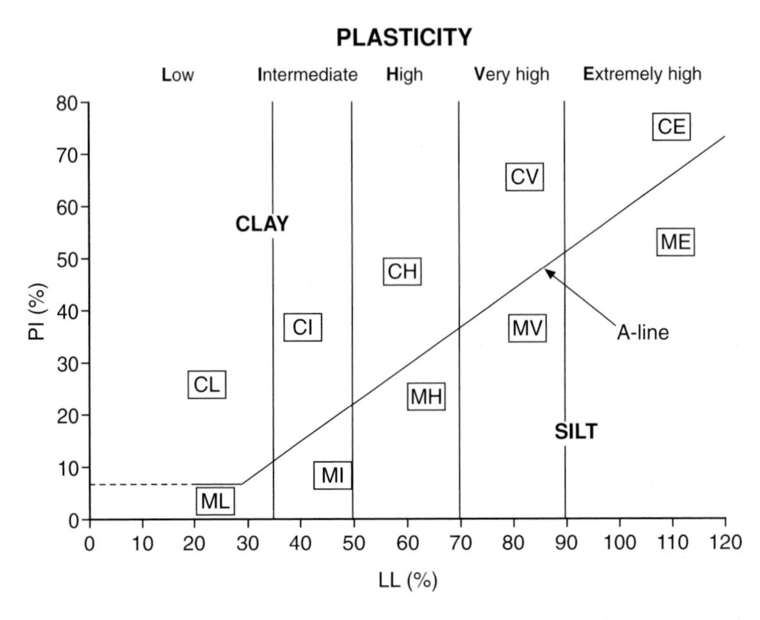

FIG 8.4 a) Standard format for the plasticity chart, with clays classified as C* (above the A-line) and silts classified as M* (below the A-line).

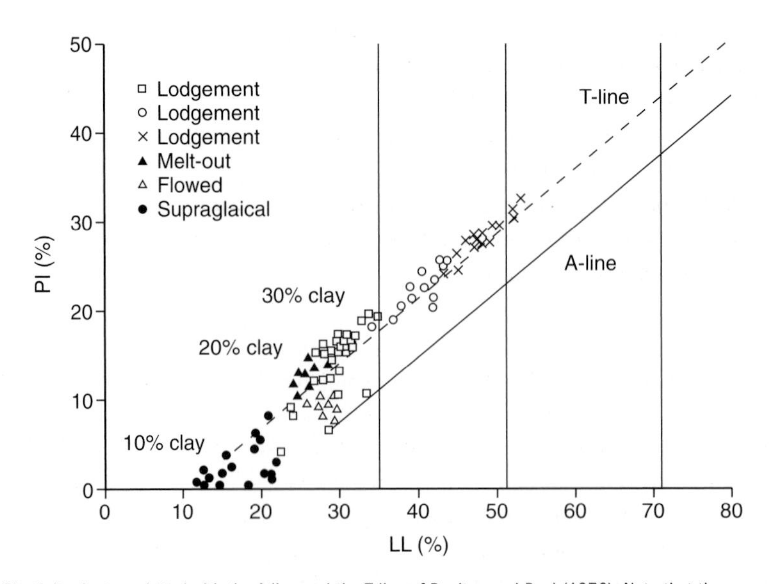

FIG 8.4 b) Plasticity chart re-plotted with the A-line and the T-line of Boulton and Paul (1976). Note that the percentage clays relates to clay-sized material only (not to clay minerals).

where m is the *in situ* moisture content of the soil. The relationships provided by the LI are as follows:

- LI<0 – sediment in a solid or pseudo-plastic state
- 0<LI<1 – sediment in a plastic state
- LI>1 – sediment in a liquid state.

For the glacial geologist this index property is not likely to be of great interest as it represents the present state of the sediment. For reconstruction of ice dynamics it is past conditions that are of interest, and even for sediments recently exposed at a glacier snout the 'present' conditions may well be significantly different from those operating while it was subglacial.

Atterberg limits provide, at a globally recognizable level, a classification of fine-grained sediments regarding the ranges of moisture contents at which they exist in various states. However, in order to design acceptable engineering solutions more detailed information, with regards to sediment strength and strain characteristics, is required. Likewise for the glaciologist or glacial geologist, an understanding of sediment strength and strain characteristics is vital to interpreting glacier dynamics, sedimentary processes and landform evolution. After the actual composition (physical size and mineralogy) the second most important factor in governing soil strength characteristics is water.

8.5 THE ROLE OF POREWATER

The role of porewater is fundamental in understanding the stress-strain behaviour of sediments. As mentioned above a sediment mass is made up of solids (which form a grain skeleton) and voids (which may be filled with air and/or water). From above, the ratio of void volume to total volume is the sediment porosity (n), and in any porous material it is the combination of the total stress (σ) and the pore fluid pressure (u) (in glacial sediments this is water) that produces the effective stress (σ'). In a saturated soil (which is a reasonable assumption for most subglacial and ice marginal sediments), the porewater pressure supports some of the total stress thus:

$$\sigma' = \sigma - u \qquad (7)$$

Any increase in total stress (this can be the result of shearing or vertical loading) is instantaneously taken-up by the porewater pressure, so there is no net increase in effective stress. However, provided there is hydraulic connection to a region of lower hydraulic head, porewater flows away along the hydraulic gradient. The result is a reduction of the void volume (e) and an increase in the load supported by the solid grain skeleton, the process known as consolidation. Eventually, if sufficient porewater drains away, the increase in total stress is taken up entirely by the solid grain skeleton and equals the increase in effective stress. The concept of effective stress is also critical in understanding the shear strength of sediments (see below). Obviously then, the rate at which excess porewater dissipates is fundamental for any treatment of stress and strain in saturated sediments.

8.5.1 Hydraulic conductivity

The rate of flow of porewater (q) through sediments can be described by Darcy's Law:

$$q = K \frac{\Delta h}{\Delta x} \quad (8)$$

where, K is the hydraulic conductivity, $\Delta h/\Delta x$ is the change in hydraulic head with length along the flow path (the hydraulic gradient). The volume of porewater discharged per unit time Q is then given by:

$$Q = q A \quad (9)$$

where, A is the cross sectional area under consideration. K represents the capacity of the sediment to allow water to pass through it. It is controlled by a number of sediment properties including the particle-size distribution, particle orientations and porosity, which in saturated soils is a function of the effective stress. So from equations 8 and 9, the higher the hydraulic conductivity of the sediment the faster water can flow through it: the faster it can dissipate elevated porewater pressures, the further σ' increases.

Many empirically derived relationships have been developed in order to make possible rapid approximations of K. One such widely applied relationship uses the effective grain size of the sediment (d_{10} – the maximum grain diameter at 10% by weight of the sample) and is given below:

$$K = B\, d_{10}^2 \quad (10)$$

where, B is an experimentally derived coefficient. This relationship was developed originally for filter sands but has been shown to provide an acceptable approximation for other sediments in the fine sand to gravel range (Freeze and Cherry 1979; Whitlow 1983). Not surprisingly, given the nature of glacial sediments in terms of particle-size range, Hubbard and Maltman (2000) report a considerable spread of data around the relationship described in equation 10. The best method for determining K is by physically measuring it, and two standard sets of apparatus are used; the falling head permeameter for fine-grained soils (Fig. 8.5a) and the constant head permeameter for coarse-grained soils (Fig. 8.5b).

Falling head permeameter

In the falling head permeameter a laterally confined sediment sample (with filter papers at either end) is connected to a standpipe of known cross-sectional area, with the bottom of the sample just in a water reservoir (Fig. 8.5a). The standpipe is filled with water to a known level and a series of readings of this level are recorded as water drains through the

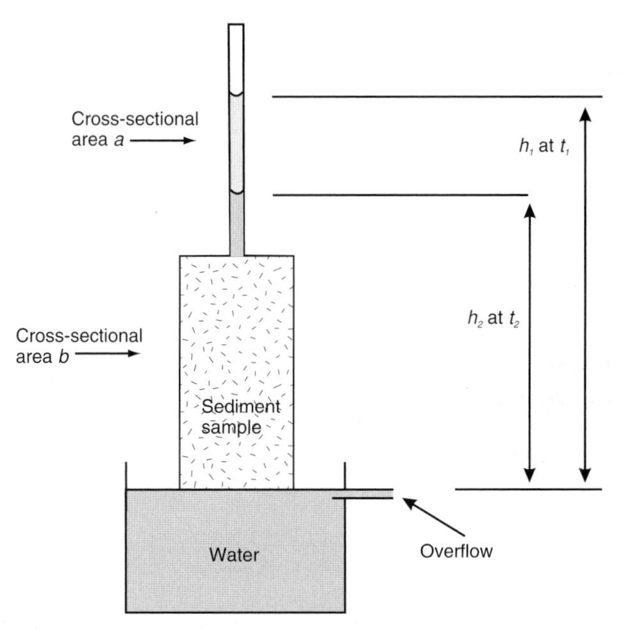

FIG 8.5 a) The falling head permeameter, best suited to testing fine-grained soils.

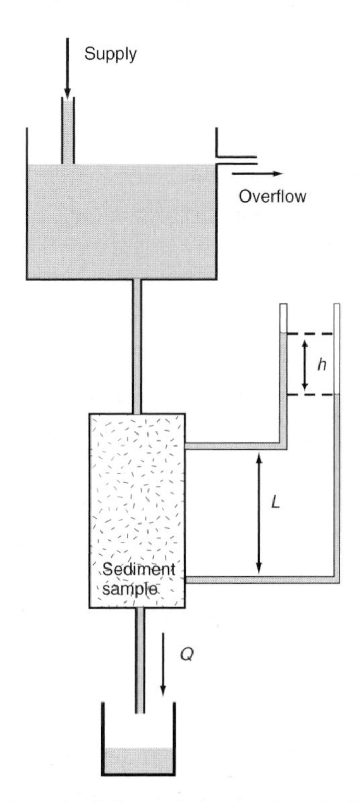

FIG 8.5 b) The constant head permeameter, best suited to testing coarser-grained soils. The Rowe test cell, (initially designed for consolidation testing) is a more sophisticated type of constant head device and can measure *K* under realistic values of effective stress.

sediment. The test is repeated for a number of standpipe diameters and an average obtained for K is calculated from:

$$K = \frac{aL\ln\left(\frac{h_1}{h_2}\right)}{A(t_1 - t_2)} \qquad (11)$$

Where $L = h_n - h_{n+1}$

An important point to note is that the falling head permeameter test may be conducted *in situ*, on undisturbed sediments. This is useful as the *in situ* properties of some sediment will be lost or adversely affected by sampling (see below).

Constant head permeameter

In the constant head permeameter a laterally confined, saturated sample (again with filter papers at either end) is attached to a constant head water supply. The fall in head pressure across the sample is indicated by the water level difference L (Fig. 8.5b) and the discharge of water is also measured. The test is repeated for a number of head pressures and an average for K is obtained from:

$$K = \frac{QL}{Aht} \qquad (12)$$

As mentioned above K will affect the shearing behaviour of sediments, but the relationship is not simple, as K has been shown to vary with effective pressure (large changes at lower {<100kPa} effective pressures) and strain (Hubbard and Maltman 2000). Values of K under typical field conditions of effective stress can be measured using more sophisticated equipment, for example, the Rowe test cell (Rowe and Barden 1966), or modified triaxial testing equipment (e.g. Hubbard and Maltman 2000). It has been shown that particle fabric and layering produce significant differences in values of K (Arch and Maltman 1990), and given the strong macro- and micro-fabrics developed in strained glacial sediments (e.g. Benn 1995; Benn and Evans 1996, 1998; see Chapters 5 and 6) it is probable that they will exhibit significant anisotropy in hydraulic conductivity (Murray and Dowdeswell 1992).

8.5.2 Consolidation

The general conceptual framework for understanding consolidation is to consider slow sedimentation, such that, as a mass of solid is gradually added, so the porewater is concomitantly expelled, and porewater pressure essentially remains hydrostatic (i.e. the porewater provides only buoyant support to the particles) and so the effective stress σ' is equal to the buoyant weight of the overlying sediment (Middleton and Wilcock 1994). In this situation the sediment is 'normally consolidated' and the current effective stress is the largest to which the sediment has been subjected. If deposition is more rapid then porewater pressures may rise, supporting some of the weight of the overlying sediments, and in this situation it is under-consolidated. Given time, if the porewater can drain away, the sediment will become normally consolidated. The usual results of consolidation are a reduction in volume, compressibility and permeability with an increase in strength and swelling potential (though not all of the deformation is recoverable). If the effective stress is reduced some of the volume reduction is recoverable, but not all, and in this condition the sediment is said to be 'over-consolidated', in other words the effective pressure at one time was greater than at the present day. This situation may be the result of, for example, ice-sheet loading, porewater pressure rise or overburden erosion. It should be noted that, while ice-sheet/glacier advance will increase the total stress, subglacial water pressures may also be very high, even approaching buoyancy, which results in sediments being only very slightly overconsolidated (e.g. Piotrowski and Kraus 1997; Tulaczyk *et al.* 2001).

The simplest engineering test, and treatment of consolidation, is the one dimensional Casagrande Oedometer Test. An undisturbed sediment sample is trimmed to fit a metal ring (usual dimensions 75mm diameter × 15–20mm thick). This is placed into the test apparatus, sandwiched between two porous plates and submerged in water (Fig. 8.6). A vertical load is applied via a hanger and a dial gauge and/or displacement transducer is used to record the change in volume (the sample is confined laterally so all the volume change is in the vertical, depth dimension). Data are recorded until the consolidation is complete and a new increment of mass added, a process repeated a number of times.

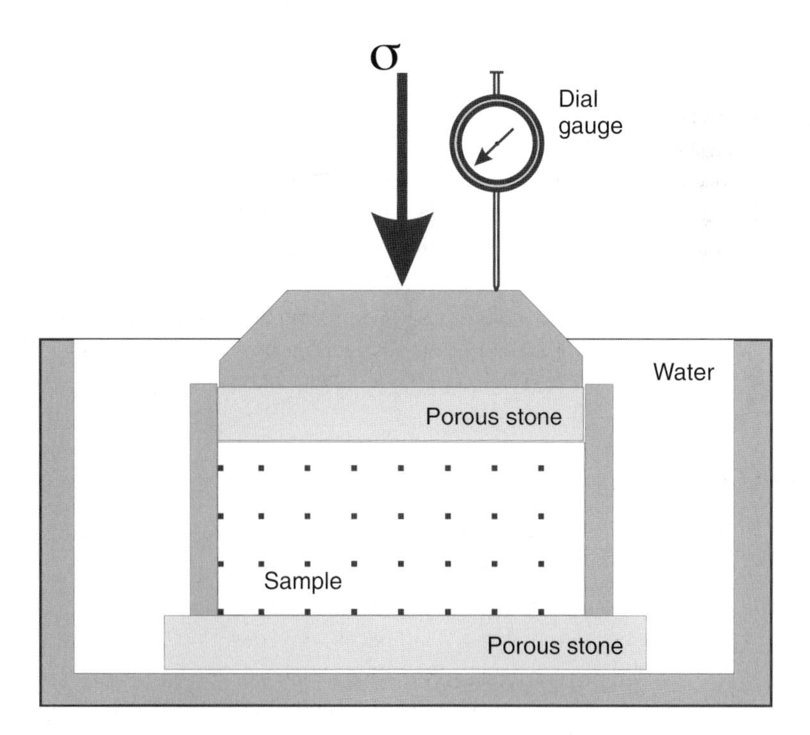

FIG 8.6 The one-dimensional Casagrande oedometer, which represents the simplest consolidation test. More complex tests involve the back pressuring of sediment samples.

It should be noted that in the one-dimensional consolidation test total stress σ is equivalent to the effective stress σ' because, for each incremental load, consolidation is completed only when the excess porewater pressure is dissipated and volume reduction has ceased. After the final consolidation phase is complete the sample is unloaded and allowed to swell to equilibrium, before removal for moisture content calculation. From these data a void ratio (*e*) vs. effective stress (σ') plot can be constructed (Fig. 8.7), and this relates the amount of consolidation (change in void ratio) due to increase in effective stress, for specific sediment. In the conceptual model described above, the plot of *e* vs. σ' represents the virgin or normal consolidation line (NCL), and for a remoulded sample a complete oedometer test will follow the NCL. If an undisturbed sediment sample is tested, the shape of the *e* vs. σ' curve can provide information on the loading history. The solid line in Figure 8.7 shows the reduction in *e* along the NCL with increasing σ' (in nature loading may be due, for example, to deposition of sediments, over-riding by ice or reduction of porewater pressure). If unloading occurs (dashed line) some, but not all, of the strain is recoverable, so the sediment is over-consolidated. Unloading may be the result of sediment erosion, ice retreat or an increase in porewater pressure. If the sample is re-loaded (dashed-dot line) it approaches the virgin compression curve at approximately the value of the previously experienced maximum 'pre-consolidation' stress (σ_{pc}'), and further loading continues along the NCL. Note also that from equation 7, fluctuations in porewater pressure will change the effective pressure.

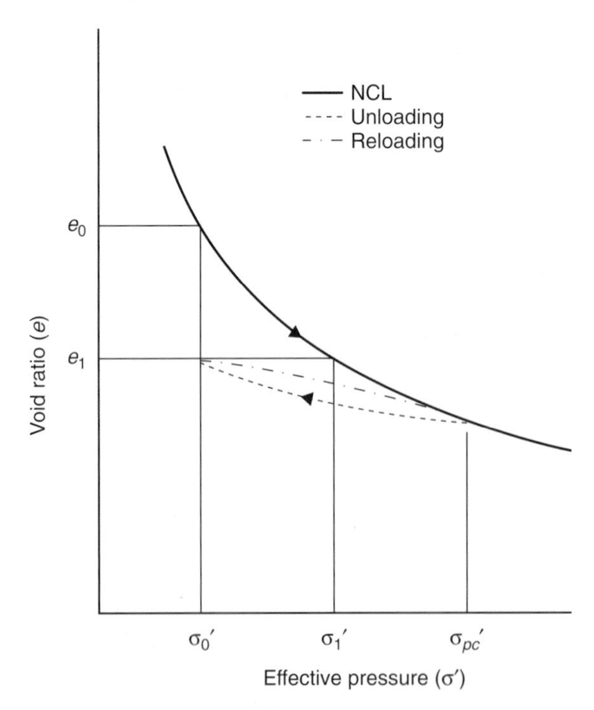

FIG 8.7 Change in void ratio (volume of voids/volume of solids) as a function of effective stress, first along the Normal Consolidation Line (solid line) to the preconsolidation stress (σ_{pc}'), then along an unloading curve (dashed line) back to a stress (σ_0') and finally re-loaded back to rejoin (dashed-dot line) and continue along the NCL.

Sediments in the subglacial environments will commonly experience porewater pressure fluctuations on a daily, seasonal and annual basis, resulting from changes in meltwater input to the ice-bed interface. In the absence of till shearing, this cyclic fluctuation in porewater pressure will lead to over-consolidation of the sediment (Clarke 1987; Iverson *et al.* 1998).

If consolidation test data are plotted as e vs. $\log\sigma'$ the NCL plots as a near straight line (CD in Figure 8.8) with any recompression following a curve (AB in Figure 8.8) until it reaches the NCL. In glacial reconstruction studies a very useful application of consolidation testing is to estimate the preconsolidation stress (σ_{pc}') i.e. the maximum effective stress that the sediment body has experienced, provided it has not been remoulded, for example, sheared. The estimate of the maximum effective stress ($\sigma - u$) if used in conjunction with other information can give a good indication of ice thicknesses and porewater pressures (Schokking 1990; Sauer and Christiansen 1993; Sauer *et al.* 1993; Larsen *et al.* 1994; Piotrowski and Kraus 1997). Indeed, where the sediment body has been sheared, provided it was initially over-consolidated, it will dilate and the effect of this 'remoulding' is to increase the natural water content, which can be used as an effective method for identifying shear zones (Sauer *et al.* 1990; Christiansen and Sauer 1993; 1997). Boulton and Dobbie (1993) used preconsolidation stress data as the basis for inferring a number of ice sheet dynamic properties for former ice sheets in England and Holland. The σ_{pc}' value may be determined using a number of methods, two of which are listed below.

The original, a graphical method, is based on an empirical relationship defined by Casagrande (1936). On a plot of e vs. $\log\sigma'$ (ABC), at the point of maximum curvature (M) both a tangent (T-T) and a stress axis parallel line (MN) are drawn (Fig. 8.8). The

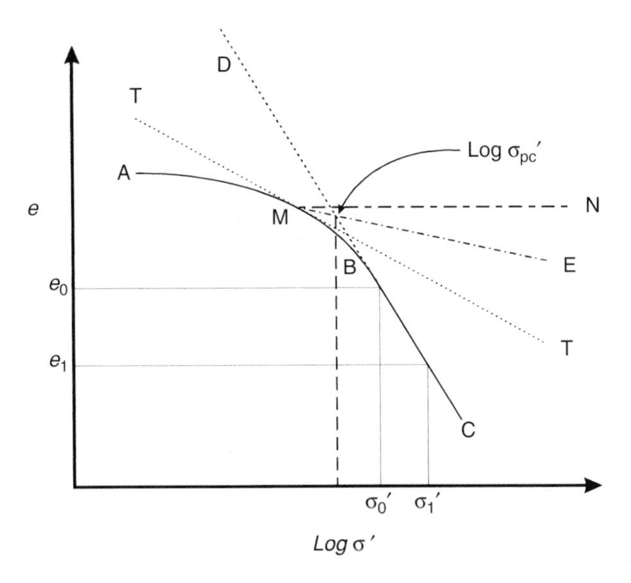

FIG 8.8 Plot of e vs. $\log\sigma'$ showing a reloading curve (AB), which joins the virgin compression line (DC). The remaining lines are used to provide an estimate following the method of Casagrande (1936) and are explained in detail in the text.

straight-line portion of the e vs. log σ' plot (NCL) is extended backward (CD) and a bisector ME (of the angle NMT) drawn. The intersection of ME and CD approximate the preconsolidation effective stress (Fig. 8.8). An alternative technique is described by Sridharan *et al.* (1991) which plots the log (1+e) vs. log σ' and is shown in Figure 8.9. The reloading line on the plot (the sub-horizontal line) is approximated by a straight line, as is the NCL (the steeper straight-line portion) obtained after the preconsolidation effective stress level is passed. The intersection of the two lines gives an approximation of the preconsolidation effective stress. In the results presented by Sridharan *et al.* (1991) this approach yielded significantly better results than the Casagrande method and given its ease of construction would appear to have major advantages over the Casagrande method.

On a plot of e vs. log σ' (Fig. 8.8) the slope of the NCL is referred to as the compression index (C_c) and is given by:

$$C_c = -\frac{\Delta e}{\Delta \log \sigma'} = \frac{e_0 - e_1}{\log\left(\frac{\sigma'_1}{\sigma'_0}\right)} \qquad (13)$$

which describes consolidation along the NCL, and can be taken as a constant for any given sediment (provided it is not over-consolidated). Further, the coefficient of volume compressibility (m_v), which is the change in volume resulting from a unit change in effective stress, is given by:

$$m_v = \frac{\Delta e}{\Delta \sigma'} \frac{1}{1+e} = \frac{\varepsilon}{\Delta \sigma'} \qquad (14)$$

where ε is strain because this is the one dimensional case. The slope of the e vs. σ' curve (Fig. 8.7) is given by $\Delta e/\Delta\sigma'$ so m_v is not constant for a given sediment but changes with

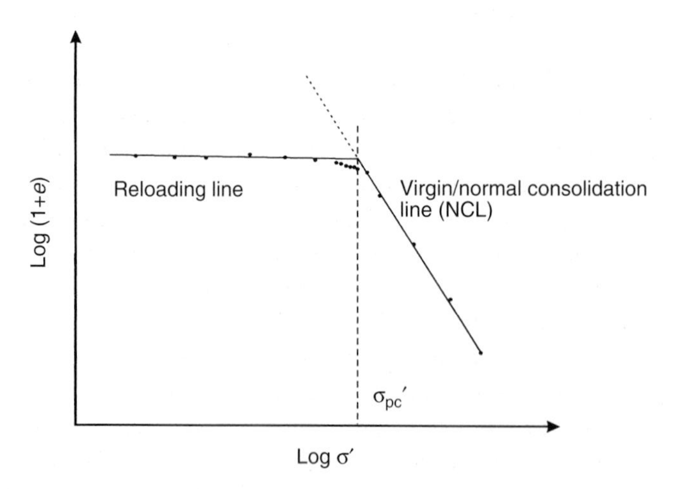

FIG 8.9 Plot of log(1+e) vs. log σ' showing the intersection between the sub-horizontal reloading line and the steeper virgin compression curve (NCL) (extended backward as the dashed line) giving the value of σ_{pc}'.

the effective stress: the higher the effective stress the lower the compressibility. The consolidation of a thickness h_0 of virgin sediment can then be predicted using:

$$\varepsilon = m_v \Delta\sigma' h_0 \qquad (15)$$

Finally, mention must be made of the coefficient of consolidation c_v which governs the rate at which sediment can expel excess porewater, which in turn controls the rate of consolidation:

$$c_v = \frac{K}{m_v \gamma_w} \qquad (16)$$

where, K is the coefficient of permeability/hydraulic conductivity and γ_w is the unit weight of water.

8.6 SHEAR STRENGTH

Since the 'paradigm shift' (Boulton 1986) in glaciology following the identification of a deforming till layer beneath Ice Stream B (Blankenship *et al.* 1986), quantifying the strain behaviour of sediments in relation to glacier-induced stresses has become fundamental to understanding and reconstructing the dynamics and sedimentary record of glaciers and ice sheets. The general and most widely used relationship to describe the strength of glacial sediments is the Mohr–Coulomb failure criterion, which describes the shear strength of the sediment as a straight-line relationship using the following equation:

$$\tau = c' + \sigma' \tan\phi' \qquad (17)$$

where τ is the shear strength (i.e. shear stress at failure), σ' is the effective stress (total normal stress (σ) minus porewater pressure (u)) and c' and ϕ' are respectively the cohesion and friction angle at the effective stress. Important points to note are that failure (deformation) of the sediment is the result of both shear (τ) and normal (σ') components, increasing normal stress (loading) increases shear strength (by increasing frictional resistance between particles), while increasing water pressure reduces effective stress (equation 7) and thus, shear strength. Rapid increases in normal stress will lead to increases in porewater pressure and as such a decrease in sediment strength. However, under sustained loading, the dissipation of the excess porewater pressures, i.e. drainage, will lead to consolidation. It will be shown below that the shearing of sediment can alter the porewater pressure, via alteration of the grain skeleton, and thus sediment strength. The classic graphical representation of Mohr–Coulomb shear strength is shown in Figure 8.10 and, following equation (17), the apparent cohesion is given by the intercept and the internal angle of friction is taken from the gradient of the line.

The response of the sediment to an applied shearing stress is dependent upon its particle-size characteristics and previous loading history, and typically follows one of two paths on the stress vs. strain (displacement) diagram (Fig. 8.11). For non-cohesive sediments the dashed line represents the stress-strain behaviour in the over-consolidated

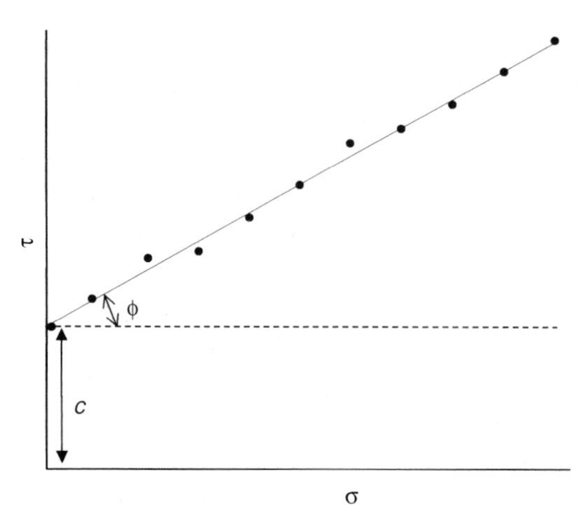

FIG 8.10 Coulomb failure envelope.

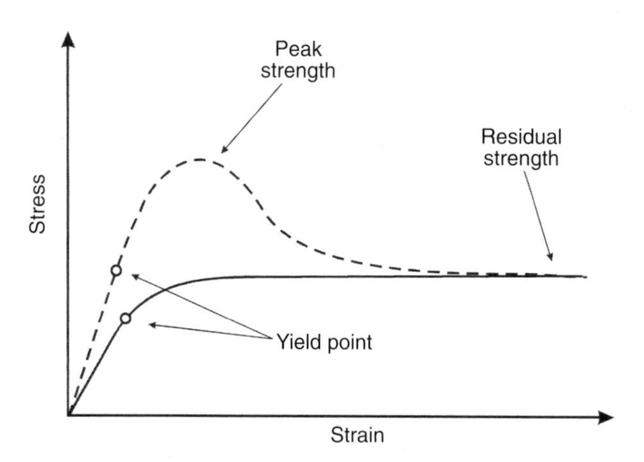

FIG 8.11 Two end-members of the stress-strain (displacement) graph for non-cohesive sediments. The dashed line represents an over-consolidated sediment and the solid line represents a normally consolidated or under-consolidated sediment. A clay-rich soil will have a relationship similar to the dashed line.

state and the solid line represents the normally consolidated state. Clay-rich soils will tend to exhibit a stress-strain curve similar to that for over-consolidated non-cohesive sediment.

8.6.1 Measurement methods

There are a number of methods for determining the shear strength of sediments; some of these are *in situ* field tests while others are conducted in the laboratory. There are advantages and disadvantages associated with both, especially in relation to engineering interests. *In situ* tests are more difficult to undertake, in that the boundary conditions are less well controlled and instrumentation, etc is more difficult. However, in terms of measuring the peak shear strength of the sediments, disturbance, resulting from sampling, can introduce significant errors into lab tests. Remoulded samples (remixed and saturated with water) may also be tested, a process that removes the strain history from the sediments. This can result in errors of estimation of factors of safety (i.e. the ratio of applied stress to *in situ* strength) if the *in situ* sediments have previously been sheared (i.e. they have been sheared through their peak strength and now can only support a lower shear stress (Fig. 8.11)). An example of this is provided by Early and Skempton (1972) where pre-existing shear surfaces developed in a palaeo-landslide were reactivated following embankment construction for the M6 motorway in England. Peak strengths of the un-sheared clays were significantly higher than those found along the shear surfaces, due to the re-orientation of the clay particles parallel to the shear plane. The sediment then only had residual strength.

Before testing begins the conditions under which the loading will occur should first be decided upon and what type of sample is to be tested, i.e. a remoulded or undisturbed sample. Tests may be *undrained*, where no drainage is allowed, so there is no loss of moisture during the test, and for a saturated soil therefore no volume change occurs. In a *consolidated undrained* test the sample is allowed to consolidate under a constant effective stress to the desired level (i.e. drainage is allowed) and then during loading and deformation undrained conditions are applied. Finally *drained* tests are where the sample is allowed to consolidate under a constant effective stress to the desired level and then full drainage is allowed during deformation, with the loading applied at a controlled rate, equal to the dissipation of excess porewater pressure. In the following section a number of the commonly used field and lab shear strength tests are described.

Direct soil shear box

A schematic of the direct soil shear box is shown in Figure 8.12. This is a relatively simple test to perform and the results are easily interpreted using the Mohr–Coulomb failure criterion (equation 17). Test sediment is prepared, by trimming an undisturbed sample or by consolidating to the desired degree a remoulded sample. Perforated, ribbed metal plates (to better transmit the stress) (Whitlow 1983) contact the sample, backed with porous plates for a drained test (excess porewater can drain away during the test) or impermeable membranes for an undrained test. The normal load (σ) is applied to the sample, via dead weights or hydraulics. It should be noted that, in the direct shear box test, porewater pressure is not measured, so only total stress measurements can be given. However, in a drained test $\sigma = \sigma'$ as the excess porewater pressures are allowed to dissipate, as in the oedometer test described above. A motor applies the shearing force at either a constant rate of strain or constant stress. The shear force (shear strength) supported across the sample is measured directly using a proving ring with a dial gauge and/or displacement tranducer, or using load cells. When the sample fails the test may be stopped if only the peak shear strength is of interest, for example, for plotting the Mohr–Coulomb failure envelope to derive c and ϕ. Otherwise, the test may be continued until the motor reaches its limit at which point it is reversed, the procedure being repeated until a constant value is obtained. This measure is the residual strength, which is an important parameter for

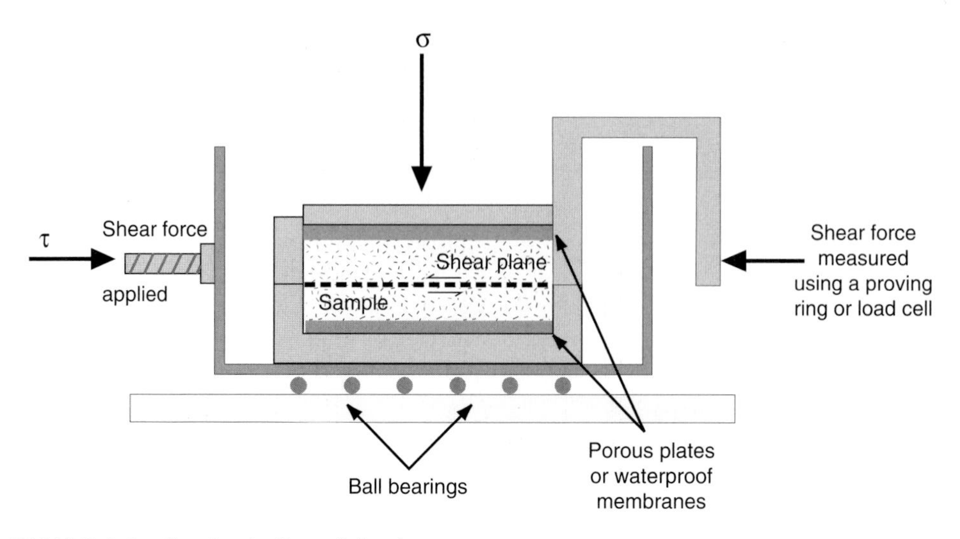

FIG 8.12 Typical configuration of a direct soil shear box.

understanding the behaviour of subglacial sediments as many will have been sheared to their residual strength. In order to determine representative values for c and ϕ (or c' and σ' in drained tests) a series of tests are run with differing normal loads and a plot of τ vs. σ can be used to derive the values of c and ϕ (Fig. 8.10). The main disadvantages of this test are that porewater pressures cannot be monitored and the location of the shear plane is pre-determined. However, it is very useful for determining the residual values of τ, c and ϕ, which is of significant interest in glacial research. Shear box tests are also useful for testing sand and gravel-rich sediments, where drainage takes place rapidly. Shear box sizes vary from the small 0.02m × 0.02m up to the larger end of the spectrum at 0.3m × 0.3m. With respect to glacial sediments the larger boxes allow undisturbed and remoulded samples with a more representative particle-size range to be tested. One important advantage of this equipment is that it is generally present in all engineering departments, and many earth science/geography/geology departments have one hiding away in a cupboard or basement.

Triaxial cell

Triaxial testing equipment provides greater control on the stress regime applied to the sample, allows measurement of porewater pressures and is the most widely used test in engineering. A cylindrical sediment sample, either undisturbed or remoulded and consolidated to the desired degree, is placed on the rig with sealed or porous end plates (Figure 8.13a). A waterproof membrane is then fitted over the sample and sealing O-rings put into place. The chamber is flooded and σ_3, the confining pressure, raised to the level required for the test. The cell is driven against a proving ring creating the deviatoric stress (σ_1 - σ_3). During the test changes in the sample length, volume (produced by dilatancy and consolidation), axial load and porewater pressure can all be recorded. This is a more

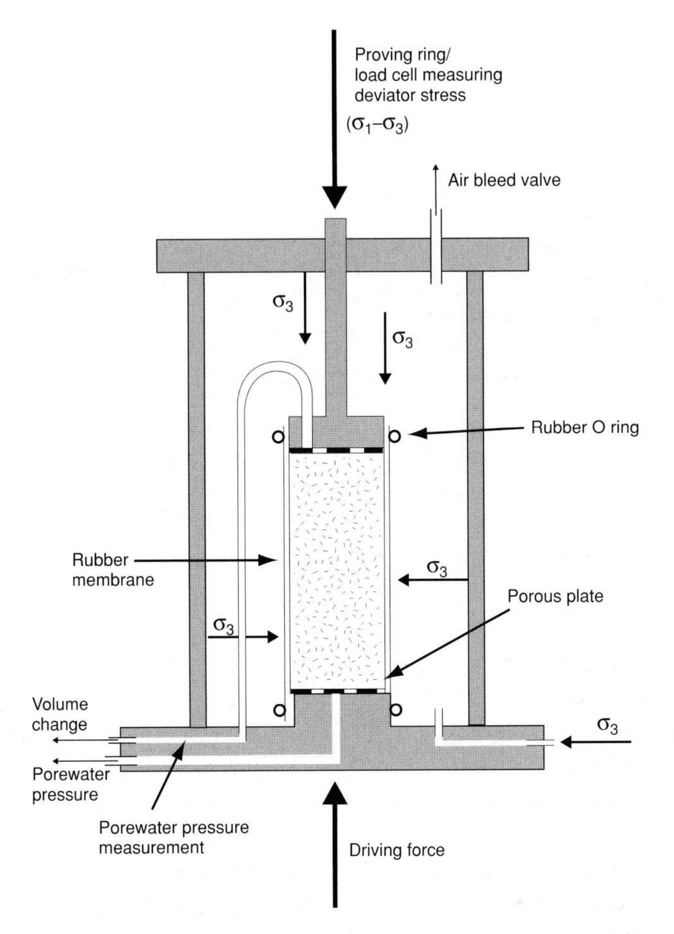

FIG 8.13 a) A typical triaxial testing rig configuration, with the ability to measure sample volume and pore water pressure changes.

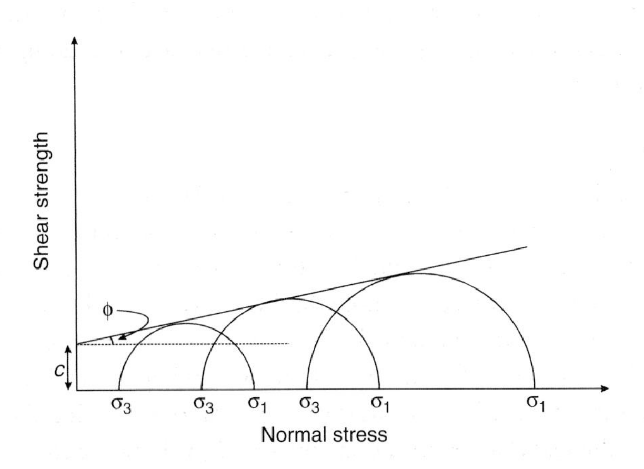

FIG 8.13 b) Graphical calculation of c and σ using a Mohr circle and Coulomb failure envelope construction.

sophisticated test than the direct shear box and, importantly, the location of the shear plane is not pre-defined. The sample size range is typically 38–100mm diameter, with a height to diameter ratio of 2:1 (length 76–200mm) although some larger rigs use samples with a ratio of 1:1 (this speeds consolidation of the sample) and diameters up to 254mm. Drained and undrained tests can be undertaken under controlled stress or strain conditions. More sophisticated triaxial devices have been developed, which provide a means of measuring, for example, the hydraulic conductivity of sediment undergoing deformation (e.g. Hubbard and Maltman 2000). Normally, a test takes place to failure, or the development of a number of sub-parallel shear failures. In tests on low-strength, saturated sediments the sample may barrel, without brittle failure occurring, so failure is assumed to occur at 0.02 axial strain. The limited axial strain to which samples can be subjected in a triaxial test (typically 0.2) is one shortcoming of this approach, as determination of the residual strength may not be possible. However, it has recently been shown that for some glacial sediments (Ice Stream B) the limited axial strain developed in a triaxial rig may be sufficient for the sediment to reach its residual strength (Tulaczyk *et al.* 2000a and b). Test results are easily analysed by plotting a Mohr Circle for each test, with a best fit tangent to the circles representing the Coulomb failure envelope (Fig. 8.13b). From this values for c' and ϕ' can be obtained.

Ring shear apparatus

The ring shear apparatus was specifically designed for measuring the residual strength of soils as it can subject a sample to large strains (Bishop *et al.* 1971). It offers significant potential for the study of glacial sediments and is becoming more widely used for this purpose (Iverson *et al.* 1997, Tulaczyk 1999, Müller and Schlüchter 2001). A diagram of a ring shear device specifically designed for shearing glacial sediments is shown in Figure 8.14, although commercial ring-shear equipment is available and can also be used (e.g. Müller and Schlüchter 2001). The general test approach is to remould a till sample and pour it into the sediment chamber (Iverson *et al.* 1997, Müller and Schlüchter 2001) and then consolidate this incrementally to the desired degree. Vertical columns of displacement markers may be inserted into the sample and are later excavated to reveal the quantity, and distribution, of strain developed in the sample during shearing (Iverson *et al.* 1997). The remoulded sediment is confined by vertical sidewalls and sandwiched between perforated ribbed plates (in order to transmit the shear stress) backed by filter papers and/or porous stones (in order to transmit water). The sediment chamber is either contained within a bath that may be filled with water (Müller and Schlüchter 2001) or has connection to a reservoir which is 'in hydraulic communication' with the sample (Iverson *et al.* 1997). Strain localization can be dictated by the configuration of the vertical sidewalls; full depth sidewalls focus strain on the upper ribbed plate (Bishop *et al.* 1971), or split sidewalls, the lower and upper (Fig. 8.14), which focus strain on the boundary between them (Iverson *et al.* 1997). The sample is sheared by rotating (generally) the lower portion of the apparatus relative the upper portion, with varying rates of rotation available. The normal stress is applied via dead weights or hydraulics. Various load cells, proving rings, torque sensors and displacement transducers may then be used to measure

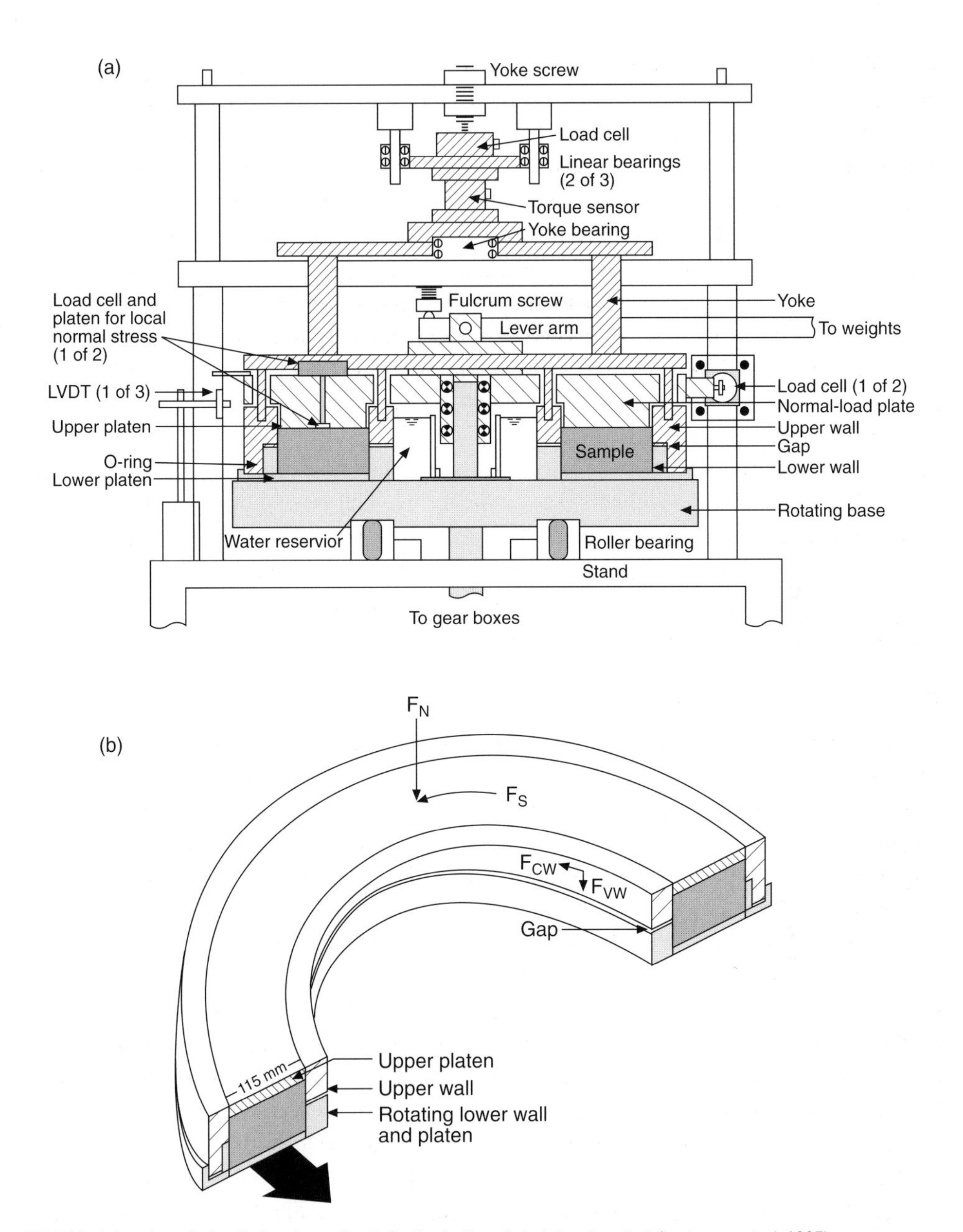

FIG 8.14 A ring shear device designed specifically for the testing of glacial sediments (after Iverson *et al*. 1997).

forces and displacements during the consolidation and shearing phases (Iverson *et al.* 1997). As with the direct shear box and triaxial rig, commercial device sizes are variable with the larger sized apparatus being most applicable for glacial sediments, as it allows more realistic particle size distributions to be tested. The ring shear apparatus shown in Figure 8.14 was designed and constructed specifically for testing glacial sediments, which is perhaps the best approach, as it accommodates larger samples (Iverson *et al.* 1997). One potential problem with the ring shear approach is the introduction of a strain gradient (lower on the inside and higher to the outside) due to the shorter diameter on the inside as opposed to the outside of the sediment chamber (Fig. 8.14), although account can be taken of this effect. As yet no ring shear device has been constructed that will allow elevated porewater pressures to be maintained. It is possible to develop elevated porewater pressures in sediments with low hydraulic conductivity under high shearing rates. However, in order to make use of this phenomenon the data must be recorded using miniature porewater pressure transducers, for example. The focus of the investigation will dictate the amount of strain required, but generally tests are run until the residual strength has been attained. Results are easily interpreted as the shear force is measured directly (Bishop *et al.* 1971).

One major disadvantage with all three of the apparatus sets described above is the particle-size range of the sediments that can be tested. The general practice is that particles larger than ~10% of the sample thickness should be excluded (Head 1989). In the smaller sizes of the apparatus described above this will make testing of undisturbed samples very difficult, as there will very likely be particles that exceed this 10% threshold. The problem is overcome by reconstituting the sample in the lab, minus the 'too large' particles (e.g. Müller and Schlüchter 2001) and/or by making the sample size larger (Iverson *et al.* 1997). However, by remoulding the sample in the lab the fabric will be lost, a factor that could significantly affect the shear strength or drainage characteristics of the sample. This can to some degree be overcome by deforming the sample until the residual strength is reached. However, reproducing the original fabric cannot be guaranteed so results may not be comparable to the field situation.

Shear vane

The shear vane is a very useful portable piece of testing equipment, which comes in both a frame-mounted variety (most suited to engineering site tests) and smaller handheld versions (of greater use in field based investigations), which are ideal for *in situ* testing (e.g. Sharp 1985; Benn 1995). The handheld shear vane has four orthogonal blades (height to diameter ratio 2:1) connected to a dial via a rod (Fig. 8.15). There are different sizes of blades depending upon the stiffness of the sample being tested: for softer sediments the larger blades are required. The vane is pushed vertically into the sediment until the desired depth of measurement is reached. The dial is rotated at constant speed and the sediment holds the vane firm until sufficient torque is developed to overcome the undrained shear strength of the sediment. At this point the sediment fails and the torque reduces, with the maximum reading obtained being recorded on the dial. The dial scale is zeroed and the test repeated, the final recorded value being an average of several readings. Extension rods allow vertical profiles of shear strength through significant depths of sediment to be obtained. The shear vane is of particular importance for testing sediments that would suffer significantly from sampling and is best used in fine-grained sediments. It provides a useful means of profiling reasonably thick sedimentary sequences where little or no exposure exists.

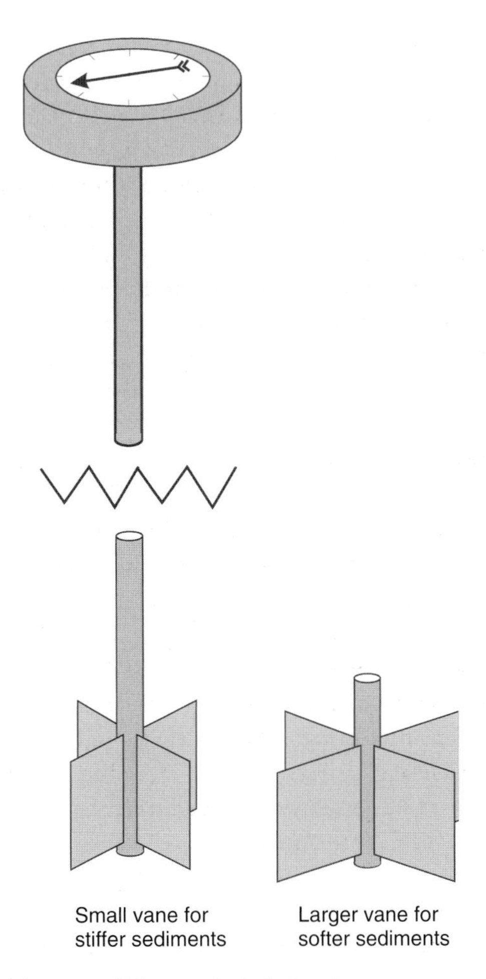

FIG 8.15 The portable, handheld shear vane, with two vane-head attachments.

8.7 *IN SITU* TESTING

Until now *in situ* has referred to undisturbed deposited sediments, but in the realm of glaciology, 'true' *in situ* conditions are really experienced at the base of glaciers and ice sheets beneath 10s, 100s even 1000s of metres of ice. The collection of data from such locations demands the use of a suite of equipment, not readily available to many researchers. Holes are generally 'drilled' to the bed using hot water drilling techniques, and instrumentation is introduced to the bed, down the borehole. The instruments required for making these *in situ* measurements thus require some ingenious engineering. Rapid developments, in both drilling and instrumentation, over the past 15–20 years have led to the collection of ever increasing quantities of *in situ* data and material, via down-borehole operations (e.g. Engelhardt *et al.* 1990; Blake *et al.* 1992; Fischer and Clarke

1994, 1997; Iverson *et al.* 1994; Hooke *et al.* 1997; Porter and Murray 2001). These types of studies have provided *in situ* estimates of till shear strengths (e.g. Iverson *et al.* 1994), hydraulic properties of subglacial tills (e.g. Hubbard *et al.* 1995; Fisher *et al.* 1998) basal sliding and till deformation rates (e.g. Engelhardt and Kamb 1998). The full details of these measurement procedures lies beyond the scope of this book, given the focus on Quaternary, i.e. deposited glacial sediments. However, the reader is encouraged to remember the value of contemporary process understanding for interpreting the sedimentary record in both recently exposed ice marginal and long deglaciated locations.

8.8 CONCLUSIONS

The application of engineering tests to the investigation of glacial sediments is becoming a more common procedure. This has been driven, in large part, by the recognition of the contribution that subglacial sediment deformation can make to ice velocity (e.g. Alley *et al.* 1986; Boulton and Hindmarsh 1987; Humphrey *et al.* 1993; Iverson *et al.* 1995; Engelhardt and Kamb 1998). So an understanding of the behaviour of subglacial sediment, 'till', has become imperative in modelling the dynamics of modern and former ice masses (e.g. Boulton and Hindmarsh 1987; Clarke 1987; Alley 1989, 1991; Kamb 1991; Tulaczyk *et al.* 2000a; 2000b). Combining a better understanding of the physical process of subglacial sediment deformation with detailed sedimentological, tectonic and micromorphological studies of glacial sediments, will lead to an improvement in our understanding of the role of subglacial sediment deformation in controlling the dynamics of glaciers and ice sheets. As more and more evidence points towards Coulomb-plastic rather than viscous fluid behaviour of till, a re-appraisal of the wholesale invoking of pervasive subglacial sediment deformation has been called for (Piotrowski *et al.* 2001). Further detailed gathering of engineering (geotechnical) properties including *in situ* and laboratory data by the glacial geologist will add significantly to the advancement of our understanding of subglacial bed deformation. This will allow better interpretation of glacial deposits, in terms of both the timing and extent of sediment deformation and its role in controlling the dynamics of glaciers and ice sheets.

The research project – a case study of Quaternary glacial sediments

Douglas I. Benn, David J.A. Evans, Emrys R. Phillips, John F. Hiemstra, John Walden and Trevor B. Hoey

9.1 INTRODUCTION

In this chapter, we show how a range of sedimentological techniques can be used in combination to build up a detailed reconstruction of former glacial environments, using a case study of sediments exposed in a quarry south of Loch Lomond, west central Scotland. The quarry is located at Drumbeg, immediately to the south-east of Drymen village (Fig. 9.1), where sections have been cut through one of a series of flat-topped ridges that lie inside the terminal moraine of the Loch Lomond Readvance. The ridges were formed at the eastern margin of the Loch Lomond glacier, where the ice terminated in a proglacial lake, known as Lake Blane. The general stratigraphic sequence in the area has been central to reconstructions of late Quaternary palaeoenvironments in west central Scotland (Rose 1981, 1989a; Browne and McMillan 1989; Gordon 1993). Detailed sedimentological studies of Drumbeg quarry have been reported by Benn and Evans (1996), Phillips *et al.* (2002), Phillips *et al.* (2003) and Hiemstra *et al.* (2003).

9.2 RESEARCH AIMS AND METHODS

During the Younger Dryas (Loch Lomond) Stade, an extensive ice cap and glacier transection complex developed in the western Scottish Highlands, and numerous outlet glaciers extended into the surrounding lowlands, including the Loch Lomond basin. Moraines and other depositional landforms deposited by these glaciers provide important records of former glacier behaviour and palaeo-climate at this time. The aim of this case study is to reconstruct former depositional processes and environments for part of the margin of the Loch Lomond glacier,

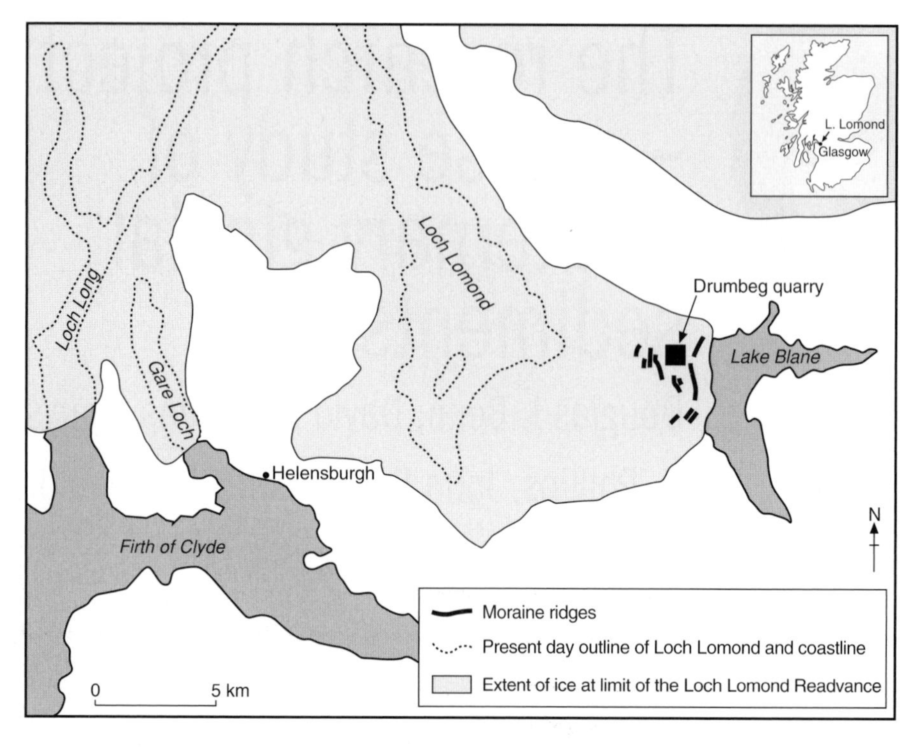

FIG 9.1 Palaeogeography of Drumbeg site during deposition of the ice-contact delta and overlying till.

where the sedimentary record is exceptionally well exposed. In particular, the study aims to demonstrate how sedimentological and structural data can be used to gain insights into former glacier behaviour, and its possible relationship with climatic forcing.

Section logs (2-D drawings and vertical profiles) are used to illustrate the lithofacies and lithofacies associations exposed in the quarry, and form the basis of a preliminary environmental interpretation. Then, specific hypotheses are developed and tested using a variety of sedimentological techniques, with particular emphasis on subglacial depositional and deformational processes. This second phase of the investigation shows how a hypothesis-testing approach can be used to arrive at a detailed understanding of sedimentary systems and their controls.

9.3 LITHOFACIES AND LITHOFACIES ASSOCIATIONS

Detailed logs were made on two occasions, in 1995 and 1999. In 1995, a 2-D drawing was made of the main quarry face showing the large-scale architecture of the lithofacies, supplemented with vertical profile logs at four representative locations (Fig. 9.2). In addition, more detailed sketches were made of the complex ice-proximal sediments

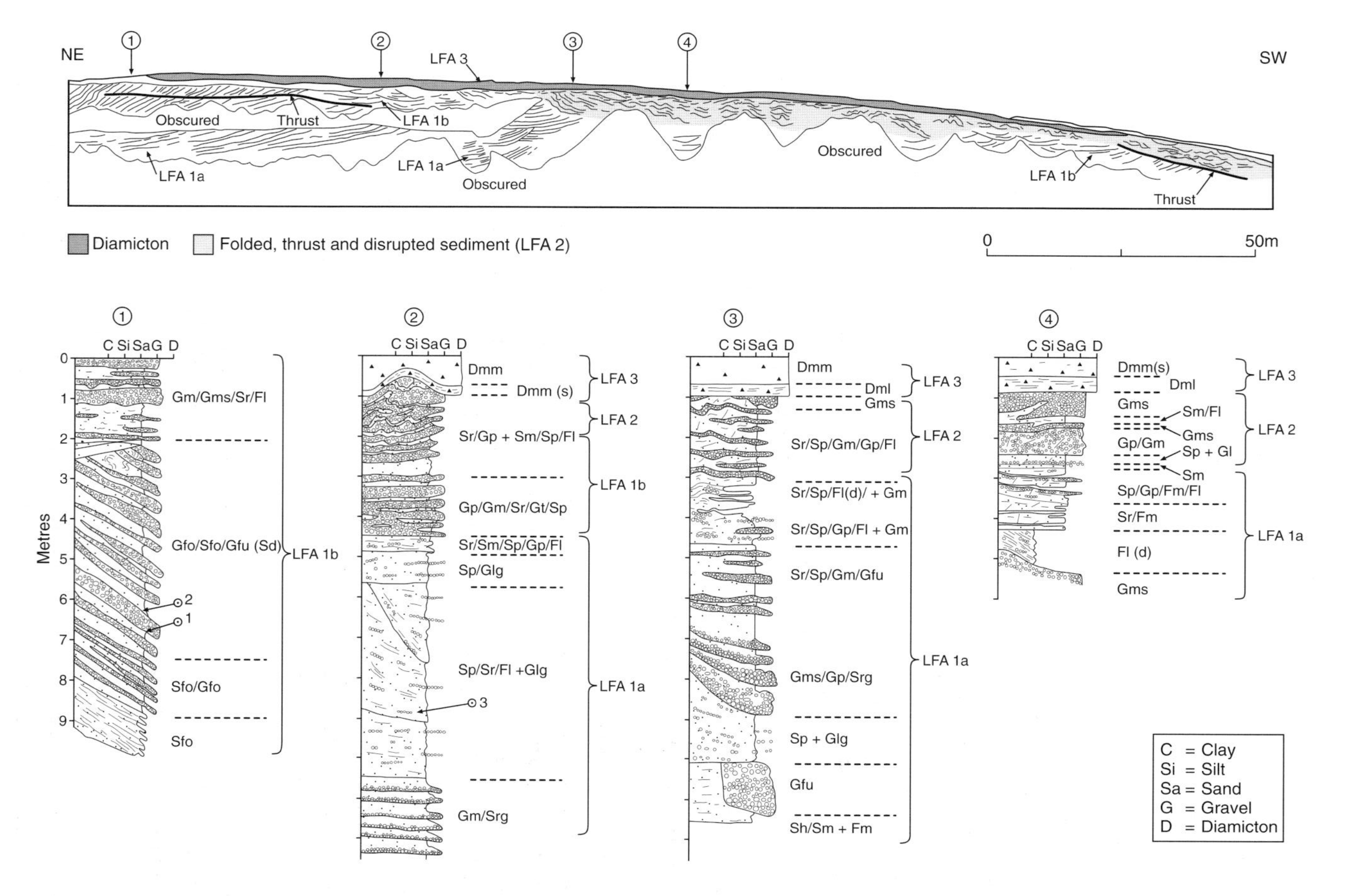

FIG 9.2 Two-dimensional section sketch of the 1995 quarry exposure showing large-scale architecture of the lithofacies together with vertical profile logs at four representative locations.

exposed in the western part of the face in 1995 (Fig. 9.3). New sections of the ice-proximal sediments were exposed in 1999, and were recorded in 2-D drawings and annotated photographs (Fig. 9.4). The lithofacies observed in the quarry are now summarized and grouped into lithofacies associations (LFAs).

9.3.1 Description

LFA-1: Gravels and sands

Gravel and sand lithofacies were exposed in much of the quarry in 1995 (Fig. 9.2). In the ice-distal (north-eastern) part of the section, the lowermost sediments consist of low-angle sandy clinoforms dipping towards the north-east. A variety of internal structures are present, including bed-parallel lamination, climbing ripple cross-lamination (Type A and Type B) and normal grading, as well as occasional thin gravel lags between beds. The sandy clinoforms are overlain by more steeply dipping gravelly clinoforms, consisting of well-sorted gravels with locally well-developed normal grading, and sand interbeds. In places, the gravel units contain soft-sediment intraclasts of silt and fine sand. Overlying the gravelly clinoforms is 1–4m of horizontally-bedded gravels, sands and silts. These include tabular beds of massive gravel, planar and trough cross-bedded gravels and sands, and ripple cross-laminated sands and silts. Lower contacts of units are conformable or form shallow, concave-up scours.

In the lower, central and south-western parts of the section, the gravel and sand facies form more complex cross-cutting units, recording repeated phases of erosion and deposition. Lithofacies include planar bedded, massive and matrix-supported gravels with discontinuous sand and silt drapes, in places interstratified with cross-bedded sands. Dropstones and sparse rafts of graded gravel were recorded within some of the sandy cross-bedded deposits. Near the top of the section, low-angle thrust faults cut through LFA-1, with displacements of several dm to the north-east. More extensive deformation occurs in the south-western part of the quarry, where LFA-1 is locally folded and faulted (Fig. 9.4).

LFA-2: Pervasively-deformed gravels, sands and silts

In the south-western (ice-proximal) parts of the quarry, pervasively-deformed gravels, sands and silts were exposed in 1995 and 1999 (Figs. 9.2, 9.3 and 9.4). The sediments have been deformed into a melange, consisting of numerous detached sand and gravel lenses and pods enclosed within an anastomosing network of laminated diamicton, sand and silt (Figs. 9.3 and 9.4). Deformation is highly heterogeneous. The most intense deformation affects the fine-grained materials, in which sedimentary lamination is transposed by a well-developed, finely-banded tectonic fabric. Deformation structures include NE-dipping shears, normal and reverse faults, water escape conduits and loading structures. This locally intense deformation is in marked contrast to the enclosed sand and gravel lenses and pods, in which primary sedimentary structures (e.g. lamination and cross-bedding) are commonly well preserved. Some lenses and pods have been partially disaggregated, and have diffuse margins.

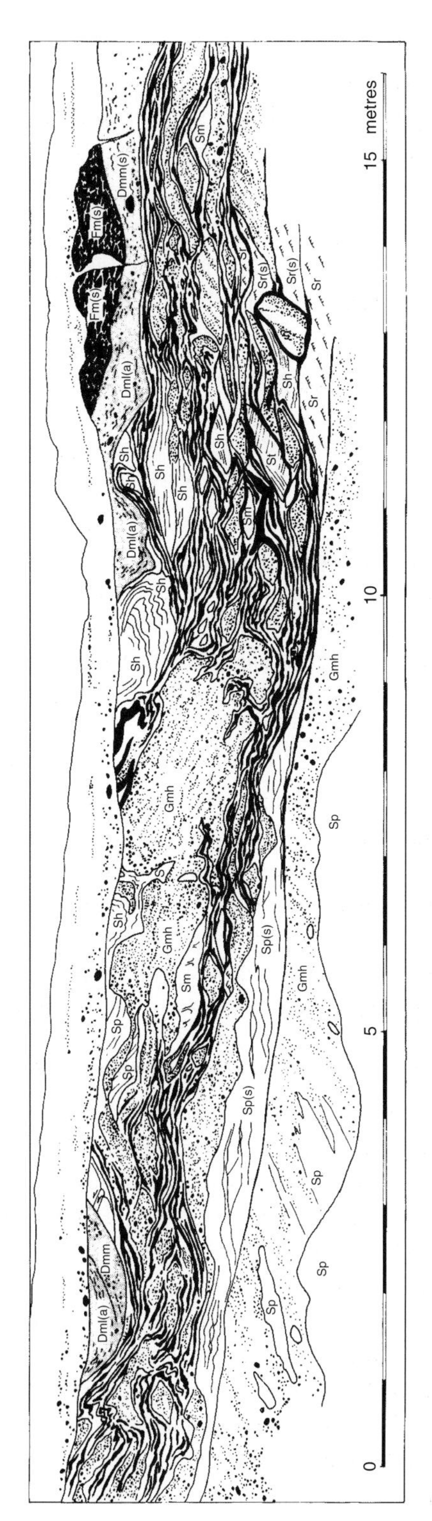

FIG 9.3 Detailed sketch of structures in the western part of the 1995 exposure, showing complex deformation structures in the gravel, sands and silts of LFA 2 (from Benn and Evans 1996).

LFA-3: Massive to laminated diamicton

LFA-1 and 2 are overlain by a laminated to massive, matrix-supported diamicton. This pinches out towards the east end of the 1995 section face (Fig. 9.2) and is generally less than 2.5m thick. The contact between the diamicton and the underlying stratified sediments is erosive and sharp. However, the base of the diamicton is typically laminated, incorporating numerous lenses of sand, fines and gravel that have been stretched and contorted. Locally, these lenses thicken into pods with tapered ends enclosed within laminated diamictons, silt and mud. Above the basal zone, much of the diamicton appears to be massive, with no apparent macroscopic structure. Careful examination of the diamicton in the 1999 section (Fig. 9.4) showed that at this locality it comprised four inclined sub-units or slabs separated by planar, sharp contacts. The contact planes between the slabs dip towards W to WNW, and the base of each slab is finely laminated, grading upward into massive diamicton. The slabs are variably tabular and wedge shaped, and are 0.18–0.59m thick.

LFA-4: Gravels, sand and diamicton

The uppermost sediments at the quarry comprise horizontally-bedded gravel and pebbly gravel fining upwards into thinly bedded sand, and overlain by a thin massive, matrix-supported diamicton. In places this diamicton lies directly over the massive to laminated diamictons of LFA-3 but is distinguishable by its more open structure and numerous sub-horizontal partings.

9.3.2 Relationships between lithofacies associations

The structural relationships between the four lithofacies associations can be seen in Figures 9.2, 9.3 and 9.4. The north-eastern (ice-distal) part of LFA 1 consists of clinoforms overlain by horizontally-bedded gravels and sands. To the south-west, the clinoforms pass laterally into more complex cross-cutting units, with multiple cut and fill structures, which are locally folded and faulted. At the south-western end of the quarry, LFA-1 is sharply overlain by the deformed sediments of LFA-2. This consists of a melange of dislocated sediment pods within laminated diamicton, sand and silt, forming a unit up to 2m thick which pinches out towards the north-east. LFA-1 and LFA-2 are erosively overlain by LFA-3 (laminated to massive diamicton), which extends further north-east than LFA-2 before pinching out. Finally, LFA-4 rests unconformably on LFA-2 and 3, its lower boundary being marked by a prominent erosion surface.

9.3.3 Initial interpretation

The vertical succession of (1) low-angle sandy clinoforms, (2) steeper gravelly/sandy clinoforms and (3) upper, horizontally-bedded gravels and sands found in the north-eastern part of LFA-1 is interpreted as the bottomset, foreset and topset components of a

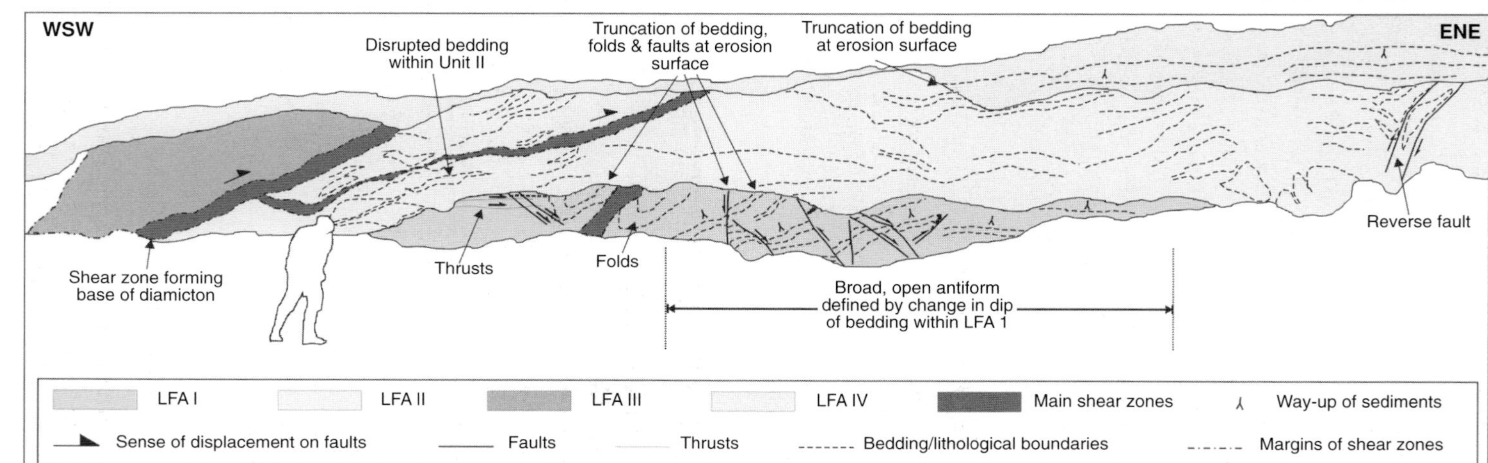

FIG 9.4a Annotated photograph of the 1999 exposure showing main lithofacies associations, and locations of grain-size samples (samples 1, 2 and 3 from foreset gravels, foreset sands and bottomset muds not shown).

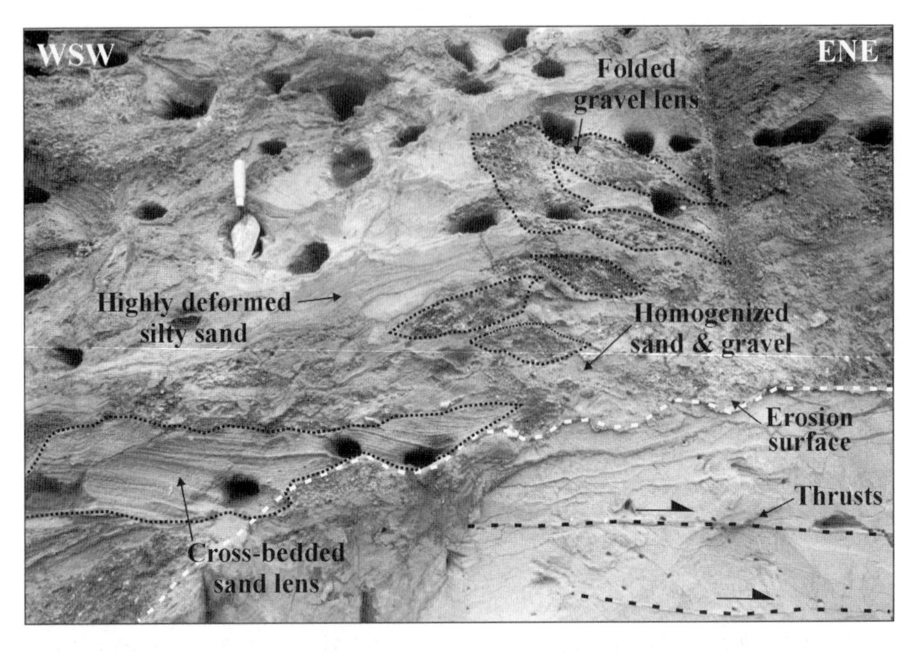

FIG 9.4b Detail of disrupted bedding at the contact between LFA's 1 and 2 (located on Fig. 9.4a). Note the variation in intensity of deformation within the sand and silty sand and the folded gravel lenses.

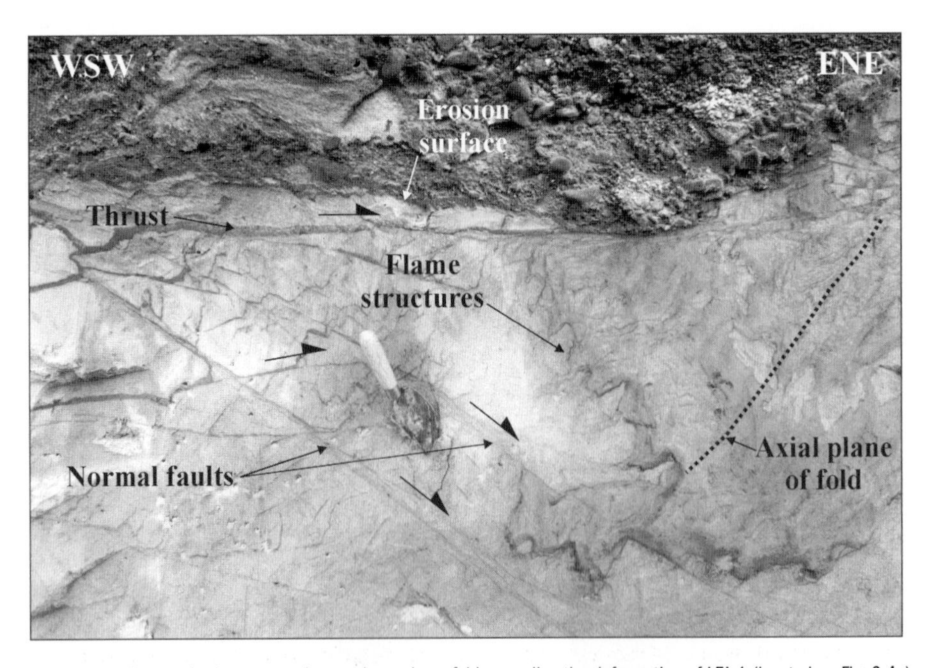

FIG 9.4c Detail of low angle thrusts superimposed on a large fold, recording the deformation of LFA 1 (located on Fig. 9.4a). The structures are truncated by an erosion surface at the base of LFA 2.

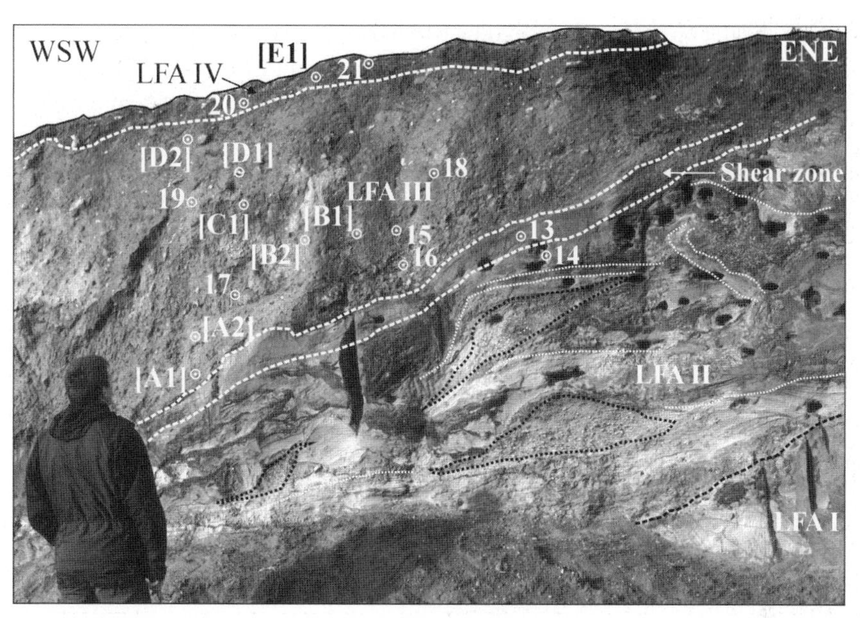

FIG 9.4d Annotated photograph of the left hand part of the section illustrated in Fig. 9.4a, showing lithofacies associations, grain-size samples (circled dots) and clast macro-fabric locations (bracketed numbers classified according to sub-units of LFA 3 (till slabs A–D; see Fig. 9.4e)).

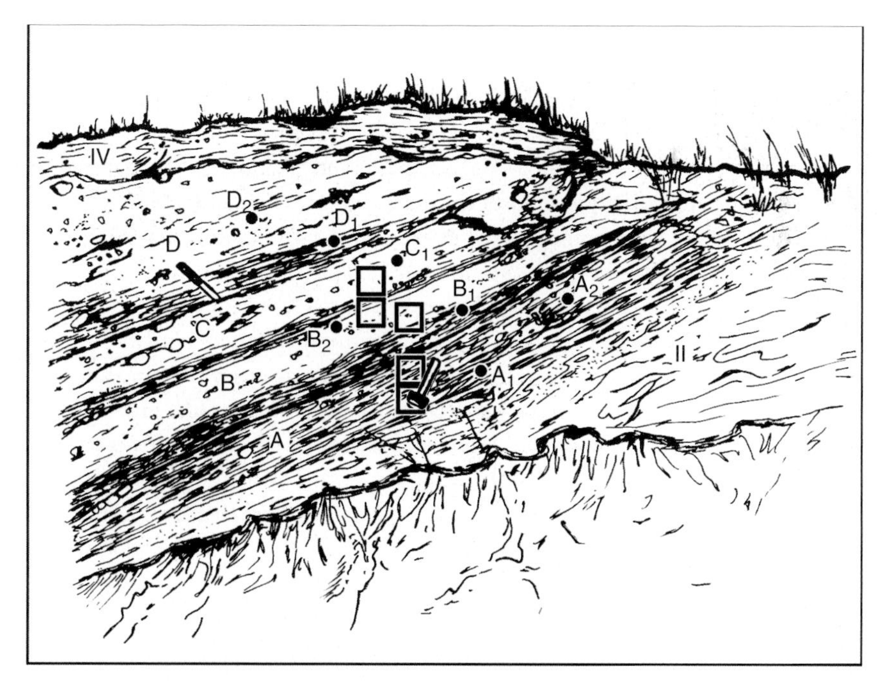

FIG 9.4e Sketch of the LFA sub-units (till slabs A–D; illustrated in Fig. 9.4d) as identified by internal, macro-scale structures. Locations of clast fabrics (same notation as in Fig. 9.4d) and thin section samples (boxes) are labelled. The boxes in sub-unit A are A_{L1} (lower laminated) and A_{L2} (upper laminated), in sub-unit B are B_L (laminated) and B_M (massive) and in sub-unit C is C_M (massive).

Gilbert-type delta. Topsets are deposited by braided streams on the subaerial delta top, foresets represent sediment that avalanches down the steep frontal slopes of the delta, and bottomsets are deposited from sediment-laden underflows (turbidity currents) which intermittently extend into deep water beyond the delta front. Progradation of the delta results in foresets being deposited on top of bottomsets, and topsets on top of foresets, thus providing an excellent example of *Walther's Law*, in which an unbroken vertical succession of sediments records the lateral migration of neighbouring sedimentary environments (see Chapter 2). In this case, the delta was fed by glacial meltwater, and prograded into ice-dammed Lake Blane (Fig. 9.1).

The more complex structure of the lower and more south-western parts of LFA-1 is typical of a more ice-proximal environment. We interpret the cross-cutting units as erosional subaqueous mass-flows (e.g. scour fills of matrix-supported and graded gravel) interspersed with the rain-out of suspended sediments and ice-rafted debris. The sedimentary architecture and abundant evidence of ice-contact deposition and iceberg rafting suggests that deposition occurred in a series of coalescent subaqueous fans fed by a subglacial drainage system. This interpretation is consistent with the evidence for a Gilbert-type delta; subaqueous fans will evolve into deltas if the glacier terminus is relatively stable and progressive deposition builds sediment up to the waterline.

LFA-2, the melange of dislocated sand and gravel lenses and pods enclosed within laminated diamicton, sand and silt, is interpreted as glacially-deformed sorted sediments. The original sediments appear to have been similar to those of LFA-1, which were sheared by advancing ice. Deformation structures penetrate and overprint the original sedimentary structures, suggesting that LFA-2 is a *Type A* or *penetrative glacitectonite* (Banham 1977; Benn and Evans 1996). The tabular geometry of the massive to laminated diamicton (LFA-3) and its locally gradational relationship with the underlying glacitectonite, indicate that it is a subglacial till. Benn and Evans (1996) suggested that the till is the highly strained equivalent of the glacitectonite; that is, it consisted of similar source sediments which were homogenized in the upper, more mobile parts of a subglacial deforming layer. This idea is tested in Section 9.4.1 below.

The gravels, sands and diamicton of LFA-4 are interpreted as sediment deposited during and following the withdrawal of glacier ice. The lowermost sediments in LFA-4 consist of sands and gravels. These must have been deposited when glacier ice remained in contact with the sediment pile, as there is no alternative stream catchment. The gravels and sands pass upward into non-fluvial sediments, which are interpreted as wind-blown sands and slope deposits (colluvium and slopewash).

From field observations of lithofacies character and architecture, we have arrived at an initial environmental interpretation of the sediments exposed in Drumbeg quarry. This suggests the following sequence of events:

1| glacier meltstreams deposited subaqueous fans at the ice margin, below the surface of Lake Blane;

2| aggradation of the subaqueous fans built sediment up to the waterline, evolving into a Gilbert-type delta, which prograded into the lake;

3| the glacier margin over-rode the south-eastern part of the fan-delta, deforming the sediments into a glacitectonite and depositing a basal till;

4| following withdrawal of the ice, the fan-delta surface was reworked, initially by glacial meltwater and then by aeolian and slope processes.

9.4 DETAILED SEDIMENTOLOGICAL STUDIES

The proposed sequence of events raises a number of interesting questions, particularly about the character of the complex sub-marginal environment. Specifically:

- what is the relationship between LFA-2 and LFA-3 (the glacitectonite and the till)?
- what is the origin of the till?
- were the deformations at the site caused by a single glacier readvance, or a more complex polyphase sequence?

These questions can be addressed by using detailed sedimentological techniques to test competing hypotheses. By taking this approach, we collect data with the explicit purpose of refining our understanding of the system, in a series of purpose-designed experiments. As we will see, this approach is considerably more powerful than the routine application of standard techniques.

9.4.1 Is the diamicton (LFA-3) the highly strained equivalent of the glacitectonite (LFA-2)?

As noted in Section 9.3.3, Benn and Evans (1996) argued that the laminated to massive diamicton (LFA-3) is a subglacial deformation till, and suggested that it is the highly-strained equivalent of the underlying glacitectonite. We can test this idea using data on particle-size distributions (Chapter 3), clast morphology (Chapter 4) and clast lithology and mineral magnetic susceptibility (Chapter 7). All samples were taken from the 1999 section (Fig. 9.4).

If the diamicton was derived from the same source sediments as the glacitectonite, we should expect clasts in both deposits to have similar *morphological characteristics* and *lithologies*, or that they should differ in ways compatible with a greater degree of modification by shearing in the till compared with the glacitectonite. To test this prediction, 150 clasts were taken from each LFA for lithological analysis, and three samples of schist and sandstone clasts were taken from each LFA for morphological analysis (n = 42–47). The samples were taken in close proximity to each other at the 1999 section.

Examination of the data shows no systematic differences in clast morphology between the two rock types. There are, however, clear differences in clast morphology between LFA-3 (diamicton) and the underlying glacitectonized gravels of LFA-2 (Fig. 9.5). LFA-

2 exhibits a broader range of clast form, with C40 values (% of clasts with c:a ratios ≤ 0.4) in the range 16–52%, in contrast with 30–38% in the till; and contains higher percentages of rounded and well-rounded clasts (28–46%) than the diamicton. Note that the percentage of rounded and well-rounded clasts is used as a summary statistic, not the RA index used by Benn and Ballantyne (1994) in their study of glacial sediment transport in Norway. This is because no angular or very angular clasts occurred in either the till or glacitectonite samples, so the RA index is useless as a discriminant statistic in this case. Examination of roundness histograms, however, indicates that there are systematic differences in the proportions of rounded and well-rounded clasts between the two sets of samples, suggesting that %R + WR is a useful statistic. There is also a good theoretical basis for this choice, because fluvial transport produces rounded pebbles, whereas subglacially modified clasts tend to be predominantly sub-angular or sub-rounded.

Both the diamicton (till) and the glacitectonite contain: a) sandstone and grit derived from Devonian strata south of the Highland Boundary Fault; b) Dalradian mica schist from north of the HBF; c) white quartz, derived from either quartz veins in the Dalradian or pebbles in conglomeratic facies in the Devonian and d) igneous rocks from minor intrusions (Table 9.1). The diamicton (till), however, contains a higher percentage of sandstone clasts, and a lower percentage of schist clasts, than the gravels.

The clast form and lithological data are not compatible with the hypothesis that LFA-3 is the highly-strained, well-mixed equivalent of the underlying glacitectonite. If the hypothesis were valid, we should expect the lithological and morphological characteristics of the two facies to be similar, or vary in a way compatible with the modification of the till clasts by shearing. Modification by shearing, however, cannot explain the excess of

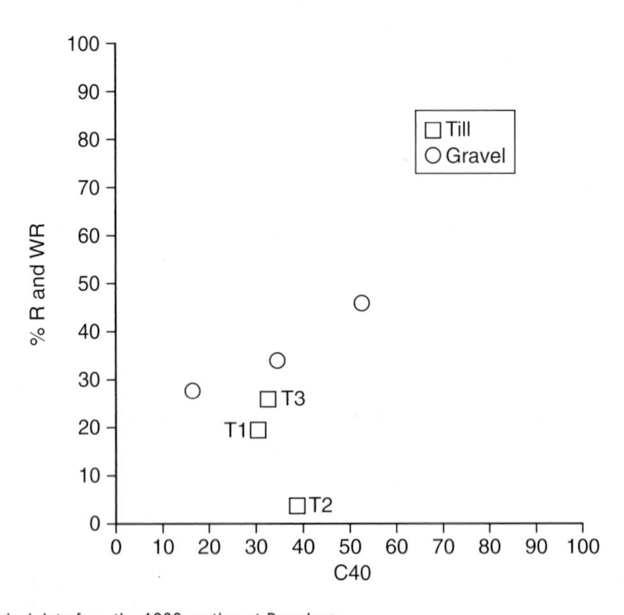

FIG 9.5 Clast morphological data from the 1999 section at Drumbeg.

Table 9.1: Clast lithology of LFA 2 and LFA 3.

	Sandstone	Schist	Vein Quartz	Igneous
Gravels (LFA-2)	47	44	8	1
Diamicton (LFA-3)	63	29	7	1

sandstone clasts in the till relative to the glacitectonite (Table 9.1). Instead, it is probable that the relatively low proportion of sandstone in the glacitectonite reflects fluvial transport processes in the parent gravels. High energy flow would rapidly have worn down the softer lithologies (sandstone), leaving the gravels relatively enriched in durable metamorphic rocks. The morphological data tell a similar story. If the till were derived from the glacitectonite, then the constituent clasts would have to have become less well-rounded in the process. This could have been accomplished by fracture, although only one broken, rounded clast was found in all three till samples, compared with three from the glacitectonite. In contrast, the greater clast roundness in the glacitectonite compared with the till can be explained if the parent gravels (and subsequently the glacitectonite) were derived from till (or basal debris with a similar composition), with additional rounding occurring during subglacial or proglacial fluvial transport. The clast lithological and morphological data thus falsify the hypothesis that the till and the glacitectonite are closely genetically related.

Mineral magnetic data provide striking support for this conclusion. Samples from the unmodified sands and gravels of LFA-1 and the glacitectonite (LFA-2) both have low concentrations of magnetic minerals (low χ, χ_{arm} and SIRM, for example) and their soft IRM, hard IRM and -40 mT backfield ratios suggest the presence of high coercivity minerals such as hematite (Table 9.2). In contrast, the massive parts of the till (LFA-3) have higher concentrations of magnetic minerals, and greater proportions of low coercivity minerals such as magnetite. These observations are consistent with derivation of the glacitectonite (LFA-2) and the till (LFA-3) from somewhat different assemblages of source rocks.

The compositional relationships between the glacitectonite and the laminated parts of the till appear more complex. Small-scale compositional variability in stratified glacigenic sediments is commonly observed (e.g. Walden *et al.* 1992a). Even so, where the laminated and massive components of the till were sampled in close proximity across a contact

Table 9.2 Summary mineral magnetic properties from LFA-1 to 3, Drumbeg.

Description	Number of samples	χ	χ_{arm}	SIRM	soft IRM	hard IRM	−40 mT IRM ratio
Massive till (LFA-3)	3	92.0	27.0	1321.5	518.2	660.8	0.22
Laminated till (LFA-3)	2	56.9	13.6	749.0	270.8	374.5	0.28
Glacitectonized sand (LFA-2)	4	55.8	19.2	968.7	353.5	484.3	0.29
Unmodified sand (LFA-1)	4	56.2	11.5	823.4	355.2	411.7	0.14

(samples 13/14 and 15/16), a marked compositional difference is observed. The laminated till samples show mineral magnetic properties much closer to those of the glacitectonite than the massive till, suggesting that they are derived by re-working of the underlying unit. In conclusion, the mineral magnetic data show that the glacitectonite is closely similar to the underlying sands and gravels of LFA-1, whereas the till (LFA-3) appears to have originated from different source materials. The basal, laminated parts of the till show intermediate properties, but have closer affinities with the underlying glacitectonite, indicating some degree of mixing at the contact.

Summary *particle size* data from the 1999 section are shown in Figure 9.6, and clearly show systematic differences between the four lithofacies associations. The whole sample data show generally coarser median grain sizes in the till (LFA-3) than the glacitectonite (LFA-2) or LFA-1, although some samples from LFA-1 and LFA-2 are as coarse as the till samples (Fig. 9.6a). The till and glacitectonite exhibit finer median matrix sizes than LFA-1 (Fig. 9.6b). Taken together, these characteristics indicate that the till is the least sorted, LFA-1 is the best sorted, and LFA-2 generally has intermediate values. This observation is confirmed by the data on the proportions of clay, silt and sand in the samples (Fig. 9.6d), and the sample standard deviations shown in Fig. 9.6e (higher standard deviations indicate poorer sorting). Skewness is generally negative in LFA-1 and LFA-2, and close to zero in the till samples. These characteristics indicate that LFA-1 and LFA-3 are dissimilar, and that LFA-2 exhibits intermediate characteristics. Taken in isolation, the particle-size data may be taken as support for the mixing model proposed by Benn and Evans (1996), as the range of grain sizes in the till is also present in the underlying glacitectonite. However, as we have seen, the clast lithological and morphological data and the mineral magnetics are incompatible with a simple mixing model. Thus, although the till has a similar range of grain sizes as the underlying glacitectonite, they do not appear to have been derived from identical sources. This result highlights a potential pitfall in the hypothesis-testing approach: some methods may yield results which appear to be compatible with our original hypotheses, whereas others may provide conclusive evidence to the contrary. Thus, hypotheses should be tested using as broad a range of techniques as possible.

9.4.2 What is the origin of the till?

The foregoing has shown that the till is not simply a highly-strained, homogenized equivalent of the underlying glacitectonite. Instead, the till appears to have been transported to the site and deposited on top of the glacitectonite with some degree of mixing only in a relatively narrow transition zone. As noted in Section 9.3.1, the till at the 1999 section consists of four inclined slabs (A-D) dipping towards the WNW, each consisting of finely-laminated diamicton grading up into massive diamicton. This structure suggests the possibility that the till was emplaced in four separate events, perhaps by the process described by Krüger (1993, 1994) for the margins of Icelandic glaciers. In Krüger's model, till beneath the margin of the glacier is frozen on to the glacier sole in the early winter, is carried forward during the winter advance, then melts out in its new

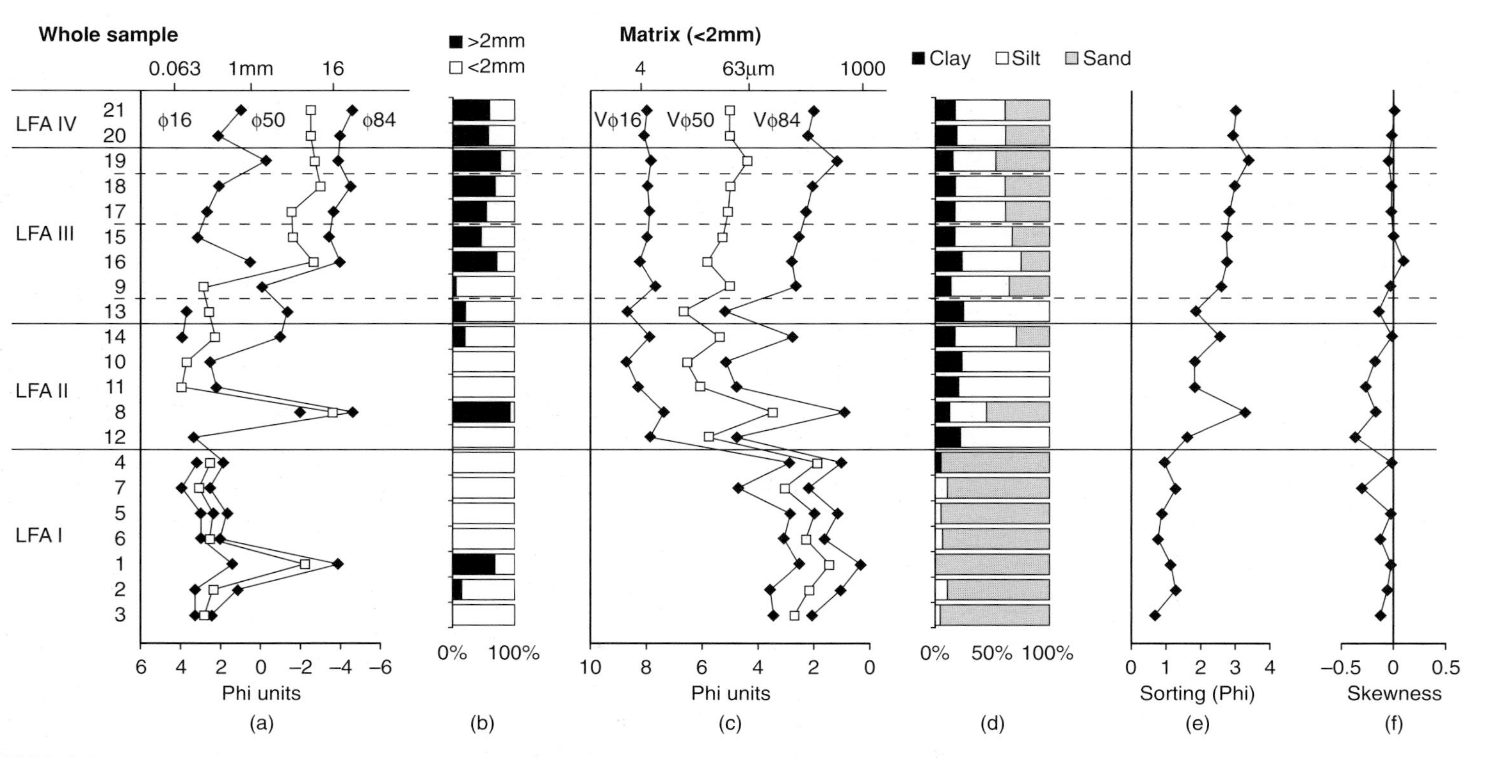

FIG 9.6 Grain-size data from bulk sediment samples. (a) + (b) show dry-sieving results; (c)–(f) are from laser diffraction analysis of the < 2mm (0 phi) fraction. Due to the fact that the two techniques measure different properties of the sediment, results from dry-sieving are not directly comparable to those from laser diffraction. (a) Median ($\phi50$), 16th and 84th percentiles of the bulk grain-size distributions; (b) Percentages of the bulk samples above and below 2mm (0 phi); (c) Median, 16th and 84th percentiles of the <2mm size fraction in each sample. These percentiles are based on the grain volume converted to a diameter, hence the notation $V\phi16$ etc.; (d) Proportions of sand, silt and clay in the matrix samples; (e) Sorting of the <2mm fraction, expressed as the graphic standard deviation (equation 5a); (f) Skewness of the <2mm size fraction expressed as the graphic skewness (equation 6a). Samples are plotted in stratigraphic order, with sample numbers shown.

position the following summer. Repetitions of the cycle produce a stacked sequence of till slabs at the glacier margin. We express this model in the form of a testable hypothesis: *LFA-3 at Drumbeg was emplaced by the melt-out of successive thrust slabs of frozen till.*

To test this hypothesis, we need to have a clear idea of its predictions. In other words, what evidence would we expect to find if this hypothesis were true, and conversely, what evidence could be used to falsify the hypothesis? Two possibilities immediately suggest themselves: (1) if the slabs were emplaced in separate events, we might expect the slabs to have different properties, particularly directional properties (macrofabric); (2) if the slabs were emplaced when frozen and then melted out *in situ*, we should expect to find extensive dewatering structures within each slab, recording the evacuation of meltwater. Since the till slabs are structureless at the macro-scale, examination of micromorphology could provide a way of testing this prediction. There is, however, a difficulty with prediction (2), which is that dewatering structures would also be expected to occur if the till was deposited in a saturated state, rather than a frozen one. It is important to make sure that the predictions of a hypothesis are unique, because if they are not, observations will not necessarily mean what we think they mean. In the present case, we should take care not to infer that the till slabs were frozen unless there is additional firm, unequivocal evidence.

Macrofabrics

The first prediction was tested by collecting seven clast fabric samples from the till slabs (Fig. 9.4). Each sample consisted of the a-axis orientations (azimuth and dip) of 50 clasts, which were plotted on equal-area stereonets using the Rockware computer programme (Fig. 9.7). The clast fabric from the laminated basal part of slab A (A1, Fig. 9.4) forms a well-developed westerly-orientated cluster, whereas that from the massive upper part of the slab (A2), is a slightly weaker cluster oriented towards the SW. Similarly, the a-axis fabrics from the basal, laminated part of slab B (B1) form a strong westerly-orientated cluster, and that from the macroscopically massive upper part of the slab has a less clustered, more girdle-like distribution with a mean orientation towards the WNW. Sample C1, from the central, macroscopically massive part of slab C, is strongly clustered towards the WNW. Sample D1, from the basal laminated zone of slab D, forms a strong W–WNW-orientated cluster, and D2 (from the massive upper part) is a strong WNW–NW-orientated cluster.

The fabric data, when plotted on a fabric-shape ternary diagram (Fig. 9.7), are comparable with those from shear zones in the glacitectonite (LFA-2) presented by Benn and Evans (1996). When considered in detail, the data reveal some interesting patterns. First, fabric maxima from the laminated basal zones of slabs A, B and D are oriented towards the W or W–WNW, closely similar to the direction of dip of the associated contact planes between the slabs. Second, in each case, fabric maxima from the macroscopically massive, upper parts of the slabs deviate from those of the underlying laminated zones. Third, in slabs A and B fabric strength decreases from bottom to top: the principal eigenvalues (S_1) are 0.68 and 0.66 at the base, and 0.59 and 0.58 in the massive parts of slabs A and B, respectively. Fabric strengths from the laminated and

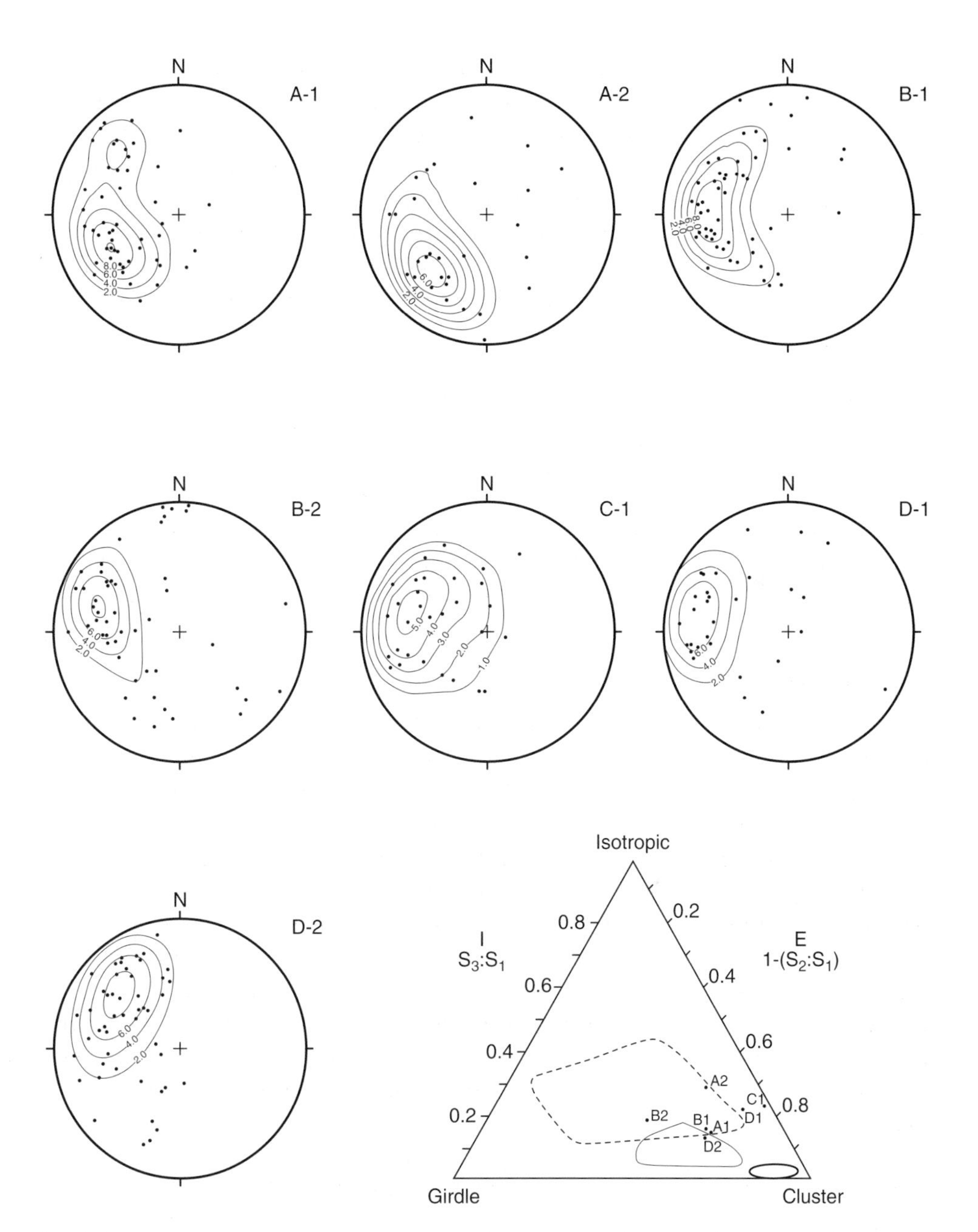

FIG 9.7 Clast macrofabrics from the till units at Drumbeg (samples located on Fig. 9.4d and e). Clast form triangle includes envelopes for previously sampled subglacial tills in Iceland (broken line = upper till horizon; thin solid line = lower till horizon; thick solid line = lodged clasts).

massive parts of slab D are similar (S_1 = 0.66 and 0.68), but there is a considerable difference in preferred orientation (~ 18°).

These patterns are compatible with the hypothesis that the slabs were emplaced in four separate events. The strong fabrics in the basal laminated zones, and the close similarity of their fabric maxima and the direction of dip of the underlying contact planes, indicate that the laminated diamictons represent shear zones over which the slabs were thrust. Conversely, the massive parts of the slabs exhibit a range of preferred orientations (differing from each other and the intermediate shear zones), suggesting that they represent slices of till with varying properties. Under this interpretation, the fabric characteristics of the massive parts of the slab predate the thrusting events, whereas those of the laminated zones were imposed during slab emplacement. The varying directions of the fabric maxima within the slabs need not imply varying ice-flow directions during formation of the parent till, because fabric maxima in tills may deviate significantly from the flow direction of overpassing ice, and may exhibit considerable variation in strength (Section 5.5.1).

Micromorphology

To further investigate the origin of the till, five Kubiëna-size (8 × 6cm) samples were taken from slabs A, B and C (Fig. 9.4e), impregnated with an unsaturated polyester resin and thin sectioned (Fig. 9.8). Two samples were taken from the laminated part of slab A (A_{L1}, A_{L2}), one each from the massive and laminated parts of slab B (B_M, B_L), and one from the massive part of slab C (C_M). The five thin sections were examined under an optical microscope using plane-polarized and cross-polarized light at up to 35 × magnification.

The matrix of the diamict appears to be uniform and has a low density (Fig. 9.8a). The lamination in samples A_{L1}, A_{L2} and B_L is evident in thin section. It comprises silty to sandy diamict laminae and relatively sorted silty to sandy bands or lenses. Locally the diamict laminae show irregular clayey to silty lenses or wisps, or thin, laminar grain concentrations, which occasionally account for an anastomosing type of internal lamination (Fig. 9.8c). Contacts between laminae and lenses may be sharp or gradational. Plasmic fabrics in the diamict matrix are poorly developed in all samples, both in laminated and massive parts of the slabs. This may be related to low overall clay content, but is more likely to reflect non-directional (i.e. random) orientation of clay particles.

None of the samples show a distinctive microfabric signal. However, arcuate and turbate (circular) silt and sand grain arrangements were observed in all samples. They are abundant in samples A_{L1} and A_{L2}, common in C_M, but relatively sparse in B_M and B_L. Diameters of the features are between 1–3mm, and occasionally they seem to be associated with short lineaments, which consist of at least three linearly aligned grains (Figs. 9.8a–b; see Hiemstra and Rijsdijk 2003). Although soft sediment intraclasts are rare, grains or clasts with diffusely bounded fine-grained casings or rims occur in all samples (Fig. 9.8d). In sample B_M, the diamict consists exclusively of a loose packing of such aggregates (up to 3mm diameter), some of which seem to have coalesced to form loosely defined clusters, with irregularly shaped, but sharply defined inter-aggregate spaces.

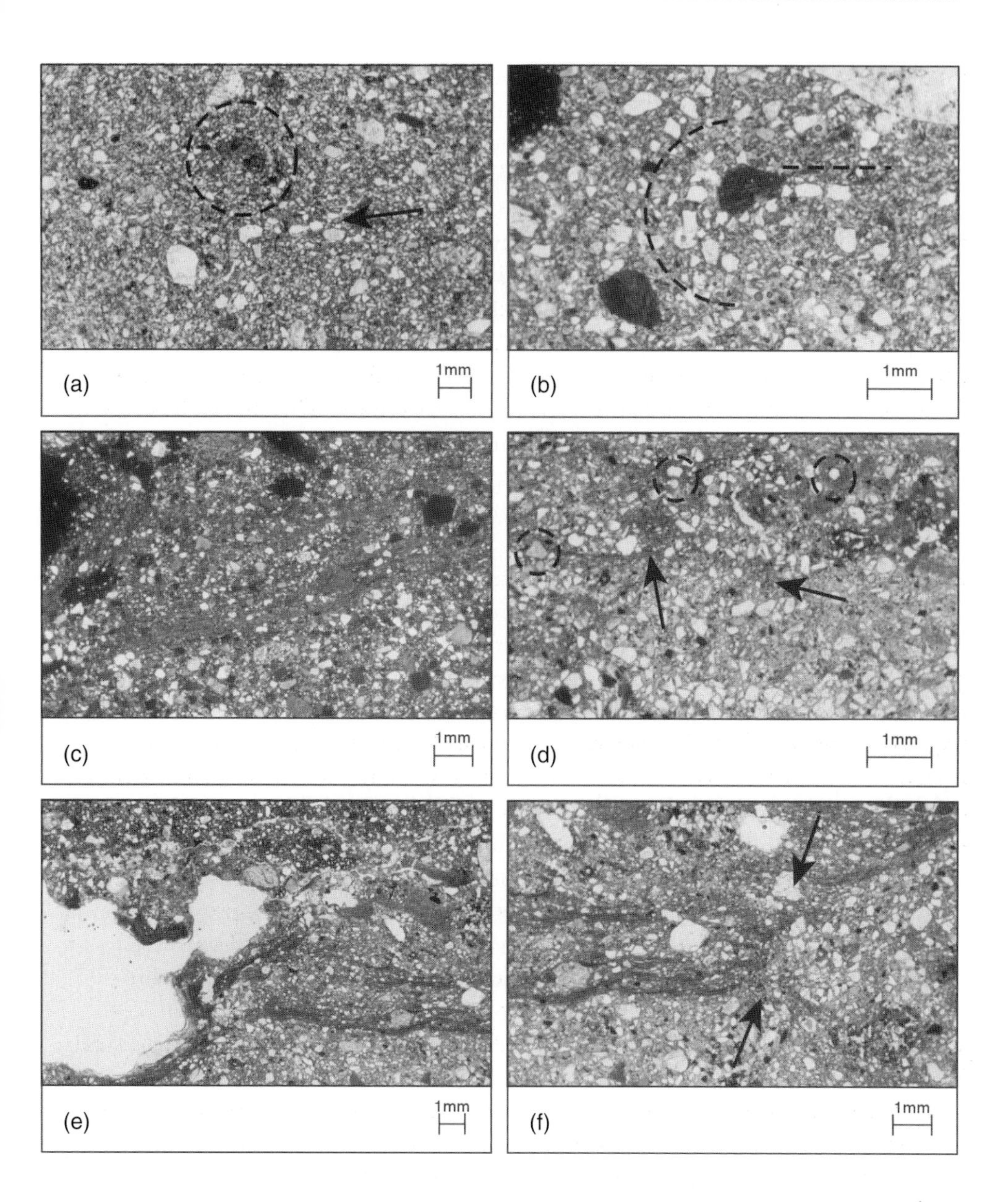

FIG 9.8 Selected photomicrographs showing details of Drumbeg tills: a) Silt-rich till with high-grain density and low matrix or clay density. Note linear arrangement of silt and sand grains (arrow). Also note diffusely bounded intraclast (circle). Slab C. Plane-polarized light; b) Turbate structure consisting of dark, fine-grained nodule (core) with a circular arrangement of silt and sand grains around it. Note that long axes of the silt and sand grains in the direct vicinity of the core have aligned to form a semi-circle. Also note associated lineament 'tail' to the right. Slab B. Plane-polarized light; c) Example of fine, anastomosing silt and clay laminae. Note that many of the silt grains have aligned parallel to the lamination. Also note that the clay laminae are highlighted by light-coloured illuviated clay-iron complexes. Slab A. Cross-polarized light; d) Muddy intraclasts (arrows) and casings around individual grains (indicated with circles). Note that grains tend to occur in clusters. Slab A. Plane-polarized light; e) Example of an irregularly shaped mammilated void, composed of intergrown more rounded pore spaces. Note associated textural separation and silt-clay lamination. Slab A. Plane-polarized light; f) Example of a water escape structure (arrows). Note associated laminae. Slab A. Plane polarized light.

Mammilated pore spaces with sharply defined walls occur in the laminated zones, in samples A_{L1}, A_{L2} and B_L (Fig. 9.8e). It is likely that these represent collapsed or intergrown circular vughs, suggestive of former water pockets in the diamict. There is a marked absence of planar fractures in all samples, but there are several examples of short, crinkly, low oblique to subvertical features, like irregular fractures, ill-defined, winnowed silty tracks and pipe-like features (Fig. 9.8f), which represent low-energy water-escape structures and which again suggest formerly abundant porewater. The occurrence of such features seems to culminate in the laminated base of slab A (samples A_{L1} and A_{L2}) and in sample B_M just above the laminated basal zone of slab B. The laminae and water-escape features are often associated with iron staining or clay-iron compound illuviation. Cutans, neocutans, detached and transported papules and clay veinlets, all exhibiting strong birefringence, tend to highlight and accentuate stratification, lamination, fractures and pores. While papules occurring in samples A_{L1} and A_{L2} may be contemporaneous with the lamination structures, most clay translocations in the other samples must be younger than the water escape or flow structures.

In summary, the abundance of interconnected and mammilated pore spaces, turbate and arcuate grain arrangements, microfractures and other water escape structures provides evidence for sorting, fluidization and illuviation by low energy water movements. Large parts of the till must have been close to saturation prior to dewatering. The observed microstructures are most abundant in the laminated zones at the base of the till slabs, indicating that laminar water flow was focused in these areas.

Reconciling the macro- and micro-scale evidence

One of the most striking features of the micromorphology is the total absence of microstructures indicative of shear in either the laminated or massive parts of the till. In contrast, at the macroscopic level, there are consistent strong preferred a-axis orientations in the laminated zones, and strong (but rather more variable) a-axis orientations in the massive parts of the till. The macroscopic evidence thus suggests that the directional properties of both the massive and laminated parts of the till were imposed during shearing (though apparently at different times), whereas the microscopic evidence shows pervasive dewatering structures and no evidence of shear. This apparent paradox can be resolved if shear and dewatering affected the sediments at different times and/or different spatial scales.

The clast macrofabric data indicate that the laminated diamicts represent shear zones over which the overlying massive till was thrust, with tectonic transport consistently towards the E/ESE. The shear zones apparently also acted as conduits for water flow, both during and following shearing, locally sorting matrix grains and destroying microscopic shear structures. However, water flow was insufficiently energetic to realign larger particles (macrofabric). The macrofabric data from the massive diamict are consistent with the hypothesis that the slabs consist of basal deformation till (*cf.* Benn 1995; Benn and Evans 1996; Hooyer and Iverson 2000a; Larsen and Piotrowski 2003), while the micromorphological data provide convincing evidence for abundant porewater and subsequent water-escape, but no evidence for shear. Significantly, the a-axis fabrics in the massive diamict show no evidence of overprinting during E/ESE tectonic transport, strongly suggesting that the slabs were not deformed when the laminated shear zones were active. Taken together, the evidence implies that (a) the a-axis fabrics in the massive diamicts are inherited properties of the till slabs, and (b) the slabs were rigid during transport. Since the inferred rigidity of the slabs is incompatible with abundant porewater (which greatly reduces frictional strength), we infer that the slabs were probably frozen during tectonic transport, and that the dewatering structures reflect loss of porewater following melt of interstitial ice.

To summarize, we propose the following sequence of events:

1| the massive diamict was formed beneath the glacier, and a-axis macrofabrics were imposed during subsole deformation, lodgement and ploughing;

2| submarginal slabs of the till were frozen on to the glacier sole, possibly in successive winters;

3| the till slabs were thrust over narrow shear zones. A-axis macrofabrics developed in the shear zones, parallel to the direction of tectonic transport, but the massive parts of the till remained rigid and older macrofabrics were preserved. The shear zones acted as conduits for escaping subglacial meltwater, which inhibited or destroyed micro-scale shear structures;

4| the till slabs melted, and during dewatering the till matrix was remoulded, destroying micro-scale evidence of shear. The escaping porewater was insufficiently energetic to realign clast macrofabrics, which thus retained a strain signature.

Thus, Krüger's (1993, 1994) model of stacking of frozen till slabs is consistent with all of the available evidence. This does not, of course, prove that the model is true, simply that it could not be falsified by the data we collected. Alternative models may be devised (although these must be compatible with the existing evidence), which could be tested in further studies.

9.4.3 Deformational sequence: single stage or polyphase?

Deformation structures: description

It is clearly of interest to determine whether the deformation observed in LFA-1 and LFA-2 at Drumbeg occurred during a single episode or multiple events, because this information will shed light on the behaviour of the ice margin during the early stages of deglaciation from the Loch Lomond Readvance maximum. Glacial deformational histories can be reconstructed using the principles of *kineto-stratigraphy* (Berthelsen 1978; Chapter 2), which examine the relationships between deformation structures affecting a sediment sequence. If the deformation at Drumbeg occurred during a single episode, we should expect to find a single suite of deformational structures, with no cross-cutting relationships. Conversely, polyphase deformation should be expected to result in varying deformational styles and structures, some of which affect or truncate others. To determine which is the case, we systematically examined the deformation structures exposed in the 1999 section, at the western (ice-proximal) part of Drumbeg quarry. More detailed discussions of the sequence of deformation at Drumbeg can be found in Philips *et al.* (2002, 2003).

At this site, LFA-1 to 4 exhibit distinct deformational styles. Throughout much of Drumbeg quarry, LFA-1 is undeformed, but it is locally folded and/or faulted. Deformation structures are particularly well exposed in the 1999 section, where the sediments consist of overall right-way up interbeds of sand and silty sand with occasional clay and gravel beds. These sediments have been folded into a broad, open antiform and two small NE-verging overfolds with weak axial planar foliation (Fig. 9.4). Flame-like soft-sediment deformation structures, which distort and locally disrupt the silty sand and clay, are aligned parallel to the axial surface of the folds. One fold axis is traversed by a 20 to 30cm thick, SW-dipping high-strain zone composed of pale blue-grey, matrix-poor sand, in which bedding has been transposed by a weak glacitectonic foliation. Additionally, the folds are cut by at least four sets of faults: normal faults dipping to the SW or NE, and reverse faults dipping to the SW or W (Fig. 9.4). Cross-cutting relationships are complex. Some normal faults cut across thrusts and *vice versa*, and no simple sequence of faulting could be determined. Most of the faults appear to be restricted to LFA-1; only in a very few cases do they propagate upwards to deform the overlying sediments of LFA-2. In general, however, the fold and fault structures are truncated by a prominent, irregular erosion surface (subsequently locally tectonized) below LFA-2, showing that most deformation of LFA-1 pre-dated erosion and the emplacement of the overlying sediments.

LFA-2 consists of a variably deformed and disrupted melange of gravel and sand lenses, cut by an anastomosing network of fine-grained, laminated sediments. The anastomosing laminae show widespread evidence of intense deformation, including boudinage and overfolds, and although some of the laminae may originally have had a sedimentary origin, most primary structures have been overprinted by a well-developed glacitectonic fabric. Other types of deformation structures affecting the laminae include NE-dipping shears, normal and reverse faults, soft-sediment deformation structures and water-escape conduits. In contrast, the lenses of sand and gravel enveloped by the laminated sediments are relatively undeformed, and primary lamination, graded bedding and/or cross-bedding are commonly preserved. Some lenses have asymmetric forms, and in places the surrounding laminae exhibit 'pressure shadow' folds, showing that shear was directed NE to E. Towards the north-eastern end of the section, LFA-2 is traversed by a prominent SW-dipping fault and an open synclinal fold. Both fault and syncline are truncated by a prominent erosion surface which separates the disrupted sediments of LFA-2 from the overlying, horizontally-bedded sands and gravels of LFA-4. At the southernmost end of the section, LFA-2 is traversed by two SW-dipping high strain zones consisting of 20–30cm thick layers of highly-deformed silty sand and clay with a SW-dipping glacitectonic fabric. The structurally higher of these shear zones forms the boundary between LFA-2 and the overlying diamicton (LFA-3).

As discussed in detail in Section 9.4.2, LFA-3 consists of four slabs of till separated by sharp bounding surfaces and laminated shear zones. These are interpreted as evidence for thrusting (deformational event D3; see below) by ice flowing towards the E or ENE.

LFA-4 is undeformed and overlies an erosional unconformity which truncates the underlying sediments of LFA-2 and 3.

Interpretation

This examination of deformational structures and their relationships allows a detailed reconstruction of events at Drumbeg. Three main deformational events (D1, D2 and D3) can be identified in the kinetostratigraphic signatures, the first two of which were separated in time by the deposition of LFA-2.

The oldest phase of deformation (D1) occurs within LFA-1, and resulted in the folding and faulting of the fan and delta sediments. Folding caused progressive shortening of the sediment pile, which culminated in thrusting during the later stages. Folding and thrusting of this type is commonly associated with proglacial tectonics and the formation of thrust-block moraines. There is no topographic expression of proglacial tectonics in the 1999 section, although there is in LFA-1 elsewhere in the quarry (Phillips *et al.* 2002, 2003) and in the surrounding area (Evans *et al.* 2003). D1 is thus interpreted as proglacial deformation associated with the advance of ice over the fan and delta sediments. The normal and reverse faults within LFA-1 cut across the folds, and thus postdate them. However, the sense of displacement along the reverse faults is compatible with the overall NE-directed sense of movement on the D1 thrusts and consequently, the normal and reverse faults are interpreted as D1 structures. Normal faults are associated with extension and reverse faults with horizontal compression. The intimate relationships between normal and reverse faults in LFA-1 indicate a complex pattern of extension and compression during the late stages of D1, and it appears likely that they formed as a conjugate set during gravitational spreading of the thickening sediment pile during D1 thrusting.

The prominent erosion surface at the base of LFA-2 truncates LFA-1 and D1 structures. Therefore, D1 deformation was followed by a period of erosion, then deposition of the sediments of LFA-2. Subsequently, these sediments were sheared during a second phase of deformation, D2. As noted above, some D1 faults penetrate into LFA-2, indicating that they may have been reactivated during D2 deformation. Kinematic indicators within LFA-2 (such as asymmetric sand and gravel lenses and overfolds in the surrounding laminated sediments) record an apparent NE/E-directed sense of shear. The deformation of LFA-2 during D2 was highly heterogeneous, with deformation being partitioned at several levels. Deformation partitioning was probably controlled by the variation in sediment composition (matrix content, grain size, porosity), water content and porewater pressure. These variations led to marked changes in the strength of the sediments within the deforming pile, and focused deformation into particular lithologies and/or horizons. This deformation partitioning resulted in the observed anastomosing network of high strain zones, separating lenticular domains of relatively low strain, including areas of essentially undeformed sediments (*cf.* Benn and Evans 1996; Evans 2000). The widespread presence of ductile deformation structures within some of the sand and silty sand lenses, together with the overall highly disrupted nature of LFA-2, indicates that these sediments were not frozen during D2. The style of deformation affecting LFA-2, in which primary structures are transposed by a pervasive sub-horizontal tectonic fabric, is typical of that produced by subglacial shear. Thus, D2 is interpreted as the result of glacial over-riding of subaqueous fan sediments.

The shear zone at the base of LFA-3 truncates D2 sediments, and so is interpreted as evidence for a third deformational phase, D3. This deformational event resulted in the stacking of four slabs of till by thrusting along narrow shear zones. The style of deformation is consistent with the process described by Krüger (1993, 1994) from the margins of Icelandic glaciers, in which frozen slabs of till are carried forward during winter readvances, then melt out during the following summer. If correct, this interpretation implies partial ice withdrawal from the site following the subglacial deformation event D2, and a series of annual ice margin oscillations. Following on from the D3 deformation event, proglacial meltwater emanating from the receding glacier eroded LFA-2 and 3, and deposited the gravels and sands of LFA-4. The erosion surface at the base of LFA-4 truncates the earlier developed D2 and D3 structures, clearly showing that glacial deformation had ceased prior to the deposition of LFA-4.

9.5 SUMMARY AND CONCLUSIONS

The ice-contact sediments exposed at Drumbeg quarry have undergone a polyphase deformational history separated by periods of erosion and deposition. The reconstructed sequence of events can be summarized as follows:

1| Following retreat from its maximum position, the Loch Lomond glacier deposited subaqueous fans and a Gilbert-type delta into proglacial Lake Blane (LFA-1). Aggradation of this ice-contact fan/delta may have encouraged temporary stabilization of the glacier margin by reducing water depth and ice losses by calving.

2| The glacier readvanced to the NE over the proximal part of the fan/delta sequence, causing proglacial glacitectonic deformation (D1). Events 1 and 2 are depicted in Figure 9.9.

3| Erosion of LFA-1 was followed by renewed deposition of sands and gravels (LFA-2).

4| A second readvance of the glacier over-rode the fan-delta causing extensive subglacial deformation (D2). This readvance progressed further towards the E/NE than the first, and affected a greater part of the fan-delta complex. The effect of D2 on the underlying deformed sediments of LFA-1 was relatively limited.

5| Slabs of till were sheared over sediments of LFA-2, possibly in four annual events (LFA-3, D3). Clast lithology and morphology, and mineral magnetic and particle-size data indicate that the till was rafted in above the glacitectonite, and that the two lithofacies associations are not closely genetically related, except near the base of the till slabs where underlying material was incorporated into, and rafted along, narrow shear zones.

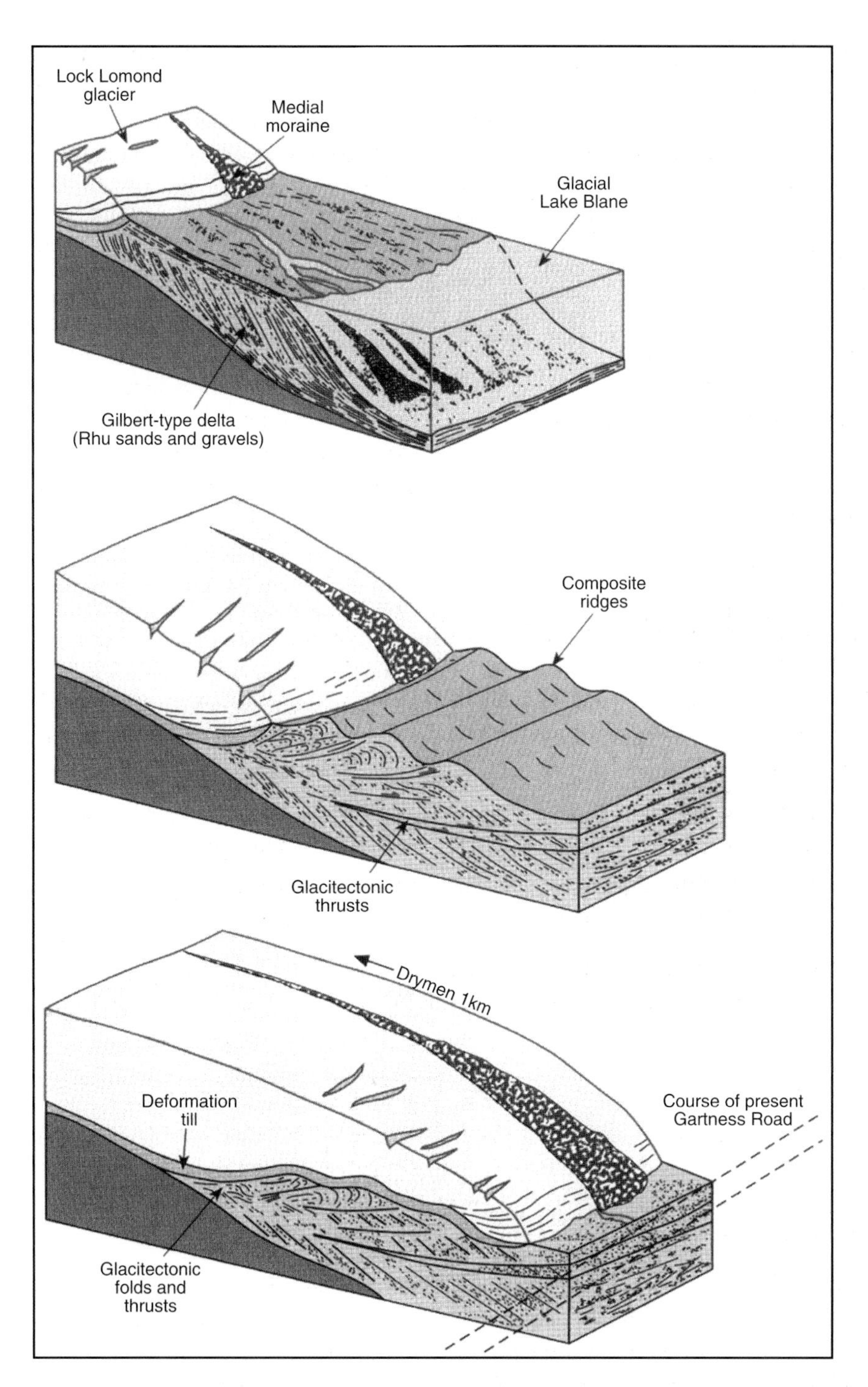

FIG 9.9 Reconstruction of early depositional events at Drumbeg (stages 1 and 2 in the text). As the Loch Lomond glacier receded its meltwaters deposited a subaqueous fan and then a Gilbert-type delta (LFA 1) into glacial Lake Blane. The glacier then readvanced over the delta sequence initiating proglacial glacitectonic deformation in LFA 1.

6| Subsequent retreat of the glacier from the over-ridden part of the ice-contact delta was accompanied by erosion and truncation of all the earlier developed deformation structures and, in some areas, the partial removal of the LFA-3 diamicton. The underlying sequence was then unconformably overlain by glacifluvial sands and gravels. Deposition of the locally channelized deltaic sequence was apparently terminated by the deposition of a thin mass flow diamicton which caps the sequence exposed in all of the sections examined.

The application of many of the sedimentological techniques reviewed in this book has allowed us to reconstruct the depositional environment at the margin of a proglacial lake (glacial Lake Blane) during the early stages of deglaciation following the Loch Lomond Readvance maximum. The complexity of the depositional and glacitectonic history in the Drumbeg area indicates that sedimentation and glacitectonism occurred in response to the oscillations of the ice margin. There is also evidence that the glacier was wet-based (unfrozen bed during D2 deformation), but that the sub-marginal zone probably underwent freezing in winter (evidence for the emplacement of frozen slabs of till in LFA-3). This condition is similar to that below Icelandic glaciers at the present day.

It is never possible to *prove* any particular interpretation of ancient glacial sediments. However, by adopting a hypothesis testing approach, it is possible to reject some models and increase our confidence in others. Central to this approach is the formulation of clear hypotheses, which allow unambiguous predictions to be made, which can then be tested against further observations. In this chapter, we have tested hypotheses using both quantitative data and systematic observations, but we have not employed any advanced statistical tests. Depending on the aims of the study and the type of available data, more or less formal approaches may be appropriate. Rigorous statistical approaches will be appropriate in some situations, whereas in others a more qualitative, inductive approach may be preferable. The important point is that observations should be collected and interpreted in as objective a way as possible, if they are to bring us closer to a balanced understanding of the Earth's glacial record.

REFERENCES

Allen J.R.L. 1963. 'The classification of cross-stratified units with notes on their origin.' *Sedimentology*, 2: 93–114.

Allen J.R.L. 1982. *Sedimentary Structures.* 2 vols. Amsterdam, Elsevier.

Allen J.R.L. 1985. *Principles of Physical Sedimentology*. London, Allen & Unwin.

Alley, R.B. 1989. 'Water-pressure coupling of sliding and bed deformation: II. Velocity-depth profiles.' *Journal of Glaciology*, 35: 119–29.

Alley, R.B. 1991. 'Deforming-bed origin for southern Laurentide till sheets?' *Journal of Glaciology*, 37: 200–7.

Alley, R.B., Blankenship, D.D., Bentley, C.R. and Rooney, S.T. 1986. 'Deformation of till beneath Ice Stream B, West Antarctica.' *Nature*, 322: 57–9.

Andrews, J.T. 1971. 'Techniques of Till Fabric Analysis.' *British Geomorphological Research Group Technical Bulletin*, Geo Abstracts.

Arch, J. and Maltman, A.J. 1990. 'Anisotropy and tortuosity in deformed wet sediments.' *Journal of Geophysical Research*, 95: 9035–46.

Ashley, G.M., Southard, J.B. and Boothroyd, J.C. 1982. 'Deposition of climbing ripple beds: a flume simulation.' *Sedimentology*, 29: 67–79.

Ballantyne, C.K. 1982. 'Aggregate clast form characteristics of deposits near the margins of four glaciers in the Jotunheimen massif, Norway.' *Norsk Geografisk Tiddskrift*, 36: 103–13.

Ballantyne, C.K. 1986. 'Protalus rampart development and the limits of former glaciers in the vicinity of Baosbheinn, Wester Ross.' *Scottish Journal of Geology*, 22: 13–25.

Ballantyne, C.K. and Benn, D.I. 1994. 'Paraglacial slope adjustment and resedimentation following recent glacier retreat, Fåbergstølsdalen, Norway.' *Arctic and Alpine Research*, 26: 255–69.

Banham, P.H. 1977. 'Glacitectonites in till stratigraphy.' *Boreas*, 6: 101–5.

Barendregt, R.W., Stalker, A. and Foster, J.H. 1976. 'Differentiation of tills in the Pakowki-Pinhorn area of southeastern Alberta on the basis of their magnetic susceptibility.' *Geological Survey of Canada Paper*, 76–1C: 189–90.

Barrett, P.J. 1980. 'The shape of rock particles, a critical review.' *Sedimentology*, 27: 291–303.

Beaumont, P. 1972. 'Clay mineralogy of glacial tills in eastern Durham, England' in: Yatsu, E. and Falconer, A. (Eds.), *Research methods in Pleistocene Geomorphology*, 2nd Guelph Symposium on Geomorphology, Geo Abstracts, 83–108.

Bell, T., Rogerson, R.J. and Mengel, F. 1989. 'Reconstructed ice-flow patterns and ice-limits using drift pebble lithology, outer Nachvak Fiord, northern Labrador.' *Canadian Journal of Earth Science*, 26: 577–90.

Benn, D.I. 1992. 'The genesis and significance of "hummocky moraine": evidence form the Isle of Skye, Scotland.' *Quaternary Science Reviews*, 11: 781–99.

Benn, D.I. 1994a. 'Fluted moraine formation and till genesis below a temperate glacier: Slettmarkbreen, Jotunheimen, Norway.' *Sedimentology*, 41: 279–92.

Benn, D.I. 1994b. 'Fabric shape and the interpretation of sedimentary fabric data.' *Journal of Sedimentary Research*, A64: 910–15.

Benn, D.I. 1995. 'Fabric signature of till deformation, Breidamerkurjokull, Iceland.' *Sedimentology*, 42: 735–47.

Benn, D.I. and Ballantyne, C.K. 1993. 'The description and representation of clast shape.' *Earth Surface Processes and Landforms*, 18: 665–72.

Benn, D.I. and Ballantyne, C.K. 1994. 'Reconstructing the transport history of glacigenic sediments: a new approach based on the co-variance of clast form indices.' *Sedimentary Geology*, 91: 215–27.

Benn, D.I. and Ballantyne, C.K. 1995. 'Grain-shape indices and isometric graphs – discussion.' *Journal of Sedimentary Research*, 65: 719–21.

Benn, D.I. and Evans, D.J.A. 1993. 'Glaciomarine deltaic deposition and ice-marginal tectonics: the "Loch Don Sand Moraine", Isle of Mull, Scotland.' *Journal of Quaternary Science*, 8: 279–91.

Benn, D.I. and Evans, D.J.A. 1996. 'The interpretation and classification of subglacially-deformed materials.' *Quaternary Science Reviews*, 15: 23–52.

Benn, D.I. and Evans, D.J.A. 1998. *Glaciers and Glaciation*. London, Edward Arnold. 734.

Benn, D.I. and Gemmell A.M.D. 2002. 'Fractal dimensions of diamictic particle-size distributions: simulations and evaluation.' *Geological Society of America, Bulletin*, 114: 528–32.

Benn, D.I. and Owen, L.A. 2002. 'Himalayan glacial sedimentary environments: a framework for reconstructing and dating former glacial extents in high mountain regions.' *Quaternary International*, 97–98: 3–25.

Benn, D.I. and Ringrose, T. 2001. 'Random variation of fabric eigenvalues: implications for the use of a-axis fabric data to differentiate till facies.' *Earth Surface Processes and Landforms*, 26: 295–306.

Bennett, M.R, Waller R.I, Glasser, N.F, Hambrey, M.J, and Huddart, D. 1999. 'Glacigenic clast fabrics: genetic fingerprint or wishful thinking?' *Journal of Quaternary Science*, 14: 125–35.

Berthelsen, A. 1978. 'The methodology of kineto-stratigraphy as applied to glacial geology.' *Bulletin of the Geological Society of Denmark*, 27: 25–38.

Berthelsen, A. 1979. 'Recumbent folds and boudinage structures formed by subglacial shear: an example of gravity tectonics.' *Geologie en Mijnbouw*, 58: 253–60.

Bertran, P. and Texier, J-P. 1999: 'Facies and microfacies of slope deposits.' *Catena*, 35: 99–121.

Beuselinck, L., Govers, G., Poesen, J., Degraer, G. and Froyen, L. (1998) 'Grain size analysis by laser diffractometry: comparison with the sieve-pipette method'. *Catena*, 32: 193–208.

Bishop, A.W., Green, G.E., Garga, V.K., Andresen, A. and Brown, J.D. 1971. 'A new ring shear apparatus and its application to residual strength.' *Geotechnique*, 21: 273–328.

Blake, E., Clarke, G.K.C. and Gérin, M.C. 1992. 'Tools for examining subglacial bed deformation.' *Journal of Glaciology*, 38: 388–96.

Blakely, R., Ackroyd, P. and Marden, M. (1981) *High Country River Processes.* Lincoln College, Tussock Grasslands & Mountain Lands Institute Special Publication, 22.

Blankenship, D.D., Bentley, C.R., Rooney, S.T. and Alley, R.B. 1986. 'Seismic measurements reveal a saturated porous layer beneath an active Antarctic Ice Stream.' *Nature*, 322: 54–7.

Bluck. B.J. 1999. 'Clast assembling, bed-forms and structure in gravel beaches.' *Transactions of the Royal Society of Edinburgh*, 89: 291–323.

Bogorodski, V.V., Bentley, C.R. and Gudmansen, P.E. 1985. *Radioglaciology*. Dordrecht, Reidel.

Boothroyd, J.C. and Ashley, G.M. 1975. 'Processes, bar morphology, and sedientary structures on braided outwash fans, northeastern Gulf of Alaska.' In A.V. Jopling and B.C. McDonald (Eds). *Glaciofluvial and Glaciolacustrine Sedimentation.* SEPM Special Publication, 23, 193–222.

Bordonau, J. and van der Meer, J.J.M. 1994. 'An example of a kinking microfabric in Upper Pleistocene glaciolacustrine deposits from Llavorsí (Central Southern Pyrenees, Spain).' *Geologie en Mijnbouw*, 73: 23–30.

Boulton, G.S. 1978. 'Boulder shapes and grain-size distributions of debris as indicators of transport paths through a glacier and till genesis.' *Sedimentology*, 25: 773–99.

Boulton, G.S. 1986. 'A paradigm shift in glaciology?' *Nature*, 322: 18.

Boulton, G.S. 1990. 'Sedimentary and sea level changes during glacial cycles and their control on glacimarine facies architecture.' In Dowdeswell J.A. and Scourse J.D. (Eds.), *Glacimarine Environments: Processes and Sediments.* Geological Society Special Publication, 53: 15–52.

Boulton, G.S. and Deynoux, M. 1981. 'Sedimentation in glacial environments and the identification of tills and tillites in ancient sedimentary sequences.' *Precambrian Research*, 15: 397–420.

Boulton, G.S. and Dobbie, K.E. 1993. 'Consolidation of sediments by glaciers: relations between sediment geotechnics, soft-bed glacier dynamics and subglacial ground-water flow.' *Journal of Glaciology*, 39 (131): 26–44.

Boulton, G.S. and Paul, M.A. 1976. 'The influence of genetic processes on some geothechnical properties of tills.' *Quarterly Journal of Engineering Geology*, 9: 159–94.

Boulton, G.S. and Hindmarsh, R.C.A. 1987. 'Sediment deformation beneath glaciers, rheology and geological consequences.' *Journal of Geophysical Research*, 92: 9059–82.

Bowles, J.W. 1981. *Engineering properties of soils and their measurement.* London, McGraw-Hill.

Brewer, R. 1976. *Fabric and Mineral Analysis of Soils.* Huntingdon, Krieger Press, 482.

Boyce, J.I. and Eyles, N. 2000. 'Architectural element analysis applied to glacial deposits: internal geometry of a late Pleistocene till sheet, Ontario, Canada.' *Geological Society of America Bulletin*, 112: 98–118.

Bridgland, D.R. 1986. *Clast Lithological Analysis.* Quaternary Research Association Technical Guide No. 3.

Brierley, G.J. 1991. 'Bar sedimentology of the Squamish River, British Columbia: definition and application of morphostratigraphic units.' *Journal of Sedimentary Petrology*, 61: 211–25.

Brindley, G.W. and Brown, G. (Eds.) 1980. *Crystal Structures of Clay Minerals and their X-ray Identification*, London, Mineralogical Society.

Bristow, C. 1996. 'Reconstructing fluvial channel morphology from sedimentary sequences.' In Carling, P.A. and Dawson, M.R. (Eds.), *Advances in Fluvial Dynamics and Stratigraphy.* Chichester, Wiley, 351–71.

Bristow, C.S. and Jol, H.M. (Eds.) 2003. *Ground Penetrating Radar in Sediments.* Geological Society Special Publication 211. 336.

Brodzikowski K. & van Loon A.J. 1991. *Glacigenic Sediments.* Amsterdam, Elsevier.

Broster, B.E. 1986. 'Till variability and compositional stratification: examples from the Port Huron lobe.' *Canadian Journal of Earth Science*, 23: 1823–41.

Browne, M.A.E. and MacMillan, A.A. 1989. *Quaternary Geology of the Clyde Valley.* British Geological Survey Research Report SA/89/1, Onshore Geology Series. London, HMSO.

Bull, P. 1981. 'Environmental reconstruction by electron microscopy.' *Progress in Physical Geography*, 5: 368–97.

Buurman, P., Paper, T. and Muggler, C.C. 1997. 'Laser-grain size determination in soil genetic studies: 1 Practical problems.' *Soil Science*, 162: 211–18.

Carr, S.J. 1998. *The Last Glacial Maximum in the North Sea Basin.* Unpublished PhD thesis, University of London.

Carr, S.J. 1999. 'The micromorphology of Last Glacial Maximum sediments in the southern North Sea.' *Catena*, 35, 123–45.

Carr, S.J. 2001. 'Micromorphological criteria for discriminating subglacial and glacimarine sediments: evidence from a contemporary tidewater glacier, Spitsbergen.' *Quaternary International*, 86, 71–9.

Carr, S.J. and Lee, J.A. 1998. 'Thin section production of diamicts: problems and solutions.' *Journal of Sedimentary Research*, 68: 217–20.

Carr, S.J., and Rose, J. 2003. 'Till fabric patterns and significance: particle response to subglacial stress.' *Quaternary Science Reviews*, 22: 1415–26.

Carr, S.J., Haflidason, H. and Sejrup, H.P. 2000. 'Micromorphological evidence supporting Late Weichselian glaciation of the Northern North Sea.' *Boreas*, 29: 315–28.

Carroll, D. 1970. *Clay Minerals: A Guide to their X-ray Identification*, Special Paper 126, Geological Society of America.

Carver, R.E. (Ed.) 1971. *Procedures in Sedimentary Petrology*, Chichester, Wiley.

Casagrande, A. 1936. 'The determination of the pre-consolidation load and its practical significance.' *First International Conference on Soil Mechanics and Foundation Engineering*, 60–4.

Cassidy, N.J., Russell, A.J., Marren, P.M., Fay, H., Rushmer, E.L., Van Dijk, T.A. and Knudsen, G.P. 2003. 'GPR-derived architecture of November 1996 jokulhlaup deposits, Skeiderarsandur, Iceland.' In Bristow, C.S. and Jol, H.M. (Eds.) *Ground Penetrating Radar in Sediments.* Geological Society Special Publication 211.

Chalmers, A.F. 1982. *What Is This Thing Called Science?* Milton Keynes, Open University, 179.

Chamberlin, T.C. 1897/1965. 'The method of multiple working hypotheses.' *Science*, 148: 745–59.

Chaolu, Y., Lun, W., Huaming, W. and Zhijiu, C. 1993. 'Thin section method for studying elongated microfabric in sediment: suitability and reliability.' *Scientia Geologica Sinica*, 2: 227–36.

Chernicof, S.E. 1983. 'Glacial characteristics of a Pleistocene ice lobe in east-central Minnesota.' *Geological Society of America Bulletin*, 94: 1401–14.

Chorley, R.J., Schumm, S.A. and Sugden, D.E. 1984. *Geomorphology*. London, Methuen, 605.

Christiansen, E.A. and Sauer, E.K. 1993. 'Red Deer Hill: a drumlinized, glaciotectonic feature near Prince Albert, Saskatchewan, Canada.' *Canadian Journal of Earth Sciences*, 30: 1224–35.

Christiansen, E.A. and Sauer, E.K. 1997. 'The Dirt Hills structure: an ice-thrust feature in southern Saskatchewan, Canada.' *Canadian Journal of Earth Sciences*, 34: 76–85.

Church, M.A., McLean, D.G. and Wolcott, J.F. 1987. 'River bed gravels: sampling and analysis.' In Thorne, C.R., Bathurst, J.C. and Hey, R.D. (Eds.), *Sediment Transport in Gravel-Bed Rivers*. Chichester, Wiley, 43–88.

Clark, P.U. 1989. 'Relative differences between glacially crushed quartz transported by mountain and continental ice- some examples from North America and East Africa: Discussion.' *American Journal of Science*, 289: 1195–98 [reply by Mahaney *et al.*, 1198–1205].

Clark, P.U. and Josenhans, H.W. 1986. 'Late Quaternary land-sea correlations, northern Labrador and Labrador Shelf.' *Current Research Part 8, Geological Survey of Canada Paper*, 86–1B, 171–8.

Clarke, G.K.C. 1987. 'Subglacial till: a physical framework for its properties and processes.' *Journal of Geophysical Research*, 92: 9023–36.

Coakley, J.P. and Syvitski, J.P.M. 1991. 'SediGraph technique.' In Syvitski J.P.M. (Ed.) *Principles, Methods, and Application of Particle Size Analysis*. Cambridge, CUP, 129–43.

Collinson, J.D. and Thompson, D.B. 1982. *Sedimentary Structures*. London, Allen & Unwin.

Collinson, D.W. 1983. *Methods in Rock Magnetism and Palaeomagnetism – Techniques and Instrumentation*, London, Chapman and Hall.

Compton, R.R. 1985. *Geology in the Field*. Chichester, Wiley.

Connally, G.G. 1960. 'Heavy minerals in the glacial drift of Western New York.' *Proceedings of the Rochester Academy of Science*, 10: 241–87.

Coutard, J.P and Mucher, H.J. 1985. 'Deformation of laminated silt loam due to repeated freezing and thawing cycles.' *Earth Surface Processes and Landforms*, 10: 309–19.

Dackombe, R.V. and Thomas, G.S.P. (Eds.) 1985. *Field guide to the Isle of Man*, Cambridge, Quaternary Research Association.

Davies, T.A., Bell, T., Cooper, A.K., Josenhans, H., Polyak, L., Solheim, A., Stoker, M.S. and Stravers J.A. 1997. *Glaciated Continental Margins: An Atlas of Acoustic Images*. London, Chapman & Hall.

Davis, J.C. 1986. *Statistics and Data Analysis in Geology.* 2nd edn, New York, Wiley, 646.

Derbyshire, E., McGown, A. and Radwan, A. 1976. ' "Total" fabric of some till landforms.' *Earth Surface Processes*, 1: 17–26.

Derbyshire, E., Love, M. and Edge, M.J. 1985. 'Fabrics of probable segregated ground-ice origin in some sediment cores from the North Sea Basin.' In Boardman, J. (Ed.) *Soils and Quaternary Landscape Evolution.* Chichester, Wiley, 261–80.

Derbyshire, E., Unwin, D.J., Fang, X.M. and Langford, M. 1992. 'The Fourier frequency-domain representation of sediment fabric anisotropy.' *Computers and Geosciences*, 18: 63–73.

Derry, D.R. 1933. 'Heavy minerals of the Pleistocene beds of the Don Valley, Toronto, Ontario.' *Journal of Sedimentary Petrology*, 3: 113–18.

Diplas, P. and Sutherland, A.J. 1988. 'Sampling techniques for gravel sized sediments.' *Journal of Hydraulic Engineering*, 114: 484–501.

Dobkins, J.E. and Folk, R.L. 1970. 'Shape development on Tahiti-nui.' *Journal of Sedimentary Petrology*, 40: 1167–203.

Domack, E.W and Lawson, D.E. 1985. 'Pebble fabric in ice-rafted diamicton.' *Journal of Geology*, 93: 577–91.

Domack, E.W., Anderson, J.B. and Kurtz, D. 1980. 'Clast shape as an indicator of transport and depositional mechanisms in glacial marine sediments: George V continental shelf, Antarctica.' *Journal of Sedimentary Petrology*, 50: 815–20.

Dowdeswell, J.A. and Sharp, M.J. 1986. 'Characterization of pebble fabrics in modern terrestrial glacigenic sediments.' *Sedimentology*, 33: 699–710.

Dowdeswell, J.A., Hambrey, M.J. and Wu, R. 1985. 'A comparison of clast fabric and shape in Late Precambrian and modern glacigenic sediments.' *Journal of Sedimentary Petrology*, 55: 691–704.

Doyle, P., Bennett, M.R. and Baxter, A.N. 1994. *The Key to Earth History: An Introduction to Stratigraphy.* Chichester, Wiley.

Dreimanis, A. 1962. 'Quantitative gasometric determination of calcite and dolomite by using Chittick apparatus.' *Journal of Sedimentary Petrology*, 32: 520–29.

Dreimanis, A. 1989. 'Tills: their genetic terminology and classification.' In Goldthwait, R.P. and Matsch, C.L. (Eds.), *Genetic Classification of Glacigenic Deposits.* Rotterdam: Balkema, 17–84.

Dreimanis, A., Reavely, G.H., Cook, R.J.B., Knox, K.S. and Moretti, F.J. 1957. 'Heavy mineral studies in tills of Ontario and adjacent areas.' *Journal of Sedimentary Petrology*, 27: 148–61.

Dreimanis, A. and Terasmae, J. 1958. 'Stratigraphy of Wisconsian glacial deposits of Toronto area, Ontario.' *Proceedings of the Canadian Geological Association*, 10, 119–36.

Droste, J.B. 1956. 'Alteration of clay minerals by weathering in Wisconsin tills.' *Geological Society of America Bulletin*, 67: 911–18.

Dunbar, C.O. and Rodgers, J. 1957. *Principles of Stratigraphy.* New York, Wiley.

Early, K.R. and Skempton, A.W. 1972. 'Investigations of the landslide at Walton's Wood, Staffordshire.' *Quarterly Journal of Engineering Geology*, 5: 19–41.

Ekes, C. and Hickin, E.J. 2001. 'Ground penetrating radar facies of the paraglacial Cheekye Fan, southwestern British Columbia, Canada.' *Sedimentary Geology*, 143: 199–217.

Engelhardt, H.F. and Kamb, B. 1998. 'Basal sliding of Ice Stream B, West Antarctica.' *Journal of Glaciology*, 44: 223–30.

Engelhardt, H.F., Humphrey, N., Kamb, B., and Fahnestock, M. 1990. 'Physical conditions at the base of a fast moving Antarctic ice stream.' *Science*, 248: 57–9.

Evans, D.J.A. (Ed.) 2003. *Glacial Landsystems*. London, Arnold.

Evans, D.J.A. 2000. 'A gravel outwash/deformation till continuum, Skalafellsjokull, Iceland.' *Geografiska Annaler*, 82A: 499–512.

Evans, D.J.A. In press. 'Glacial depositional processes and forms.' In Burt, T.P., Chorley, R.J., Brunsden, D., Goudie, A.S. and Cox, N.J. (Eds.), *The History of the Study of Landforms: Volume 4 – Quaternary and Recent Processes and Forms (1890–1965) and the Mid-Century Revolutions*. London, Routledge.

Evans, D.J.A., Owen, L.A. and Roberts, D. 1995. 'Stratigraphy and sedimentology of Devensian (Dimlington Stadial) glacial deposits, east Yorkshire, England.' *Journal of Quaternary Science*, 10: 241–65.

Evans, D.J.A., Wilson, S.B. and Rose, J. 2003. 'Glacial geomorphology of the Western Highland Boundary.' In Evans, D.J.A. (ed.), *The Quaternary of the Western Highland Boundary, Field Guide*. London, Quaternary Research Association, 5–20.

Evenson, E.B. 1970. 'A method for 3-dimensional microfabric of tills obtained from exposures and cores.' *Journal of Sedimentary Petrology*, 40: 762–4.

Evenson, E.B. 1971. 'The relationship of macro- and microfabrics of till and the genesis of glacial landforms in Jefferson County, Wisconsin.' In Goldthwait, R.P. (Ed.) *Till – A Symposium*. Columbia, Ohio State University Press, 345–64.

Eyles, N. 1993. 'Earth's glacial record and its tectonic setting.' *Earth Science Reviews*, 35: 1–248.

Eyles, N. and Eyles, C.H. 1992. 'Glacial depositional systems.' In Walker R.G. and James N.P. (Eds.), *Facies Models: Response to Sea Level Change*. Geological Association of Canada, Toronto, 73–100.

Eyles, N. and Miall A.D. 1984. 'Glacial facies.' In Walker R.G. (Ed.), *Facies Models*. Geoscience Canada Reprint Series 1, 15–38.

Eyles N., Eyles, C.H. and Miall A.D. 1983. 'Lithofacies types and vertical profile models: an alternative approach to the description and environmental interpretation of glacial diamict and diamictite sequences.' *Sedimentology*, 30: 393–410.

Eyles, N. and Eyles, C.H. 1992. 'Glacial depositional systems.' In Walker, R.G. and James, N.P. (Eds.) *Facies Models: Response to Sea-level Change*. Toronto, Geological Association of Canada, 73–100.

Fairchild, I., Hendry, G., Quest, M. and Tucker, M.E. 1988. 'Chemical analysis of sedimentary rocks.' In Tucker, M.E. (Ed.), *Techniques in Sedimentology*, Oxford, Blackwells, 274–354.

Federoff, N., Courty, M.A. and Thompson, M.L. 1990. 'Micromorphological evidence of palaeoenvironmental change in Pleistocene and Holocene palaeosols.' In Douglas, L.A. (Ed.) *Soil Micromorphology: A Basic and Applied Science*. Amsterdam, Elsevier, 653–65.

Ferguson R.I. and Paola C. 1997. 'Bias and precision of percentiles of bulk grain size distributions.' *Earth Surface Processes and Landforms*, 22: 1061–78.

Fichter, L.S. and Poché, D.J. 1993. *Ancient Environments and the Interpretation of Geologic History*. 2nd edn, Macmillan Publishing Company, 269.

Fichter, L.S., Farmer, G.T. and Clay, J.S. 1991. *Earth Materials and Earth Processes: an Introduction*. 3rd edn, London, Macmillan Publishing Company, 280.

Fischer, U.H. and Hubbard, B. 1999. 'Subglacial sediment textures: character and evolution at Haut Glacier d'Arolla, Switzerland.' *Annals of Glaciology*, 28: 241–6.

Fischer, U.H. and Clarke, G.K.C. 1994. 'Ploughing of subglacial sediment.' *Journal of Glaciology*, 40: 97–106.

Fischer, U.H. and Clarke, G.K.C. 1997. 'Clast collision frequency as an indicator of glacier sliding rate.' *Journal of Glaciology*, 43: 460–6.

Fischer, U.H., Iverson, N.R., Hanson, B., Hooke, R. LeB. and Jansson, P. 1998. 'Estimation of hydraulic properties of till from ploughmeter measurements.' *Journal of Glaciology*, 38: 51–64.

Fitzpatrick, E.A. 1984. *Micromorphology of Soils*. London, Chapman & Hall, 433.

Folk, R.L. and Ward, W.C. 1957. 'Brazos River bar: a study in the significance of grain size parameters.' *Journal of Sedimentary Petrology*, 27: 3–26.

Foscolos, A.E. and Barefoot, R.R. 1970. *Rapid determination of total organic and inorganic carbon in shales and carbonates*. Geological Survey of Canada, Paper 70–11, 14 pp.

Fraccarollo, L. and Marion, A. 1995. 'Statistical approach to bed-material surface sampling.' *Journal of Hydraulic Engineering*, 121(7): 1–6.

Francus, P. 1998. 'An image-analysis technique to measure grain-size variations in thin sections of soft clastic sediments.' *Sedimentary Geology*, 121: 289–98.

Freeze, R.A. and Cherry, J.A. 1979. *Groundwater*. Englewoord Cliffs, N.J., Prentice-Hall, 604.

Fritz, W.J. and Moore, J.N. 1988. *Basics of Physical Stratigraphy and Sedimentology*. New York, Wiley.

Gale, S.J. and Hoare, P.G. 1991. *Quaternary Sediments: Petrographic Methods for the Study of Unlithified Rocks*. London, Belhaven Press, 323.

Gardiner, V. and Dackombe, R.V. 1983. *Geomorphological Field Manual*. London, Allen and Unwin, 254.

Ghibaudo, G. 1992. 'Subaqueous sediment gravity flow deposits: practical criteria for their field description and classification.' *Sedimentology*, 39: 423–54.

Gibbs, R.J. 1968. 'Clay mineral mounting techniques for X-ray diffraction analysis: a discussion.' *Journal of Sedimentary Petrography*, 38: 242–4.

Gibbs, R.J. 1971. 'X-ray diffraction mounts.' In Carver, R.E (Ed.) *Procedures in Sedimentary Petrology*, Chichester, Wiley, 531–69.

Glasser, N.F. and Hambrey, M.J. 2002. 'Sedimentary facies and landform genesis at a temperate outlet glaciër: Soler Glacier, North Patagonian Icefield.' *Sedimentology*, 49: 43–64.

Glen, J.W., Donner, J.J. and West, R.G. 1957. 'On the mechanism by which stones in till become oriented.' *American Journal of Science*, 255: 194–205.

Glentworth, R., Mitchell, W.A. and Mitchell, B.D. 1964. 'The red glacial drift deposits of north-east Scotland.' *Clay Minerals Bulletin*, 5: 373–81.

Gordon, J.E. 1993. 'Gartness.' In Gordon, J.E. and Sutherland, D.G. (Eds.), *Quaternary of Scotland.* London, Chapman and Hall, 444–8.

Goudie, A.S. (Ed.) 1981. *Geomorphological Techniques*, London, Allen and Unwin.

Gozdzick, J.S. 1973. 'Geneza i pozycja stratygraficzna struktur peryglacjalnych w srodkowej Polsce.' *Acta Geographica Lodzienska*, 31: 119.

Graham, D.J. and Midgley, D.G. 2001. 'Graphical representation of clast shape using triangular diagrams: an Excel spreadsheet method.' *Earth Surface Processes and Landforms*, 25: 1473–7.

Gravenor, C.P. and Stupavsky, M. 1974. 'Magnetic susceptibility of the surface tills of southern Ontario.' *Canadian Journal of Earth Science*, 11: 658–63.

Gwyn, Q.H.J. and Dreimanis, A. 1979. 'Heavy mineral assemblages in tills and their use in distinguishing glacial lobes in the Great Lakes region.' *Canadian Journal of Earth Science*, 16: 2219–35.

Haines-Young, R. and Petch J. 1986. *Physical Geography: Its Nature and Methods*. London, Harper & Row.

Hallam, A. 1981. *Facies Interpretation and the Stratigraphic Record.* Oxford, Freeman.

Hambrey, M.J. 1994. *Glacial Environments.* London, UCL Press.

Hardy, R. and Tucker, M. 1988. 'X-ray powder diffraction of sediments.' In Tucker, M.E. (Ed.), *Techniques in Sedimentology*. Oxford, Blackwells, 191–228.

Harms J.C. 1979. 'Primary sedimentary structures.' *Annual Review of Earth and Planetary Science*, 7: 227–48.

Harms, J.C., Southard, J.B. and Walker, R.B. 1982. *Structures and Sequences in Clastic Rocks.* Society of Economic Paleontologists and Mineralogists, Lecture Notes for Short Course No.9.

Harris, C. Williams, G., Brabham, P., Eaton, G. and McCarroll, D. 1997. 'Glaciotectonized Quaternary sediments at Dinas Dinlle, Gwynedd, North Wales, and their bearing on the style of deglaciation in the Eastern Irish Sea.' *Quaternary Science Reviews*, 16: 109–27.

Harrison, P.W. 1957. 'A clay-till fabric: its character and origin.' *Journal of Geology*, 65: 275–307.

Harry, D.G. and Gozdzick, J.S. 1988. 'Ice wedges: growth, thaw transformation and palaeoenvironmental significance.'*Journal of Quaternary Science*, 3: 39–55.

Hart, J.K. 1994. 'Till fabric associated with deformable beds.' *Earth Surface Processes and Landforms*, 19: 15–32.

Hart, J.K., Hindmarsh, R.C.A. and Boulton, G.S. 1990. 'Styles of subglacial glaciotectonic deformation within the context of the Anglian Ice Sheet.' *Earth Surface Processes and Landforms*, 15: 227–41.

Hart, J.K. and Roberts, D.H. 1994. 'Criteria to distinguish between subglacial glaciotectonic and glaciomarine sedimentation, I. Deformation styles and sedimentology.' *Sedimentary Geology*, 91: 191–213.

Head, K.H. 1989. *Soil Technician's Handbook.* New York, John Wiley and Sons.

Hein, F.J. 1984. 'Deep-sea and fluvial braided channel conglomerates: a comparison of two case studies.' In Koster, E.H. and Steel, R.J. (Eds.) *Sedimentology of Gravels and Conglomerates*, Canadian Society of Petroleum Geologists Memoir, 10, 33–49.

Hicock, S.R. 1991. 'On subglacial stone pavements in till.' *Journal of Geology*, 99: 607–19.

Hicock, S.R. and Dreimanis, A. 1992. 'Deformation till in the Great Lakes region: implications for rapid flow along the south-central margin of the Laurentide Ice Sheet.' *Canadian Journal of Earth Sciences*, 29: 1565–79.

Hiemstra, J.F. and van der Meer, J.J.M. 1997. 'Pore water controlled grain fracturing as an indicator for subglacial shearing in tills.' *Journal of Glaciology*, 43: 446–54.

Hiemstra, J.F. 2001. *Dirt Pictures Reveal the Past Extent of the Grounded Antarctic Ice Sheet*. PhD Thesis, Universteit van Amsterdam, 231.

Hiemstra, J.F. and Rijsdijk, K.F. 2003. 'Observing artificially induced strain: implications for subglacial deformation.' *Journal of Quaternary Science*, 18: 373–83.

Hiemstra, J.F., Evans, D.J.A. and Benn, D.I. 2003. 'Drumbeg: zooming in on Unit III (an alternative scenario for deformation phase D2).' In Evans, D.J.A. (Ed.) *The Quaternary of the Western Highland Boundary: Field Guide.* London, Quaternary Research Association, 104–10.

Hindson, R. and Andrade, C. 1999. 'Sedimentation and hydrodynamic processes associated with the tsunami generated by the 1755 Lisbon earthquake.' *Quaternary International*, 56: 27–38.

Hirono, T. 2000. 'Determination of slip sense along a fault in unlithified sediment.' *Journal of Structural Geology*, 22: 537–41.

Hockey, B. 1970. 'An improved co-ordinate system for particle shape representation.' *Journal of Sedimentary Petrology*, 40: 1054–6.

Hoey T.B. and Bluck B.J. 1999. 'Identifying the controls over downstream fining of river gravels.' *Journal of Sedimentary Research*, 69A: 52–62.

Hofmann, H.J. 1994. 'Grain-shape indexes and isometric graphs.' *Journal of Sedimentary Research*, A 64: 916–20.

Holmes, C.D. 1941. 'Till fabric.' *Bulletin of the Geological Society of America*, 52: 1299–354.

Hooke, R. LeB., and Iverson N. 1995. 'Grain-size distribution in deforming subglacial tills: role of grain fracture.' *Geology*, 23: 57–60.

Hooke, R. LeB., Hanson, B., Iverson, N.R. Jansson, P. and Fischer, U.H. 1997. 'Rheology of till beneath Storglaciären, Sweden.' *Journal of Glaciology*, 443: 172–9.

Hooyer, T.S. and Iverson, N.R. 2000a. 'Clast-fabric development in a shearing granular material: Implications for subglacial till and fault gouge.' *Geological Society of America Bulletin*, 112: 683–92.

Hooyer, T.S. and Iverson, N.R. 2000b. 'Diffusive mixing between shearing granular layers: constraints on bed deformation from till contacts.' *Journal of Glaciology*, 46: 641–51.

Hubbard, B and Maltman, A. 2000. 'Laboratory investigations of the strength, static hydraulic conductivity and dynamic hydraulic conductivity of glacial sediments.' In Maltman, A.J., Hubbard, B. and Hambrey, M.J. (Eds.) *Deformation of Glacial Materials*, Geology Society of London Special Publication, No 176, 231–42.

Hubbard, B.P., Sharp, M.J., Willis, I.C., Nielsen, M.K. and Smart, C.C. 1995. 'Borehole water-level variations and the structure of the subglacial hydrological system of Haut Glacier d'Arolla, Valais, Switzerland.' *Journal of Glaciology*, 41: 572–83.

Humphrey, N., Kamb, B., Fahnestock, M. and Engelhardt, H. 1993. 'Characteristics of the bed of the Lower Columbia Glacier, Alaska.' *Journal of Geophysical Research*, 98: 837–46.

Illenberger, W.K. 1991. 'Pebble shape (and size!).' *Journal of Sedimentary Petrology*, 61: 756–67.

Ingram R.L. 1954. 'Terminology for thickness of stratification and parting units in sedimentary rocks.' *Bulletin of the Geological Society of America*, 65: 937–8.

Iverson N., Hooyer T., and Hooke R.LeB. 1996. 'A laboratory study of sediment deformation: stress heterogeneity and grain-size evolution.' *Annals of Glaciology*, 22: 167–75.

Iverson, N.R. Jansson, P. and Hooke, R. LeB. 1994. 'In-situ measurement of the strength of deforming subglacial till.' *Journal of Glaciology*, 40: 497–503.

Iverson, N.R., Baker, R.W. and Hooyer, T.S. 1997. 'A ring-shear device for the study of till deformation: tests on tills with contrasting clay contents.' *Quaternary Science Reviews*, 16: 1057–66.

Iverson, N.R., Hanson, B., Hooke, R. LeB. and Jansson, P. 1995. 'Flow mechanism of glaciers on soft beds.' *Science*, 267: 80–1.

Iverson, N.R., Hooyer, T.S. and Baker, R.W. 1998. 'Ring-shear studies of till deformation: Coulomb-plastic behaviour and distributed strain in glacier beds.' *Journal of Glaciology*, 44: 634–42.

Jenkins, R. and de Vries, J.L. 1970. *Practical X-ray Spectrometry*, London, Macmillan.

Johns, W.D., Grim, R.E. and Bradley, W.F. 1954. 'Quantitative estimates of clay minerals by diffraction methods.' *Journal of Sedimentary Petrology*, 24: 242–51.

Johnson, W.M. and Maxwell, J.A. 1981. *Rock and Mineral Analysis*, New York, Wiley.

Johnson, M.D. 1983. 'The origin and microfabric of Lake Superior Red Clay.' *Sedimentary Petrology*, 53: 859–73.

Jones, A.P., Tucker, M.E. and Hart, J.K. (Eds). 1999. 'The description and analysis of Quaternary stratigraphic field sections.' Technical Guide No. 7, London, Quaternary Research Association, 293.

Josenhans, H.W., Zevenhuizen, J. and Klassen, R.A. 1986. 'The Quaternary geology of the Labrador Shelf.' *Canadian Journal of Earth Science*, 23: 1190–213.

Kamb, B. 1991. 'Rheological nonlinearity and flow instability in the deforming-bed mechanism of ice-stream motion.' *Journal of Geophysical Research*, 96B: 16585–95.

Kellerhals R. and Bray D.I. 1971. 'Sampling procedures for coarse fluvial sediments.' *Proceedings, American Society of Civil Engineers, Journal of the Hydraulics Division*, 97: 1165–79.

Kemmis, T.J. 1996. 'Lithofacies associations for terrestrial glacigenic successions.' In J. Menzies (Ed.) *Past Glacial Environments: Sediments, Forms and Techniques*, Oxford, Butterworth-Heinemann, 285–300.

Kemp, R.A. 1985. *Soil Micromorphology and the Quaternary*. Quaternary Research Association Technical Guide No. 2, Cambridge, 80.

Kemp, R.A. 1998. 'Role of micromorphology in paleopedological research.' *Quaternary International*, 51/52: 133–41.

Kemp, R.A. 1999. 'Soil micromorphology as a technique for reconstructing palaeoenvironmental change.' In Singhvi, A.K. and Derbyshire, E. (Eds.) *Paleoenvironmental Reconstruction in Arid Lands*. New Delhi, Oxford and IBH Publishing Co. PVT, 41–71.

Kettles, I.M. and Shilts, W.W. 1989. 'Geochemistry of drift over the Precambrian Grenville Province, southeasten Ontario and southwestern Quebec', In DiLabio, R.N.W. and Coker, W.B. (Eds) *Drift prospecting*, Geological Survey of Canada, Paper 89–20, 97–112.

Khatwa, A. and Tulaczyk, S. 2001. 'Microstructural interpretations of modern and Pleistocene subglacially deformed sediments: the relative role of parent material and subglacial process.' *Journal of Quaternary Science*, 16: 507–17.

Kirkbride, M.P. and Spedding, N.F. 1996. 'The influence of englacial drainage on sediment-transport pathways and till texture of temperate valley glaciers.' *Annals of Glaciology*, 22: 160–6.

Kjaer, K.H. and Krüger, J. 1998. 'Does clast size influence fabric strength?' *Journal of Sedimentary Research*, 68: 746–9.

Klug, H.P. and Alexander, L.E. 1974. *X-ray Diffraction Procedures for Polycrystalline and Amorphous Material*, New York, Wiley.

Krinsley, D.H. and Doornkamp, J.C. 1973. *Atlas of Quartz Sand Surface Textures*. Cambridge, Cambridge University Press, 91.

Krüger, J. 1970. 'Till fabric in relation to direction of ice movement. A study from the Fakse Banke, Denmark.' *Geografisk Tidskrift*, 69: 133–70.

Krüger, J. 1979. 'Structures and textures in till indicating subglacial deposition.' *Boreas*, 8: 323–40.

Krüger, J. 1983. 'Glacial morphology and deposits in Denmark.' In Ehlers J. (Ed.), *Glacial Deposits in North-West Europe*. Rotterdam, Balkema, 181–91.

Krüger, J. 1984. 'Clasts with stoss-lee form in lodgement tills: a discussion.' *Journal of Glaciology*, 30: 241–3.

Krüger, J. 1993. 'Moraine-ridge formation along a stationary ice front in Iceland.' *Boreas*, 22: 101–9.

Krüger, J. 1994. 'Glacial processes, sediments, landforms, and stratigraphy in the terminus region of Myrdalsjökull, Iceland.' *Folia Geographica Danica*, 21: 1–233.

Krüger J. and Kjaer K.H. 1999. 'A data chart for field description and genetic interpretation of glacial diamicts and associated sediments – with examples from Greenland, Iceland and Denmark.' *Boreas*, 28: 386–402.

Krumbein W.C. 1938. 'Size frequency distributions of sediments and the normal phi curve.' *Journal of Sedimentary Petrology*, 8: 84–90.

Krumbein, W.C. 1939. 'Preferred orientation of pebbles in sedimentary deposits.' *Journal of Geology*, 47: 673–706.

Krumbein, W.C. 1941. 'Measurement and geological significance of shape and roundness of sedimentary particles.' *Journal of Sedimentary Petrology*, 11: 64–72.

Krumbein, W.C. and Pettijohn, F.J. 1938. *Manual of Sedimentary Petrography*. Appleton-Centuary Company.

Kujansuu, R. and Saarnisto, M. (Eds). 1990. *Glacier Indicator Tracing*. Rotterdam, Balkema, 252.

Lake, R.D., Ellison, A. and Moorlock, B.S.P. 1977. 'Middle Pleistocene stratigraphy in southern East Anglia.' *Nature*, 265: 663.

Lane S.N. 2001. 'The measurement of gravel-bed river morphology.' In Mosley M.P. (Ed.) *Gravel-Bed Rivers V*, Wellington, New Zealand Hydrological Society, 291–338.

Larsen, E., Sandven, R., Heyerdahl, H. and Hernes, S. 1994. 'Glacial geological implications of preconsolidation values in sub-till sediments at Skorgenes, western Norway.' *Boreas*, 24: 37–46.

Larsen, N.K. and Piotrowski, J.A. 2003. 'Fabric pattern in a basal till succession and its significance for reconstructing subglacial processes.' *Journal of Sedimentary Research*, 73: 725–34.

Lawson, D.E. 1979. 'A comparison of pebble orientations in ice and deposits of the Matanuska Glacier, Alaska.' *Journal of Geology*, 87: 629–45.

Lawson, D.E. 1982. 'Mobilisation, movement and deposition of subaerial sediment flows, Matanuska Glacier, Alaska.' *Journal of Geology*, 90: 279–300.

Lee, J.A. and Kemp, R.A. 1992. *Thin Sections of Unconsolidated Sediments: A Recipe*. CEAM report #2, London Royal Holloway, University of London, 32.

Lee, J.R. 2001. 'Genesis and palaeogeographical significance of the Corton Diamicton (basal member of the North Sea Drift Formation), East Anglia, U.K.' *Proceedings of the Geologists Association*, 112: 29–43.

Leeder, M.R. 1982. *Sedimentology: Process and Product*. London, Allen & Unwin.

Levson, V.M. and Rutter, N.W. 1988. 'A lithofacies analysis and interpretation of depositonal environments of montane glacial diamictons, Jasper, Alberta, Canada.' In Goldthwait, R.P. and Matsch, C.L. (Eds.), *Genetic Classification of Glacigenic Deposits*. Rotterdam, Balkema, 117–42.

Lindholm R.C. 1987. *A Practical Approach to Sedimentology*. London, Allen & Unwin.

Loizeau J-L., Arbouille D., Santiago S. and Vernet J-P. 1994. 'Evaluation of a wide range laser diffraction grain size analyser for use with sediments.' *Sedimentology*, 41: 353–61.

Lowe D.R. 1982. 'Sediment gravity flows II. Depositional models with special reference to the deposits of high-density turbidity currents.' *Journal of Sedimentary Petrology*, 52: 279–97.

Lowe J.J. and Walker M.J.C. 1997. *Reconstructing Quaternary Environments*. Harlow, Longman.

Mahaney, W.C. 1995. 'Glacial crushing, weathering and diagenetic histories of quartz grains inferred from scanning electron microscopy.' In Menzies, J. (Ed.) *Modern Glacial Environments: Processes, Dynamics and Sediments*. Oxford, Butterworth-Heinemann, 487–506.

Mahaney, W.C., Claridge, G. and Campbell, I. 1996. 'Microtextures on quartz grains in tills from Antarctica.' *Palaeogeography, Palaeoclimatology, Palaeoecology*, 121: 89–103.

Mahaney, W.C., Vortisch, W. and Julig, P. 1988. 'Relative differences between glacially crushed quartz transported by mountain and continental ice- some examples from North America and East Africa.' *American Journal of Science*, 288: 810–26.

Mahaney, W.C. and Kalm, V. 2000. 'Comparative scanning electron microscopy study of oriented till blocks, glacial grains and Devonian sands in Estonia and Latvia.' *Boreas*, 29: 35–51.

Mahaney, W.C., Stewart, A. and Kalm, V. 2001. 'Quantification of SEM microtextures useful in sedimentary environmental discrimination.' *Boreas*, 30: 165–71.

Maher, B.A. 1986. 'Characterisation of soil by mineral magnetic measurement.' *Physics of the Earth and Planetary Interiors*, 42: 76–92.

Maher, B.A. and Thompson, R. (Eds). 1999. *Quaternary Climates, Environments and Magnetism*. Cambridge, Cambridge University Press, 390.

Maizels, J.K. 1993. 'Lithofacies variations within sandur deposits: the role of runoff regime, flow dynamics and sediment supply characteristics.' *Sedimentary Geology*, 85: 299–325.

Mange, M.A. and Maurer, H.F.W. 1992. *Heavy Minerals in Colour*. London, Chapman and Hall, 147.

Mark, D.M. 1973. 'Analysis of axial orientation data, including till fabrics.' *Geological Society of America Bulletin*, 84: 1369–74.

Mark, D.M. 1974. 'On the interpretation of till fabrics.' *Geology*, 2: 101–4.

Martin, J.H. 1980. 'The classification of till: a sedimentologist's viewpoint.' *Quaternary Newsletter*, 32: 1–10.

Martini, I.P. and Brookfield, M.E. 1995. 'Sequence analysis of upper Pleistocene (Wisconsinan) glaciolacustrine deposits of the north-shore bluffs of Lake Ontario, Canada.' *Journal of Sedimentary Research*, B65: 388–400.

Matthews, M.D. 1991. 'The effect of grain shape and density on size measurement.' In Syvitski J.P.M. (Ed.) *Principles, Methods, and Application of Particle Size Analysis*. Cambridge, CUP, 22–33.

Matthews, J.A. 1987. 'Regional variation in the composition of Neoglacial end moraines, Jotunheimen, Norway: an altitudinal gradient in clast roundness and its possible palaeoclimatic significance.' *Boreas*, 16: 173–88.

Matthews, J.A. and Petch, J.R. 1982. 'Within-valley asymmetry and related problems of Neoglacial lateral moraine development at certain Jotunheimen glaciers, southern Norway.' *Boreas*, 11: 225–47.

McCave, I.N. and Syvitski, J.P.M. 1991. 'Principles and methods of geological particle size analysis.' In Syvitski, J.P.M. (Ed.) *Principles, Methods, and Application of Particle Size Analysis.* Cambridge, CUP, 3–21.

McManus J. 1988. 'Grain size determination and interpretation.' In Tucker M. (Ed.) *Techniques in Sedimentology.* Oxford, Blackwell Scientific: 63–85.

Melkerud, P. 1984. 'Distribution of clay minerals in soil profiles – a tool in chronostraigraphical and lithostratigraphical investigations of tills.' *Striae*, 20: 31–7.

Menzies, J. and Maltman, A.J. 1992. 'Microstructures in diamictons: evidence of subglacial bed conditions.' *Geomorphology*, 6: 27–40.

Menzies, J. 1998. 'Microstructures within subglacial diamictons.' In Kostrzewski, A. (Ed.) *Relief and Deposits of Present-day and Pleistocene Glaciation of the Northern Hemisphere – Selected Problems.* Poznan, Adam Michiewicz University Press, Geography Series 58, 153–66.

Menzies, J. 2000. 'Micromorphological analyses of microfabrics and microstructures indicative of deformation processes in glacial sediments.' In Maltman, A.J., Hubbard, B. and Hambrey, M.J., (Eds.) *Deformation of Glacial Materials.* Geological Society of London, Special Publications, 176, 245–57.

Miall, A.D. 1977. 'A review of the braided river depositional environment.' *Earth Science Reviews*, 13: 1–62.

Miall, A.D. 1978. 'Lithofacies types and vertical profile models in braided river deposits: a summary.' In Miall A.D. (Ed.), *Fluvial Sedimentology.* Canadian Society of Petroleum Geologists Memoir 5: 597–604.

Miall, A.D. 1985. 'Architectural-element analysis: a new method of facies analysis applied to fluvial deposits.' *Earth Science Reviews*, 22: 261–308.

Miall, A.D. 1988. 'Facies architecture in clastic sedimentary basins.' In Kleinspehn, K. and Paola, C. (eds), *New Perspectives in Basin Analysis.* New York, Springer-Verlag, 63–81.

Miall, A.D. 1992. *Principles of Sedimentary Basin Analysis.* New York, Springer-Verlag, 668.

Middleton, G.V. and Hampton, M.A. 1976. 'Subaqueous sediment transport and deposition by sediment gravity flows.' In Stanley, D.J. and Swift, D.J.P. (Eds.), *Marine Sediment Transport and Environmental Management.* New York, Wiley, 197–218.

Middleton, G.V. and Wilcock, P.R. 1994. *Mechanics in the Earth and Environmental Sciences.* Cambridge, Cambridge University Press, 459.

Miller, H. 1884. 'On boulder-glaciation.' *Royal Physical Society, Edinburgh Proceedings*, 8: 156–89.

Milner, H.B. (Ed.) 1962. *Sedimentary Petrography*, London, Allen and Unwin.

Mondadori, A. (Ed.) 1983. *The MacDonald Encyclopedia of Rocks and Minerals*, London, MacDonald and Co., 607.

Müller, B.U. and Schlüchter, C. 2001. 'Influence of the glacier bed lithology on the formation of a subglacial till sequence – ring-shear experiments as a tool for the classification of subglacial tills.' *Quaternary Science Reviews*, 20: 1113–25.

Murphy, C.P. 1986. *Thin Section Preparation of Soils and Sediments*. Berkhampstead, AB Academic Press, 149.

Murrary, T. and Dowdeswell, J.A. 1992. 'Water throughflow and the physical effects of deformation on sedimentary glacier beds.' *Journal of Geophysical Research*, 97 (B6): 8993–9002.

Nemec W. 1990. 'Aspects of sediment movement on steep delta slopes.' In Colella, A. and Prior, D. (Eds.), *Coarse Grained Deltas*. International Association of Sedimentologists, Special Publication 10: 29–73.

North American Commission on Stratigraphic Nomenclature. 1983. 'North American Stratigraphic Code.' *Bulletin of the American Association of Petroleum Geologists*, 67: 841–75.

Oldfield, F., Maher, B.A., Donoghue, J. and Pierce, J. 1985. 'Particle-size related mineral magnetic source-sediment linkages in the Rhode River catchment, Maryland, U.S.A..' *Journal of the Geological Society of London*, 142: 1035–46.

Oldfield, F., Thompson, R. and Dickson, D.P.E. 1981. 'Artificial enhancement of stream bedload: a hydrological application of superparamagnetism.' *Physics of the Earth and Planetary Interiors*, 26: 107–24.

Olsen, L. 1983. 'A method for determining total clast roundness in sediments.' *Boreas*, 12: 17–21.

Orford, J.D. and Whalley, W.B. 1991. 'Quantitative grain form analysis.' In Syvitski, J. (Ed) *Principles, Methods and Applications of Particle Size Analysis*, Cambridge, Cambridge University Press.

Ostry, R.C. and Deane, R.E. 1963. 'Microfabric analyses of Till.' *Geological Society of America Bulletin*, 74: 165–8.

Owen, L.A. 1991. 'Mass movement deposits in the Karakoram Mountains.' *Zeitschrift für Geomorphologie*, 35: 401–24.

Owen, L.A. 1994. 'Glacial and non-glacial diamictons in the Karakoram Mountains and Western Himalayas.' In Warren, W.P. and Croot, D.G. (Eds.) *The Formation and Deformation of Glacial Deposits*. Balkema, Rotterdam, 9–28.

Owen, L.A. and Derbyshire, E. 1988. 'Glacially-deformed diamictons in the Karakoram Mountains, northern Pakistan.' In Croot, D.G. (Ed.) *Glacitectonics: Forms and Processes*. Rotterdam, Balkema, 149–76.

Owen, L.A. and Derbyshire, E. 1989. 'The Karakoram glacial depositional system.' *Zeitschrift für Geomorphologie*, Supp.-Bd. 76: 33–73.

Park, R.G. 1983. *Foundations of Structural Geology*. London, Blackie.

Parker, G. and Andrews, E.D. 1985. 'Sorting of bedload sediment by flow in meander bends.' *Water Resources Research*, 21: 1361–73.

Peachey, D., Roberts, J.L., Vickers, B.P., Zalasiewicz, J.A. and Mathers, S.J. 1985. 'Resistate geochemistry of sediments – a promising tool for provenance studies.' *Modern Geology*, 9: 145–57.

Peachey, D., Roberts, J.L, Vickers, B.P., Zalasiewicz, J.A. and Mathers, S.J. 1986. *The Resistate Geochemistries of Crags and Other Quaternary Sediments from NE and SW Suffolk Compared*, British Geological Survey, ACRG Report No. 86/2.

Pedersen, S.A.S. 1993. 'The glaciodynamic event and glaciodynamic sequence.' In Aber J.S. (Ed.), *Glaciotectonics and Mapping Glacial Deposits*. Canadian Plains Research Center, University of Regina: 67–85.

Perrin, R.M.S. 1957. 'The clay mineralogy of some tills in the Cambridge district.' *Clay Mineral Bulletin*, 3: 193–205.

Pettijohn, F.J. 1957. *Sedimentary Rocks*. 2nd edn. London, Harper and Row, 718.

Pettijohn, F.J. 1975. *Sedimentary Rocks*, 3rd edn, London, Harper and Row.

Phillips, E.R and Auton, C.A. 1998. 'Micromorphology and deformation of a Quaternary glaciolacustrine deposit, Speyside, Scotland.' British Geological Survey Technical Report WG/98/15.

Phillips, E.R and Auton, C.A. 2000. 'Micromorphological evidence for polyphase deformation of glaciolacustrine sediments from Strathspey, Scotland.' In Maltman, A.J., Hubbard, B. and Hambrey, M.J., (Eds.) *Deformation of Glacial Materials*. Geological Society of London, Special Publications, 176, 279–92.

Phillips, E.R., Auton, C.A., Evans, D.J.A. and Benn, D.I. 2003. 'Drumbeg: glacial sedimentology and tectonostratigraphy.' In Evans, D.J.A. (Ed.) *The Quaternary of the Western Highland Boundary: Field Guide*. London, Quaternary Research Association, 88–103.

Phillips, E.R., Evans, D.J.A. and Auton, C.A. 2002. 'Polyphase deformation at an oscillating ice margin following the Loch Lomond Readvance, central Scotland, UK.' *Sedimentary Geology*, 149: 157–82.

Piotrowski, J.A. and Kraus, A. 1997. 'Response of sediment to ice-sheet loading in northwestern germany: effective stresses and glacier-bed stability.' *Journal of Glaciology*, 43: 495–502.

Piotrowski, J.A., Mickelson, D.M., Tulaczyk, S., Krzyszkowski, D. and Junge, F.W. 2001. 'Were deforming subglacial beds beneath past ice sheets really widespread?' *Quaternary International*, 86: 139–50.

Plumley W.J. 1948. 'Black Hills terrace gravels: a study in sediment transport.' *Journal of Geology*, 56: 526–77.

Popper, K.R. 1972. *Objective Knowledge*. Oxford, Oxford University Press.

Popper, K.R. 1974. *Conjectures and Refutations*. London, Routledge & Kegan Paul.

Porter, P.R. and Murray, T. 2001. 'Mechanical and hydraulic properties of till beneath Bakaninbreen, Svalbard.' *Journal of Glaciology*, 47: 167–75.

Posamentier, H.W., Summerhayes, C.P., Haq, B.U. and Allen, G.P. (Eds.) 1993. *Sequence Stratigraphy and Facies Associations*. International Association of Sedimentologists, Special Publication 18.

Potts, P.J. 1987. *A Handbook of Silicate Rock Analysis*, Glasgow, Blackie.

Powell, R.D. 1984. 'Glacimarine processes and inductive lithofacies modelling of ice shelf and tidewater glacier sediments based on Quaternary examples.' *Marine Geology*, 57: 1–52.

Powell, R.D. 1990. 'Glacimarine processes at grounding-line fans and their growth to ice-contact deltas.' In Dowdeswell, J.A. and Scourse, J.D. (Eds.), *Glacimarine Enviroments: Processes and Sediments*. Geological Society Special Publication 53: 53–73.

Powell, R.D. and Domack, E. 1995. 'Modern glaciomarine environments.' In Menzies, J. (Ed.), *Glacial Environments Volume 1 – Modern Glacial Environments: Processes, Dynamics and Sediments*. Oxford, Butterworth-Heinemann, 445–86.

Powers, M.C. 1953. 'A new roundness scale for sedimentary particles.' *Journal of Sedimentary Petrology*, 23: 117–19.

Prins, M. and Weltje, G. 1999. 'End-member modeling of siliclastic grain-size distributions: the Late Quaternary record of eolian and fluvial sediment supply to the Arabian Sea and its paleoclimatic significance.' In *Numerical Experiments in Stratigraphy: Recent Advances in Stratigraphic and Sedimentologic Computer Simulation,* SEPM. Special Publication No. 63: 91–111.

Prothero, D.R. and Schwab, F. 1996. *Sedimentary Geology*. New York, W.H. Freeman & Co.

Ramsay, J.G. and Huber, M.I. 1987. *The Techniques of Modern Structural Geology*. Vol 2: *Folds and Fractures*. London, Academic Press.

Rappol, M. 1985. 'Clast-fabric strength in tills and debris flows compared for different environments.' *Geologie en Mijnbouw*, 64: 327–32.

Reading, H.G. (Ed.) 1986. *Sedimentary Environments and Facies*. Oxford, Blackwell.

Reinick H.E. and Singh I.B. 1980. *Depositional Sedimentary Environments*. Berlin, Springer-Verlag.

Reiter, S. 1991. *New Directions for Teaching and Learning*. Vol 45. San Francisco, Jossey-Bass Inc.

Reynolds, J.M. 1997. *An Introduction to Applied and Environmental Geophysics*. Chichester, Wiley, 796.

Rice, S. and Church, M. 1996. 'Sampling surficial fluvial gravels: the precision of size distribution percentile estimates.' *Journal of Sedimentary Research*, 66: 654–65.

Rijsdijk, K.F., Owen, G., Warren, W.P., McCarroll, D. and van der Meer J.J.M. 1999. 'Clastic dykes in over-consolidated tills: evidence for subglacial hydrofracturing at Killiney Bay, eastern Ireland.' *Sedimentary Geology*, 129: 111–26.

Rinbgerg, B. and Erlstrom, M. 1999. 'Micromorphology and petrography of Late Weichselian glaciolacustrine varves in southeastern Sweden.' *Catena*, 35: 147–77.

Ringrose, T. and Benn, D.I. 1997. 'Confidence regions for fabric shape diagrams.' *Journal of Structural Geology*, 19: 1527–36.

Roberts, D.H. 1995. 'The development of criteria to distinguish glaciotectonic and glaciomarine sedimentary environments.' Unpublished PhD thesis, University of Southampton.

Roberts, J.L. and Peachey, D. 1983. *The Use of XRF Data for Correlating Quaternary Sediments*, British Geological Survey, ACRG Report No. 129.

Robin, G.de.Q., Evans, E. and Bailey, J.T. 1969. 'Interpretation of radio echo sounding in polar ice sheets.' *Philosophical Transactions of the Royal Society of London Ser. A*, 265: 437–505.

Rose, J. 1981. 'Field guide to the Quaternary geology of the south eastern part of the Loch Lomond basin.' *Proceedings of the Geological Society of Glasgow*, 122/123, 12–28.

Rose, J. 1989a. 'Stadial type sections in the British Quaternary.' In Rose, J. and Schluchter, C. (Eds.), *Quaternary Type Sections: Imagination or Reality?* Rotterdam, Balkema, 45–67.

Rose, J. 1989b. 'Glacier stress patterns and sedimnet transfer associated with superimposed flutes.' *Sedimentary Geology*, 62: 151–76.

Rose, J. 1992. 'Boulder clusters in glacial flutes.' *Geomorphology*, 6: 51–8.

Rose, J. and Allen, P. 1977. 'Middle Pleistocene stratigraphy in south-east Suffolk.' *Journal of the Geological Society of London*, 133: 83–102.

Rose J. and Menzies J. 1996. 'Glacial stratigraphy.' In Menzies J. (Ed.), *Past Glacial Environments: Sediments, Forms and Techniques*. Oxford, Butterworth-Heinemann, 253–84.

Rose, J., Lee, J.A., Moorlock, B.S.P. and Hamblin, R.J.O. 1999. 'The origin of the Norwich Brickearth: micromorphological evidence for pedological alteration of sandy Anglian Till in northeast Norfolk.' *Proceedings of the Geologists Association*, 110: 1–8.

Rowe, P.W. and Barden, L. 1966. 'A new consolidation cell.' *Geotechnique*, 16: 162–70.

Sambrook Smith, G.H., Nicholas, A.P. and Ferguson, R.I. 1997. 'Measuring and defining bimodal sediments: problems and implications.' *Water Resources Research*, 33: 1179–85.

Sauer, K.E. and Christiansen, E.A. 1993. 'Preconsolidation pressures in the Battleford Formation, southern Saskatchewan, Canada.' *Canadian Journal of Earth Sciences*, 28: 1613–23.

Sauer, K.E., Egeland, A.K. and Christiansen, E.A. 1993. 'Preconsolidation of tills and intertill clays by glacial loading in southern Saskatchewan, Canada.' *Canadian Journal of Earth Sciences*, 30: 420–33.

Sauer, K.E., Gareau, L.F. and Christiansen, E.A. 1990. 'Softening of overconsolidated Cretaceous clays by glacial erosion.' *Quarterly Journal of Engineering Geology*, 23: 307–24.

Saunderson H.C. 1975. 'Sedimentology of the Brampton esker and its associated deposits: an empirical test of theory.' In Jopling A.V. and McDonald B.C. (Eds.), *Glaciofluvial and Glaciolacustrine Sedimentation*. SEPM Special Publication 23: 155–76.

Scheidegger, A.E. 1965. 'On the statistics of the orientation of bedding planes, grain axes, and similar sedimentological data.' *United States Geological Survey Professional Paper*, 525-C, 164–7.

Schokking, F. 1990. 'On estimating the thickness of the Saalian ice sheet from a vertical profile of preconsolidation loads of a lacustro-glacial clay.' *Geologie en Mijnbouw*, 69: 305–12.

Shakesby, R.A. 1979. 'A simple device for measuring the primary axes of clasts.' *British Geomorphological Research Group, Technical Bulletin*, No. 24, 11–13.

Sharp, M., Jouzel, J., Hubbard, B. and Lawson, W. 1994. 'The character, structure and origin of the basal ice layer of a surge-type glacier.' *Journal of Glaciology*, 40(135): 327–40.

Sharp, M.J. and Gomez, B. 1986. 'Processes of debris comminution in the glacial environment and implications for quartz sand-grain micromorphology.' *Sedimentary Geology*, 46: 33–47.

Sharp, M.J. 1982. 'Modification of clasts in lodgement tills by glacial erosion.' *Journal of Glaciology*, 28: 475–81.

Sharp, M.J. 1985. ' "Crevasse-fill" ridges – a landform type characteristic of surging glaciers?' *Geografiska Annaler*, 67A: 213–20.

Shaw, J. 1996. 'A meltwater model for Laurentide subglacial landscapes.' In McCann, S.B. and Ford, D.C. (Eds.), *Geomorphology Sans Frontières*, Chichester, Wiley, 181–236.

Shaw J. 2002. 'The meltwater hypothesis for subglacial bedforms.' *Quaternary International*, 90: 5–22.

Sheridan, M., Wohletz, K., and Dehn, J. 1987. 'Discrimination of grain-size subpopulations in pyroclastic deposits.' *Geology*, 15: 367–70.

Shetsen, I. 1984. 'Application of till pebble lithology to the differentiation of glacial lobes in southern Alberta.' *Canadian Journal of Earth Science*, 21: 920–33.

Shilts, W.W. 1993. 'Geological Survey of Canada's contributions to understanding the composition of glacial sediments.' *Canadian Journal of Earth Science*, 30: 333–53.

Shilts, W.W. and Kettles, I.M. 1990. 'Geochemical-mineralogica profiles through fresh and weathered till', In Kujansuu, R. and Saarnisto, M. (Eds.), *Glacier indicator tracing*. Balkema, Rotterdam, 187–216.

Sime, L.C. and Ferguson, R.I. 2003. 'Information on grain sizes in gravel-bed rivers by automated image analysis.' *Journal of Sedimentary Research, Section B: Stratigraphy and Global Studies*, 73: 630–6.

Sitler, R.F. and Chapman, C.A. 1955. 'Microfabrics of Till from Ohio and Pennsylvania.' *Journal of Sedimentary Petrology*, 25: 262–9.

Sladen, J.A. and Wrigley, W. 1983. 'Geotechnical properties of lodgement tills – a review.' In Eyles, N. (Ed.), *Glacial Geology: an Introduction for Engineers and Earth Scientists*, Pergamon Press, Oxford, 184–212.

Smart, P. and Tovey, N.K. 1982. *Electron Microscopy of Soils and Sediments: Techniques*. Clarendon Press, Oxford, 264.

Smith, J.P. 1985. *Mineral Magnetic Studies on Two Shropshire-Cheshire Meres*, Unpublished Ph.D. thesis, University of Liverpool.

Smith, N.D. and Ashley, G.M. 1985. 'Proglacial lacustrine environments.' In Ashley, G.M., Shaw, J. and Smith, N.D., (Eds.), *Glacial Sedimentary Environments*. SEPM Short Course 16, 135–215.

Snäll, S. 1985. 'Weathering in till indicated by clay mineral distribution.' *Geologiska, Foreningens i Stockholm Förhandlinger*, 107: 315–22.

Sneed, E.D. and Folk, R.L. 1958. 'Pebbles in the lower Colorado River, Texas, a study in clast morphogenesis.' *Journal of Geology*, 66: 114–50.

Spedding, N.F. 2000. 'Hydrological controls on sediment transport pathways: implications for debris-covered glaciers.' In Nakawo, M., Fountain, A. and Raymond, C. (Eds.) *Debris-Covered Glaciers.* IAHS Publication 264, 133–42.

Sridharan, A., Abraham, B.M. and Jose, BT. 1991. 'Improved technique for estimation of presoncolidation pressure.' *Géotechnique*, 41: 263–8.

Starkey, H.C., Blackmon, P.D. and Hauff, P.L. 1984. *The Routine Mineralogical Analysis of Clay Bearing Samples*, U.S. Geological Survey Bulletin 1563, U.S. Government Printing Office, Washington.

Sugai, T. 1993. 'River terrace development by concurrent fluvial processes and climatic changes.' *Geomorphology*, 6: 243–52.

Syvitski J.P.M. (Ed.) 1991. *Principles, Methods and Application of Particle Size Analysis.* Cambridge, CUP.

Thomas, G.S.P. and Connell, R.J. 1985. 'Iceberg drop, dump and grounding structures from Pleistocene glacio-lacustrine sediments, Scotland.' *Journal of Sedimentary Petrology*, 55: 243–9.

Thompson, R. and Oldfield, F. 1986. *Environmental Magnetism*, Thompson, Allen and Unwin.

Tiljander, M.S., Ojala, A.E.K., Saarinen, T.J. and Snowball, I.F. 2001. 'Documentation of the physical properties of annually laminated (varved) sediments at a sub-annual to decadal resolution for environmental interpretation.' In Ojala, A.E.K. (Ed.) *Varved Lake Sediments in Southern and Central Finland: Long Varve Chronologies as a Basis for Holocene Palaeoenvironmental Reconstructions.* Espoo, Geological Survey of Finland Publication, 1–10.

Tippkötter, R., and Ritz, K. 1996. 'Evaluation of polyester, epoxy and acrylic resins for suitability in preparation of soil thin-sections for in situ biological studies.' *Geoderma*, 69: 31–57.

Toyoshima, M. 1987. 'Low downstream decrease rate of particle size in latest Pleistocene fluvial terrace deposits in the Dewa Mountains, Northeastern Japan.' *Science Reports of the Tohoku University, 7th Series (Geography)*, 37: 174–86.

Truffer, M., Harrison, W.D. and Echelmeyer. K.A. 2000. 'Glacier motion dominated by processes deep in underlying till.' *Journal of Glaciology*, 46(153): 213–21.

Tucker, M. 1982. *The Field Description of Sedimentary Rocks.* Milton Keynes, Open University Press.

Tucker, M. (Ed.) 1988. *Techniques in Sedimentology*, Oxford, Blackwell Scientific, 394.

Tulaczyk, S. 1999. 'Ice sliding over weak, fine-grained tills: dependence of ice-till interactions on till granulometry. In Mickelson, D.M. and Attig, J.W. (eds), *Glacial Processes Past and Present.* Geological Society of America Special Paper 337, Boulder, 159–177.

Tulaczyk, S., Kamb, W.B. and Engelhardt, H.F. 2000a. 'Basal mechanics of Ice Stream B, West Antarctica. 1. Till Mechanics.' *Journal of Geophysical Research*, 105 (B1): 463–81.

Tulaczyk, S., Kamb, W.B. and Engelhardt, H.F. 2000b. 'Basal mechanics of Ice Stream B, West Antarctica. 2. Undrained plastic bed model.' *Journal of Geophysical Research*, 105 (B1): 483–94.

Tulaczyk, S., Kamb, W.B. and Engelhardt, H.F. 2001. 'Estimates of effective stress beneath a modern West Antarctic ice stream from till preconsolidation and void ratio.' *Boreas*, 30: 101–14.

Twiss, R.J. and Moores, E.M. 1992. *Structural Geology*. New York, Freeman, 532.

Vandenberghe, J. 1988. 'Cryoturbations'. In Ed. M.J. Clark, *Advances in Periglacial Geomorphology*, Chichester, Wiley, 179–98.

van der Meer, J.J.M. and Laban, C. 1990. 'Micromorphology of some North Sea till samples, a pilot study.' *Journal of Quaternary Science*, 5: 95–101.

van Andel T.H. 1985. *New Views on an Old Planet*. Cambridge, Cambridge University Press.

van den Berg E.H., Bense V.F. and Schlager W. 2003. 'Assessing textural variation in laminated sands using digital image analysis of thin sections.' *Journal of Sedimentary Research, Section A: Sedimentary Petrology and Processes*, 73: 133–43.

van der Meer, J.J.M. 1987. 'Micromorphology of glacial sediments as a tool in distinguishing genetic varieties of till.' In Kujansuu, R. and Saarniston, M. (Eds.), *INQUA Till Symposium*. Geological Survey of Finland, Special Paper 3, 77–90.

van der Meer, J.J.M. 1993. 'Microscopic evidence of subglacial deformation.' *Quaternary Science Reviews*, 12: 553–87.

van der Meer, J.J.M. 1996. 'Micromorphology.' In Menzies, J. (Ed.) *Past Glacial Environments – Sediments, Forms, Techniques*. Oxford, Pergamon, 335–55.

van der Meer, J.J.M. 1997. 'Particle and aggregate mobility in till: microscopic evidence of subglacial processes.' *Quaternary Science Reviews*, 16: 827–31.

van der Meer, J.J.M. and Laban, C. 1990. 'Micromorphology of some North Sea till samples, a pilot study.' *Journal of Quaternary Science*, 5: 95–101.

van der Meer, J.J.M., Rabassa, J.O. and Evenson, E.B. 1992a. 'Micromorphological aspects of glaciolacustrine sediments in northern Patagonia, Argentina.' *Journal of Quaternary Science*, 7, 31–44.

van der Meer, J.J.M., Mucher, H.J. and Hofle, H.C. 1992b. 'Micromorphological observations on till samples from the Shackleton Range and North Victoria Land, Antarctica.' *Polarforschung*, 62: 57–65.

van der Wateren F.M. 1995. 'Processes of glaciotectonism.' In Menzies J. (Ed.), *Modern Glacial Environments: Processes, Dynamics and Sediments*. Oxford, Butterworth-Heinemann, 309–35.

van der Wateren, F.M. 1999. 'Structural geology and sedimentology of Saalian tills near Heiligenhafen, Germany.' *Quaternary Science Reviews*, 18: 1625–39.

van der Wateren, F.M., Kluiving, S.J. and Bartek, L.R. 2000. 'Kinematic indicators of subglacial shearing.' In Maltman, A., Hubbard, B. and Hambrey, M.J. (Eds.) *Deformation of Glacial Materials*. Geological Society of London, Special Publications 176, 259–78.

van Vliet-Lanoe, B. 1998. 'Frost and soils: implications for paleosols, paleoclimates and stratigraphy.' *Catena*, 34: 157–83.

van Wagoner, J.C., Posamentier, H.W., Mitchum, R.M., Vail, P.R., Sarg, J.F., Loutit, T.S. and Hardenbol J. 1988. 'An overview of the fundamentals of sequence stratigraphy and key definitions.' In Wilgus C.K. et al. (Eds.), *Sea-Level Changes: An Integrated Approach*. Society of Economic Paleontologists and Mineralogists, Special Publication 42: 39–46.

Vere, D.M. and Benn, D.I. 1989. 'Structure and debris characteristics of medial moraines in Jotunheimen, Norway: implications for moraine classification.' *Journal of Glaciology*, 35: 276–80.

Vickers B. 1983. *Laboratory Work in Soil Mechanics.* London, Granada Publishing Ltd., 170.

Visser, J.N.J., Colliston, W.P. and Terblanche, J.C. 1984. 'The origin of soft-sediment deformation structures in Permo-Carboniferous glacial and proglacial beds, South Africa.' *Journal of Sedimentary Petrology*, 54: 1183–96.

Von der Haar, S.P. and Johnson, W.H. 1973. 'Mean magnetic susceptibility: a useful parameter for stratigraphic studies of glacial till.' *Journal of Sedimentary Petrology*, 43: 1148–51.

Walden, J. and Slattery, M.C. 1993. 'Verification of a simple gravity technique for separation of particle size fractions suitable for mineral magnetic analysis.' *Earth Surface Processes and Landforms*, 18: 829–33.

Walden, J. 1999. 'Sample collection and preparation.' In Walden, J., Oldfield, F. and Smith, J.P. (Eds) *Environmental Magnetism: A Practical Guide,* Technical Guide Series, No. 6, London, Quaternary Research Association, 26–34.

Walden, J. and Addison, K. 1995. 'Mineral magnetic analysis of a "weathering" surface within glacigenic sediments at Glanllynnau, North Wales.' *Journal of Quaternary Science*, 10: 367–78.

Walden, J. Smith, J.P. and Dackombe, R.V. 1996. 'A comparison of mineral magnetic, geochemical and mineralogical techniques for compositional studies of glacial diamicts.' *Boreas*, 25: 115–30.

Walden, J., Oldfield, F. and Smith, J.P. (Eds) 1999. *Environmental Magnetism: A Practical Guide*, Technical Guide Series, No. 6, London, Quaternary Research Association, 250.

Walden, J., Smith, J.P. and Dackombe, R.V. 1987. 'The use of mineral magnetic analyses in the study of glacial diamicts.' *Journal of Quaternary Science*, 2: 77–80.

Walden, J., Smith, J.P. and Dackombe, R.V. 1992a. 'Mineral magnetic analyses as a means of lithostratigraphic correlation and provenance indication of glacial diamicts: Intra- and inter-unit variation.' *Journal of Quaternary Science*, 7 (3): 257–70.

Walden, J., Smith, J.P. and Dackombe, R.V. 1992b. 'The use of simultaneous R- and Q-mode factor analysis as a tool for assisting interpretation of mineral magnetic data.' *Mathematical Geology*, 24 (3): 227–47.

Walden, J., Smith, J.P., Dackombe, R.V. and Rose, J. 1995. 'Mineral magnetic analyses of glacial diamicts from the Midland Valley of Scotland.' *Scottish Journal of Geology*, 31 (1): 79–89.

Walker, R.G. 1992. 'Facies, facies models and modern stratigraphic concepts.' In Walker, R.G. and James, N.P. (Eds.) *Facies Models: Response to Sea-level Change*, Toronto, Geological Association of Canada, 1–14.

Walther, J. 1894. *Enleitung in die Geologie als historische Wissenschaft, Bd 3: Lithogenesis der Gegenwart.* Jena, Fischer Verlag.

Watson, G.S. 1966. 'The statistics of orientation data.' *Journal of Geology*, 74: 786–97.

Wentworth, C.K. 1922. 'A scale of grade and class terms for clastic sediments.' *The Journal of Geology*, 30: 377–92.

Whalley, W.B. 1978. 'An SEM examination of quartz grains from sub-glacial and associated environments and some methods for their characterization.' *Scanning Electron Microscopy*, 1: 355–8.

Whalley, W.B. 1996. 'Scanning Electron Microscopy.' In Menzies, J. (Ed.) *Past Glacial Environments: Sediments, Forms and Techniques.* Oxford, Butterworth-Heinemann, 357–75.

Whalley, W.B. and Orford, J. 1982. 'Analysis of SEM images of sedimentary particle form by fractal dimension and Fourier analysis methods.' *Scanning Electron Microscopy*, 2: 639–47.

Whitaker, S.H. and Christiansen, E.A. 1972. 'The Empress Group in southern Saskatchewan.' *Canadian Journal of Earth Sciences*, 9: 353–60.

Whitlow, R. 1983. *Basic Soil Mechanics.* London, Longman Group Limited, 439.

Whittaker, A., Cope, J.C.W., Cowie, J.W., Gibbons, W., Hailwood, E.A., House, M.R., Jenkins, D.G., Rawson, P.F., Rushton, A.W.A., Smith, D.G., Thomas, A.T. and Wimbledon, W.A. 1991. 'A Guide to Stratigraphical Procedure.' *Journal of the Geological Society of London*, 148: 813–24.

Wilcock, P.R. 1993. 'Critical shear stress of natural sediments.' *Journal of Hydraulic Engineering*, 119: 491–505.

Wilson, M.J. (Ed.) 1987. *A Handbook of Determinative Methods in Clay Mineralogy.* London, Blackie.

Wohl, E.E., Anthony, D.J., Madsen, S.W. and Thompson, D.M. 1996. 'A comparison of surface sampling methods for coarse fluvial sediments.' *Water Resources Research*, 32: 3219–26.

Wolcott, J. and Church, M. 1991. 'Strategies for sampling spatially heterogeneous phenomena: the example of river gravels.' *Journal of Sedimentary Petrology*, 61: 534–43.

Wolman, M.G. 1954. 'A method for sampling coarse river-bed material.' *American Geophysical Union Transactions*, 35: 951–6.

Wolpert, L. 1992. *The Unnatural Nature of Science.* London, Faber and Faber, 191.

Woodcock, N.H. and Naylor, M.A. 1983. 'Randomness testing in three-dimensional orientation data.' *Journal of Structural Geology*, 5: 539–48.

Woodcock, N.H. 1977. 'Specification of fabric shapes using an eigenvalue method.' *Geological Society of America Bulletin*, 88: 1231–6.

Zaniewski, K. 2001. 'Plasmic fabric analysis of glacial sediments using quantitative image analysis methods and GIS techniques.' Ph.D thesis, Universiteit van Amsterdam.

FIG G1 Gravel-filled channel between two basal tills, Tierra del Fuego. Photo: DIB.

FIG G2 Scour and fill structures in glacifluvial sands and gravels, southern Scotland. Photo: DJAE.

FIG G3 Normally graded gravel with welded basal contact, glacilacustrine delta, Scotland. Photo: DIB.

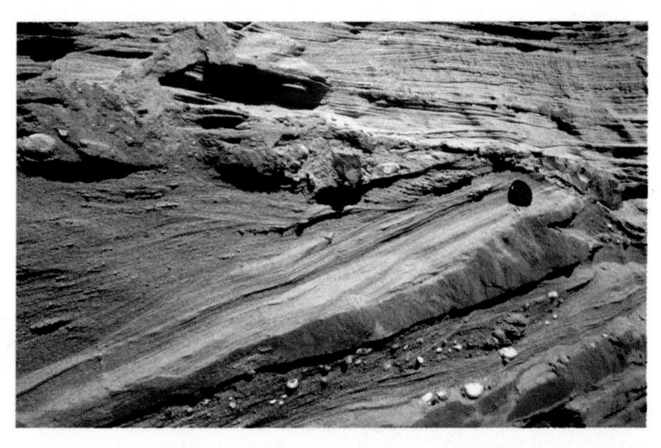

FIG G4 Sandy clinoforms showing outsized clasts, reactivation surface and diamicton intraclasts, Alberta, Canada. Photo: DJAE.

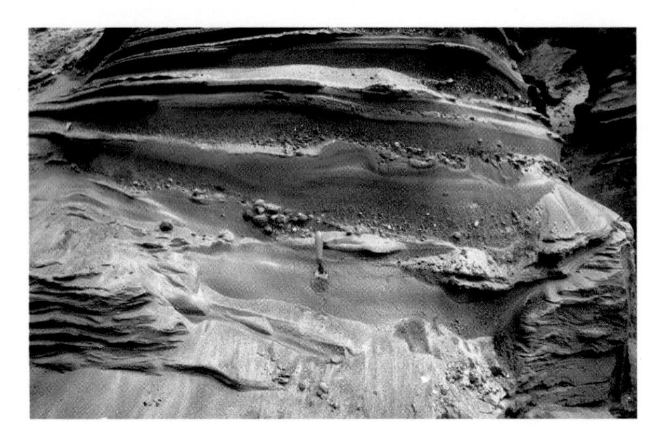

FIG G5 Normal grading in low angle clinoforms, central Scotland. Photo: DJAE.

FIG G6 Gravelly clinoforms (delta foresets), central Scotland. Photo: DJAE.

FIG G7 Climbing ripples, gravelly-mass flows and rhythmites (ice-proximal subaqueous fan), Iceland. Photo: DJAE.

FIG G8 Climbing ripple cross-lamination, Patagonia. Photo: DIB.

FIG G9 Dropstone in laminated silts, Ellesmere Island, Canada. Photo: DJAE.

FIG G10 Dropstone-rich laminated silts interbedded with stratified mass flow diamictons, south-east Ireland. Photo: DJAE.

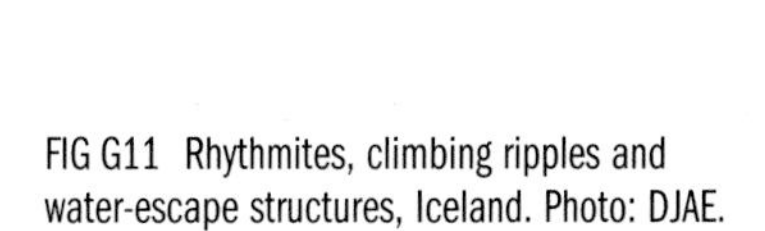

FIG G11 Rhythmites, climbing ripples and water-escape structures, Iceland. Photo: DJAE.

FIG G12 Laminated silt and mud (varves), Iceland. Photo: DJAE.

FIG G13 Normal fault in ice-contact outwash, Wales. Photo: DJAE.

FIG G14 Convolute lamination (syn-depositional soft-sediment deformation) Patagonia. Photo: DIB.

FIG G15 Sand intraclasts in gravel (rip-up clasts), central Scotland. Photo: DJAE.

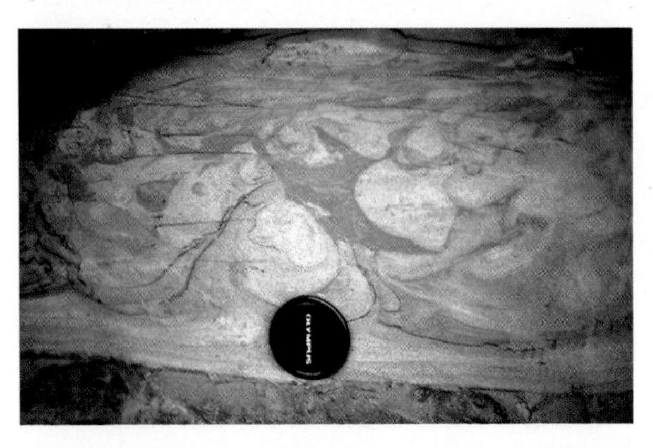

FIG G16 Ball and pillow structures in glacilacustrine silt and sand, Wales. Photo: DJAE.

FIG G17 Gravel lag between till and fluvial sand, Alberta, Canada. Photo: DJAE.

FIG G18 Boulder pavement between basal tills, Tierra del Fuego. Photo: DIB.

FIG G19 Stratified diamicton with clast-rich layers (subaerial debris flow deposits), Skye, Scotland. Photo: DIB.

FIG G20 Matrix-supported gravel (hyper-concentrated flow deposit, proglacial lake outlet), Canadian Rocky Mountains. Photo: DJAE.

FIG G21 Laminated matrix-supported to gravelly diamicton (thin subaqueous debris flow) in laminated silts, Ellesmere Island, Canada. Photo: DJAE.

FIG G22 Interbedded sand and laminated diamicton (alternating waterlaid and mass flow deposits), Alberta, Canada. Photo: DJAE.

FIG G23 Stratified to massive matrix-supported diamicton (subaqueous mass flow deposits), Scotland. Photo: DIB.

FIG G24 Matrix-supported sandy to gravelly clinoforms (subaqueous fan), Scotland. Photo: DIB.

FIG G25 Contorted sand and gravel lenses in matrix-supported diamicton (glacitectonized subglacial canal fills), eastern England. Photo: DJAE.

FIG G26 Glacitectonized laminated diamicton, western Scotland. Photo: DIB.

FIG G27 Chalk laminae and augen structures, showing pressure-shadow folds. Ice flow from right to left. South-east England. Photo: D. Roberts.

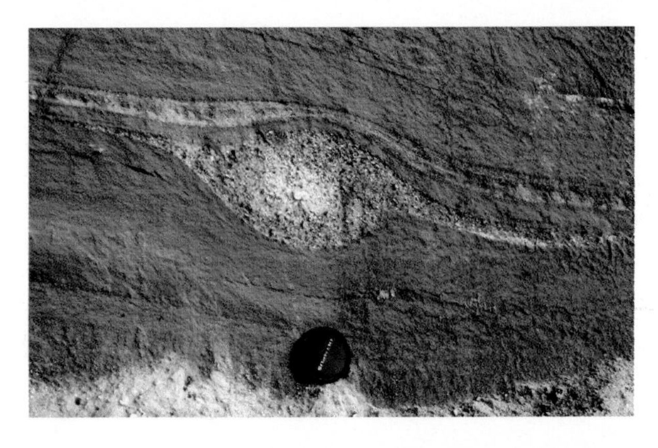

FIG G28 Gravel augen structure in glacitectonized sands and gravels, Denmark. Photo: DJAE.

FIG G29 Irregular injected clay bodies in glacitectonized sands, central Scotland. Photo: DJAE.

FIG G30 Liquefied silts and fine sands following thrust faults, Iceland. Photo: DJAE.

FIG G31 Laminated diamicton (subglacially deformed slope deposits), Ireland. Photo: DJAE.

FIG G32 Laminated diamicton overlying laminated muds, Alberta, Canada. Photo: DJAE.

FIG G33 Double stoss-lee clast in basal till, Patagonia. Photo: DIB.

FIG G34 Massive, porous diamicton overlying fissile, compact diamicton (A and B Horizon tills), Breidamerkurjokull, Iceland. Photo: DIB.

FIG G35 Fissile basal till, Scotland. Photo: DIB.

FIG G36 Matrix-supported diamicton with anastomosing silt stringers (basal till), Ireland. Photo: DIB.

FIG G37 Lower and upper laminated diamictons, Alberta, Canada. Photo: DJAE.

FIG G38 Glacitectonized sands overlain by laminated and massive diamicton, Alberta, Canada. Photo: DJAE.

FIG G39 Hydrofracture filled with gravel, Ireland. Photo: DIB.

FIG G40 Massive diamicton overlying gravels, Alberta, Canada. Photo: DJAE.

FIG G41 Laminated sediments in thin section, Marks Tey, Essex. Paired image with plane-polarized (left) and cross-polarized (right) light. Field of view 20mm. Thin sections allow the examination of laminated sediments in considerable detail. Images courtesy of Adrian Palmer.

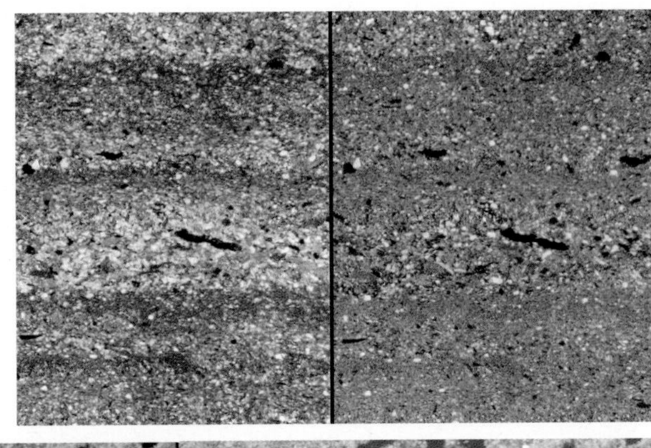

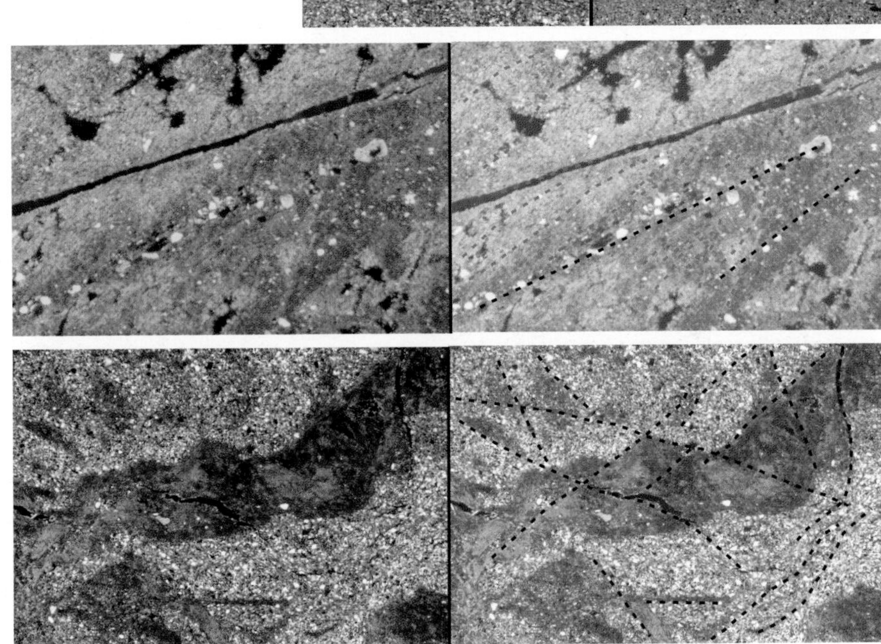

FIG G42 (above) Planar deformation structures in thin section. Note each image is paired, with original and annotated versions. Top: lineations of particles (black lines), and shear structures (red lines) reflecting zones of discrete shear, glacitectonized marine sediments, North Sea (cross-polarized light, field of view 8mm). Bottom: discrete shears (black lines) in a clay bed undergoing brecciation as a consequence of glacitectonics, Weichselian Till, Norwegian Channel (cross-polarized light, field of view 16mm). The fine-grained bed displays locally strong omnisepic plasmic fabric, reflecting an earlier stage of more pervasive deformation.

FIG G43 (right) Cyclopels from glacimarine sediments, Alaska. Photo: J. Hiemstra.

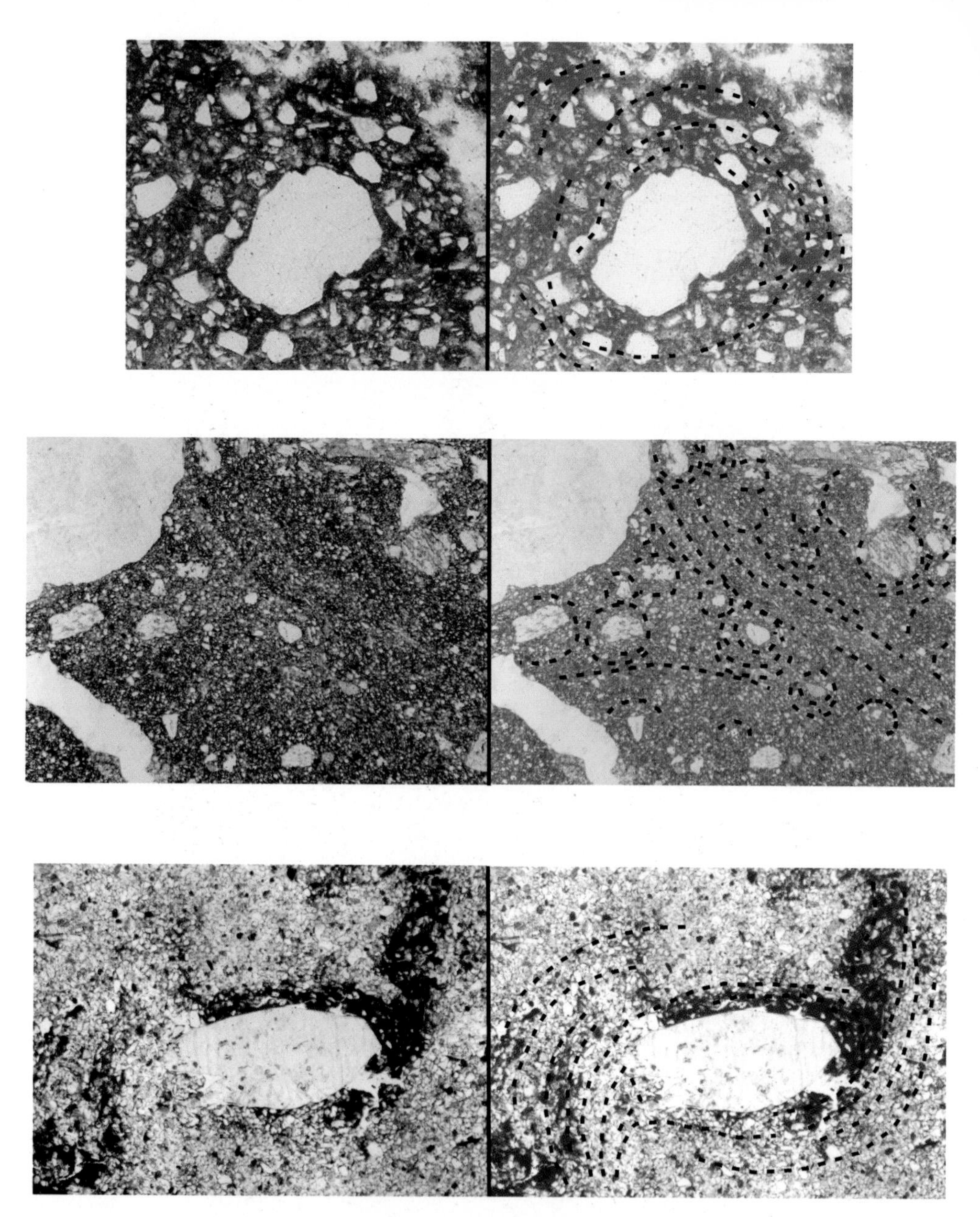

FIG G44 Rotational deformation structures in thin section. Note each image is paired, with original and annotated versions. Top: individual rotation 'galaxy' structure, with circular alignment of grains and matrix around a larger particle. Weichselian till, southern North Sea (plane-polarized light, field of view 6mm). Middle: associations of rotation features and a pressure shadow, subglacial till, Svalbard (Carr 2001). Bottom: rotational structure, highlighted by a tail of fine-grained sediment and grain orientations, Weichselian till, southern North Sea (plane-polarized light, field of view 9mm).

FIG G45 Water escape structure, Weichselian 'red-series' till, Scotland (plane-polarized light, field of view 18mm). The arrow identifies the point at which the water escape feature has propagated from. Also note the circular structure near the centre-left of the image: this is an air bubble trapped between the sample and slide coverslip.

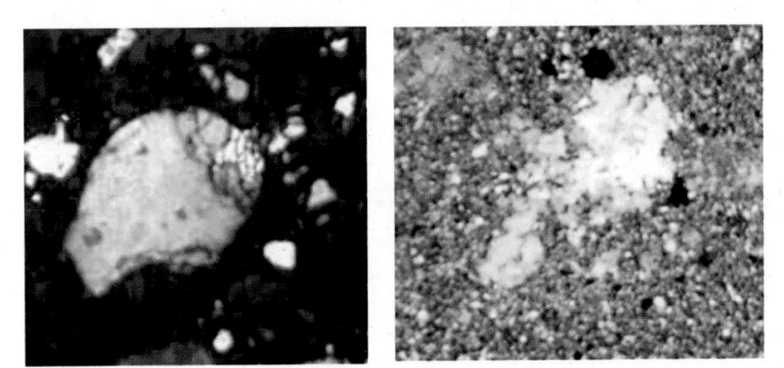

FIG G46 Crushed quartz grains, Weichselian subglacial tills. Left: *in situ* crushed grain, fractured through tangential loading (Hiemstra and van der Meer 1997). Right: fractured quartz grain with fragments being translocated (top left of main grain).

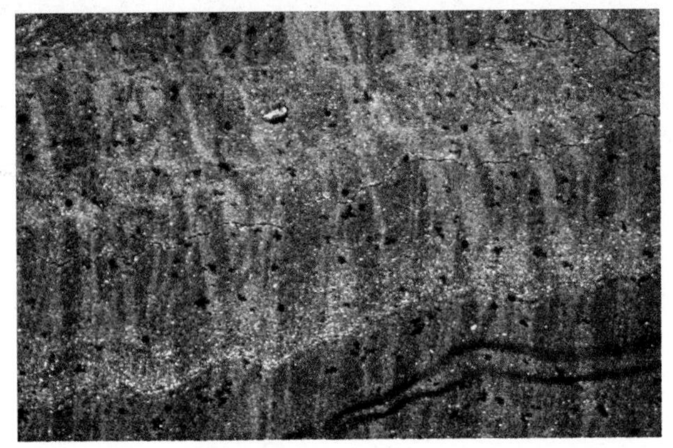

FIG G47 Kinking plasmic fabric, produced as a consequence of proglacial compressive glacitectonic deformation in glacilacustrine sediments from Llavorsí, southern Pyrenees (see Bordonau and van der Meer 1994). Cross-polarized light, field of view 18mm. Micrograph courtesy of Jaap van der Meer.

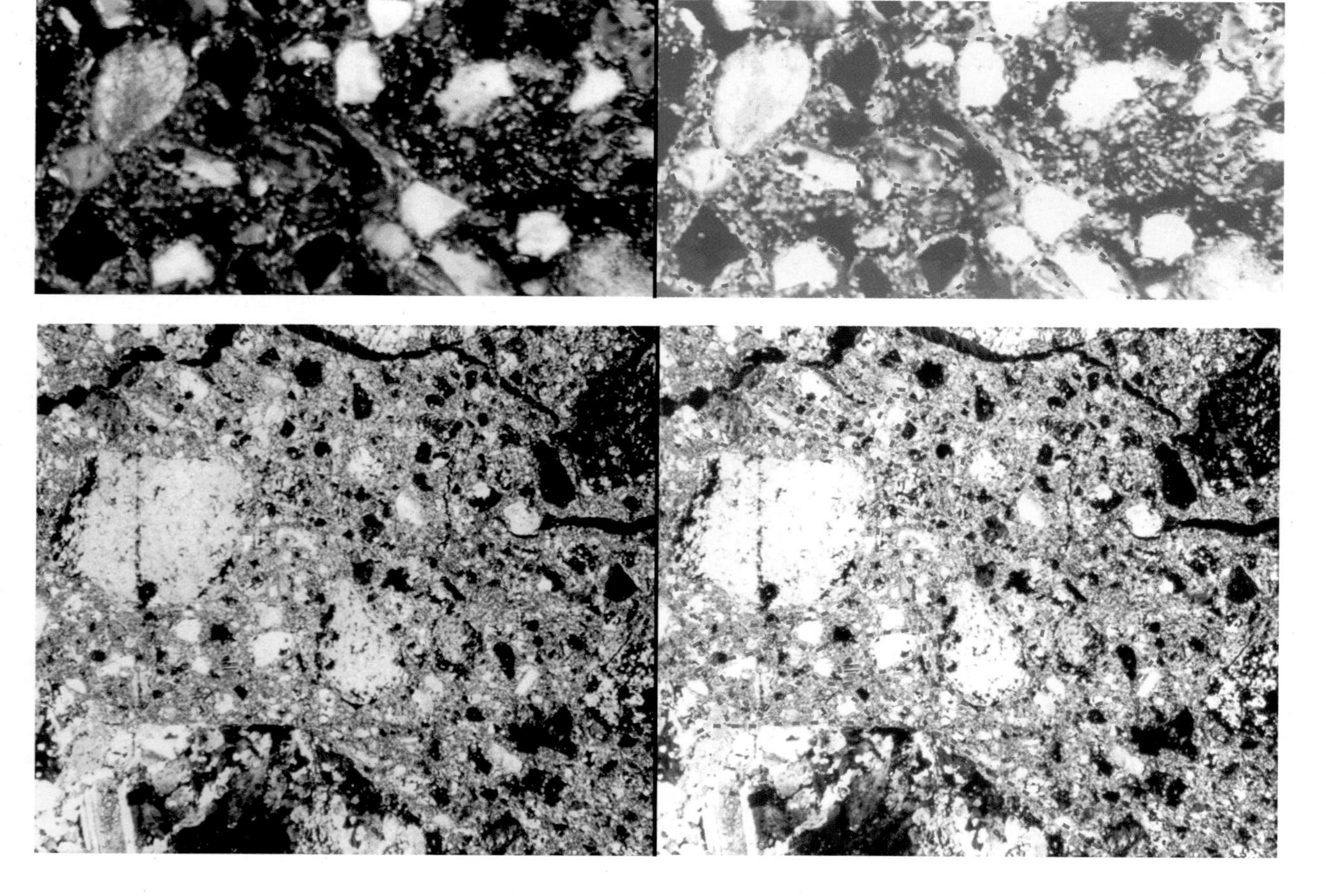

FIG G48 Examples of rotational plasmic fabrics. Note each image is paired with original and annotated versions. Top: skelsepic plasmic fabric, displaying characteristic 'extinction bands' due to the orientation of clays around each particle, Weichselian till, southern North Sea (cross-polarized light, field of view 8mm). Bottom: skelsepic and lattisepic plasmic fabrics, subglacial till, Svalbard (cross-polarized light, field of view 16mm). Whilst the skelsepic plasmic fabric clearly indicates rotation, the development of a lattice-structure may suggest a degree of strain hardening of the sediment, and the onset of planar deformation.

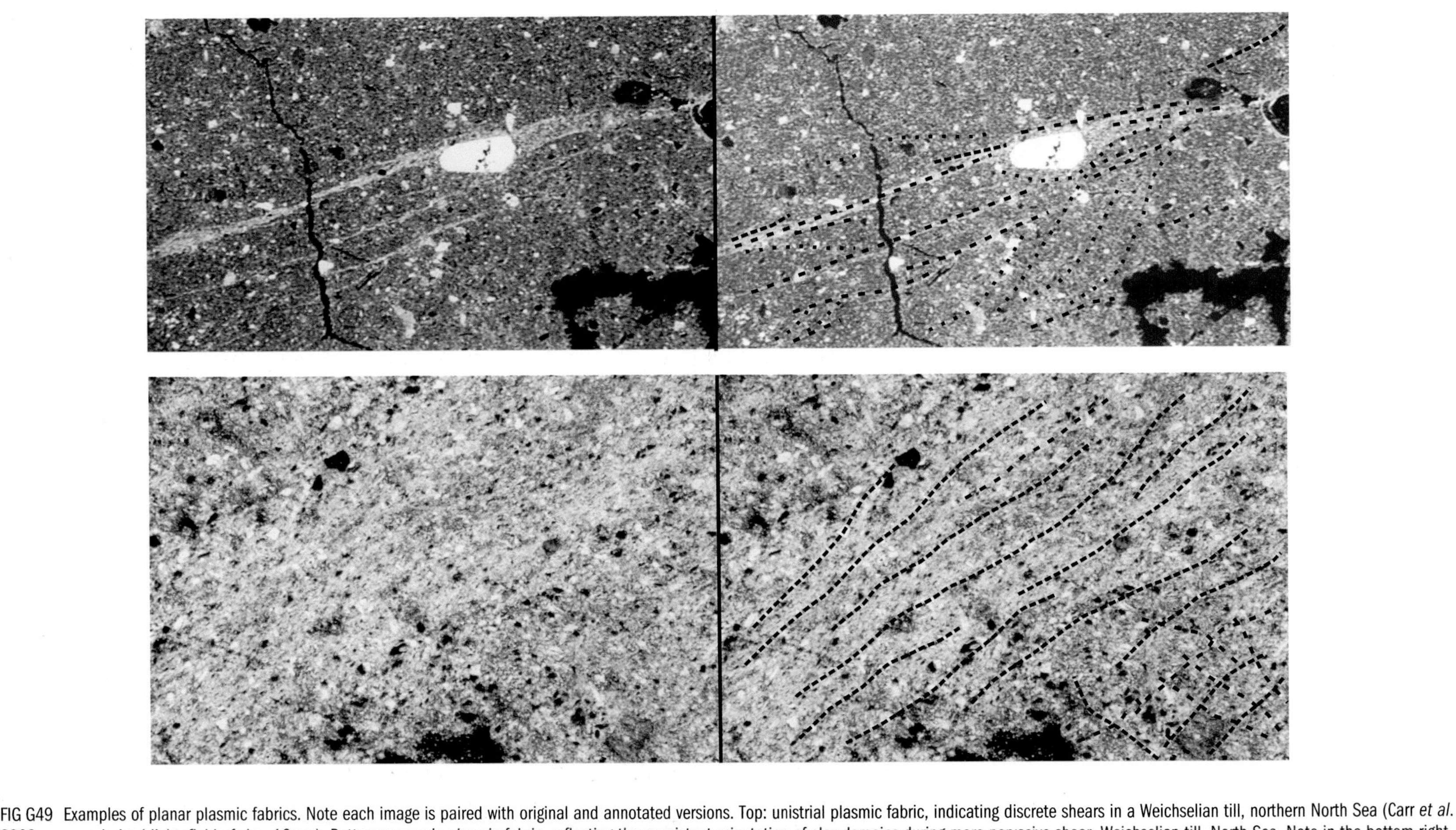

FIG G49 Examples of planar plasmic fabrics. Note each image is paired with original and annotated versions. Top: unistrial plasmic fabric, indicating discrete shears in a Weichselian till, northern North Sea (Carr *et al*. 2000: cross-polarized light, field of view 16mm). Bottom: masepic plasmic fabric, reflecting the consistent orientation of clay domains during more pervasive shear, Weichselian till, North Sea. Note in the bottom-right corner of the image, there is some development of a lattisepic plasmic fabric, possibly reflecting the transition from rotational to planar deformation (cross-polarized light, field of view 12mm).

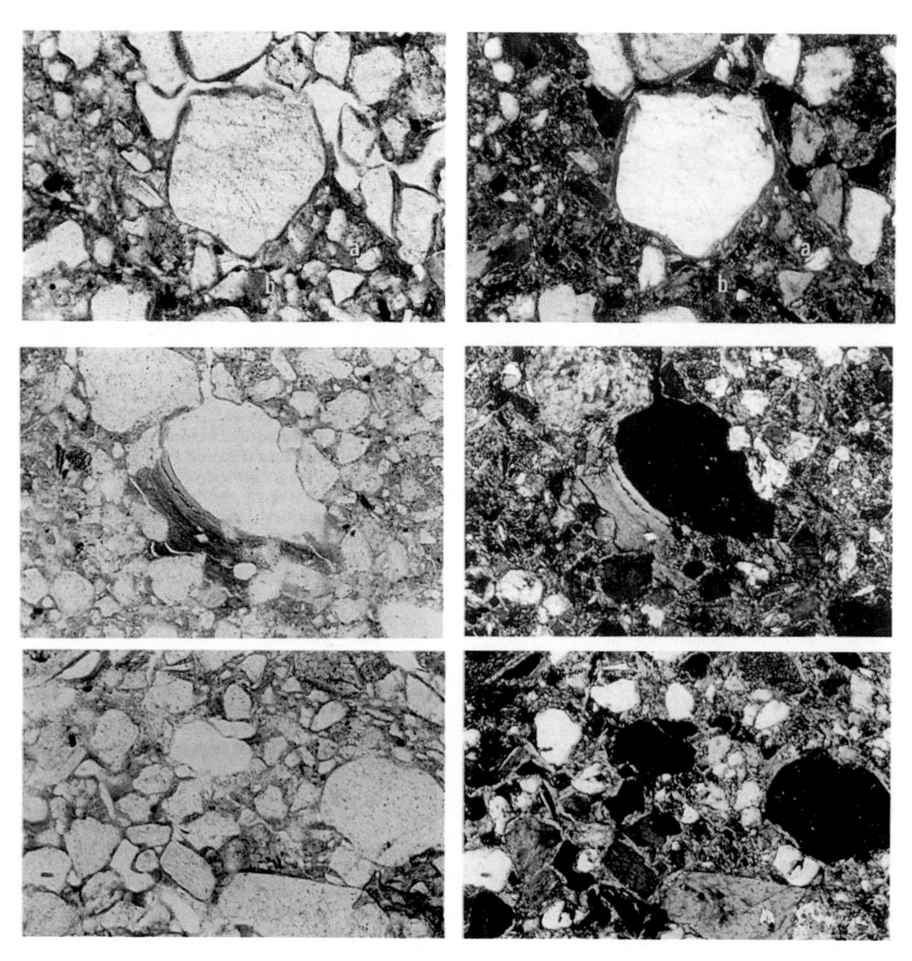

FIG G50 Micromorphological features associated with weathering and soil development in Anglian till from How Hill, Norfolk (Rose *et al.* 1999). Paired images in plane and cross-polarized light are shown. Top: clay coatings around individual grains (field of view 1.2mm). Middle: undisturbed void coating of clay (field of view 1.2mm). Bottom: development of skelsepic plasmic fabric through subglacial deformation or cryostatic pressure (field of view 1.2mm).

INDEX